Immune RNA in Neoplasia

Edited by

MARY A. FINK

Division of Cancer Research Resources and Centers
National Cancer Institute

1976

ACADEMIC PRESS, INC. New York San Francisco London
A Subsidiary of Harcourt Brace Jovanovich, Publishers

COPYRIGHT © 1976, BY ACADEMIC PRESS, INC.
ALL RIGHTS RESERVED.
NO PART OF THIS PUBLICATION MAY BE REPRODUCED OR
TRANSMITTED IN ANY FORM OR BY ANY MEANS, ELECTRONIC
OR MECHANICAL, INCLUDING PHOTOCOPY, RECORDING, OR ANY
INFORMATION STORAGE AND RETRIEVAL SYSTEM, WITHOUT
PERMISSION IN WRITING FROM THE PUBLISHER.

ACADEMIC PRESS, INC.
111 Fifth Avenue, New York, New York 10003

United Kingdom Edition published by
ACADEMIC PRESS, INC. (LONDON) LTD.
24/28 Oval Road, London NW1

Library of Congress Cataloging in Publication Data

Main entry under title:

Immune RNA in neoplasia.

Proceedings of a symposium held at the Marine Biological Laboratory, Woods Hole, Mass., Oct. 8–11, 1975.
1. Cancer–Immunological aspects–Congresses.
2.   Ribonucleic acid–Congresses.   3.   Viral carcinogenesis–Congresses.   4.   Immunotherapy– Congresses.   I.   Fink, Mary A.
RC268.3.I45   616.9'92'06   76-49873
ISBN 0–12–256940–7

PRINTED IN THE UNITED STATES OF AMERICA

Contents

List of Participants	ix
Preface	xiii
An Introduction to Immune RNA *Robert J. Crouch*	xv

Session I: Molecular Biology of RNA
HUGH ROBERTSON, Chairman

Chemical Studies on RNA *P. T. Gilham*	3
The Fractionation and Characterization of the Low Molecular Weight RNA of RNA Tumor Viruses *Gordon G. Peters*	13
Detection of Globin Messenger RNA Sequences by Molecular Hybridization *Jeffrey Ross*	27
Ribonucleases and Factors Influencing Their Activities *Robert J. Crouch*	37
Discussion *Hugh D. Robertson*	45

Session II: What Is Immune RNA?
FRANK L. ADLER, Chairman

Current Status of Immune RNA *M. Fishman and F. L. Adler*	53
Transfer of Cellular Immunity with Immune RNA Extracts and Tumor Immunotherapy *Sheldon Dray*	61
Transfer Factor: Is It Related to Immune RNA? *Fred T. Valentine*	75
Molecular Studies of Immunopathology in Plasmacytoma. Possible Role of Intracisternal A Particles *Dario Giacomoni and Jerry Katzmann*	85
I-RNA: Synthesis and Mechanisms of Action *D. Jachertz*	95

Informational RNA Directed Synthesis of Virus Neutralizing
Proteins in Cell-Free Extracts Prepared from Rat Spleen and
Mouse L Cells   113
  G. Koch, P. Bilello, M. Fishman, R. Mittelstaedt, and E. Borriss

Isolation, Purification, and Cell-Free Translation of Immunoglobulin
Messenger RNAs from Immunoglobulin Producing Cells and Variants   121
  Sidney Pestka, Laura Bailey, Roderick Brandsch, Peter Graves,
  Michael Green, Robert Jilka, James McInnes, Akira Okuyama,
  Philip Tucker, David Weiss, Lynn Yaffe, and Tova Zehavi-Willner

Discussion   145
  F. L. Adler

Session III: Mediation of Antitumor Immune Responses by I-RNA
A. ARTHUR GOTTLIEB, Chairman

The Mediation of Immune Responses by I-RNA to Animal and
Human Tumor Antigens   149
  Yosef H. Pilch, Dieter Fritze, Kenneth P. Ramming,
  Jean B. deKernion, and David H. Kern

Mediation of Antitumor Immune Responses with I-RNA   177
  David H. Kern and Yosef H. Pilch

Augmentation of the Efferent Arc of a Tumor Specific Immune
Response   201
  Peter J. Deckers, Bosco S. Wang, and John A. Mannick

The Relationship of Myeloma "RNA" to the Immune Response   223
  Paul Heller, N. Bhoopalam, Y. Chen, and V. Yakulis

An *in Vitro* Model for Transfer of Tumor Specific Sensitivity
with "Tumor Immune" RNA Extracts and Localization of
Immunologically Active RNA Fractions   235
  Ronald E. Paque

Immunotherapy of a Guinea Pig Hepatoma with Tumor-Immune RNA   245
  Seymour I. Schlager and Sheldon Dray

The Clinical Experience in the Treatment of Renal Adenocarcinoma
with Immune RNA   259
  Jean B. deKernion, Kenneth P. Ramming, Donald G. Skinner,
  and Yosef H. Pilch

Discussion   273
  A. Arthur Gottlieb

Session IV: Basic Concepts: How Does I-RNA Work?
ROBERT J. CROUCH, Chairman

Discussion   279
  Robert J. Crouch

Session V: Applications to Tumor Immunology and Immunotherapy
PETER ALEXANDER, Chairman

Discussion 287
Peter Alexander

General Discussion

Immune RNA—Questions Not Answered 293
Stewart Sell

Microdetection and Differentiation of Free and Intraviral
Nucleic Acids Using Laser Fluorescence Microscopy and
Fluorochrome Mixture Staining 303
M. A. Apple, M. Hercher, T. Hirschfeld, and T. Keefe

List of Participants

Frank L. Adler, St. Jude Children's Research Hospital, 332 North Lauderdale, Memphis, Tennessee 38101

Peter Alexander, Chester Beatty Research Institute, Clifton Avenue, Belmont, Sutton, Surrey, England

Martin A. Apple, Cancer Research Institute, University of California Medical School, San Francisco, California 94143

Frances Cohen, Division of Cancer Research Resources and Centers, National Cancer Institute, National Institutes of Health, Westwood Building, Room 848, Bethesda, Maryland 20014

Robert J. Crouch, Laboratory of Molecular Genetics, National Institute of Child Health and Human Development, National Institutes of Health, Building 6, Room 320, Bethesda, Maryland 20014

Peter J. Deckers, Department of Surgery, Boston University Medical Center, Boston University School of Medicine, 80 East Concord Street, Boston, Massachusetts 02118

Jean B. deKernion, Division of Urology, Room 66-118, Center for the Health Sciences, University of California, Los Angeles, California 90024

Sheldon Dray, Department of Microbiology, University of Illinois at the Medical Center, P. O. Box 6998, Chicago, Illinois 60612

Mary A. Fink, Division of Cancer Research Resources and Centers, National Cancer Institute, National Institutes of Health, Westwood Building, Room 848 Bethesda, Maryland 20014

Marvin Fishman, Division of Immunology, St. Jude Children's Research Hospital, 332 North Lauderdale, Memphis, Tennessee 38101

Herman Friedman, Microbiology Department, Albert Einstein Medical Center, York and Tabor Road, Philadelphia, Pennsylvania 19141

Dario Giacomoni, Department of Microbiology, University of Illinois at the Medical Center, 835 South Wolcott, Chicago, Illinois 60612

Peter T. Gilham, Department of Biological Sciences, Purdue University, Lafayette, Indiana 47907

A. Arthur Gottlieb, Department of Microbiology and Immunology, Tulane University Medical School, 1430 Tulane Avenue, New Orleans, Louisiana 70112

A. Grassmann, Institut für Molekularbiologie und Biochemie, der Freien Universität Berlin, ARNIMALIE 22, D-1 Berlin 33, West Germany

LIST OF PARTICIPANTS

Tin Han, Roswell Park Memorial Institute, 666 Elm Street, Buffalo, New York 14203

Paul Heller, West Side Veterans Administration Hospital, Chicago, Illinois 60611

D. Jachertz, University Bern, Friedbuhlstrasse 51, 3000 Bern, den, Switzerland

David H. Kern, Department of Surgery, Harbor General Hospital, 1000 West Carson Street, Torrance, California 90509

Gebhard Koch, Roche Institute of Molecular Biology, Nutley, New Jersey 07110

John C. Lee, Department of Biochemistry, The University of Texas Health Science Center at San Antonio, 7703 Floyd Curl Drive, San Antonio, Texas 78284

Ulrich Leoning, Department of Zoology, University of Edinburgh, Edinburgh, Scotland EH93JT

John A. Mannick, Department of Surgery, Boston University Medical Center, Boston University School of Medicine, 80 East Concord Street, Boston, Massachusetts 02118

Carl F. Nathan, Immunology Branch, Division of Cancer Biology and Diagnosis, National Cancer Institute, National Institutes of Health, Building 10, Room 4B08, Bethesda, Maryland 20014

Ronald E. Paque, Department of Microbiology, The University of Texas Health Science Center at San Antonio, 7703 Floyd Curl Drive, San Antonio, Texas 78284

Sidney Pestka, Roche Institute of Molecular Biology, Nutley, New Jersey 07110

Gordon G. Peters, Department of Physiological Chemistry, University of Wisconsin, 671 Medical Sciences Building, Madison, Wisconsin 53706

Yosef H. Pilch, Department of Surgery, Harbor General Hospital, 1000 West Carson Street, Torrance, California 90509

Kenneth P. Ramming, Division of Oncology, Department of Surgery, University of California School of Medicine, 924 Westwood Boulevard, Los Angeles, California 90024

Hugh D. Robertson, The Rockefeller University, Box 18, 1230 York Avenue, New York, New York 10021

Robert G. Roeder, Department of Biological Chemistry, Washington University School of Medicine, St. Louis, Missouri 63110

Martin Rosenberg, Laboratory of Molecular Biology, National Cancer Institute, National Institutes of Health, Building 37, Room 4E14, Bethesda, Maryland 20014

Jeffrey Ross, McArdle Laboratory for Cancer Research, University of Wisconsin, Madison, Wisconsin 53706

Seymour Schlager, Immunochemistry Section, Biology Branch, National Cancer Institute, National Institutes of Health, Building 37, Bethesda, Maryland 20014

Stewart Sell, Department of Pathology, School of Medicine, University of California, San Diego, La Jolla, California 92037

Ursula Storb, Department of Microbiology, University of Washington, Seattle, Washington 98195

Fred T. Valentine, Department of Medicine, New York University Medical Center, 550 First Avenue, New York, New York 10016

Preface

The idea that ribonucleic acid (RNA) from an immunized host might incite a specific immune response in another host has been with us for some time. Over the last decade an active attempt has been made in several laboratories to demonstrate that this concept applies in neoplasia and that it holds promise of being a reliable immunotherapeutic approach. In order to accomplish this, however, a concerted effort needs to be made to understand exactly what immune RNA is—how to isolate, purify, and characterize it—and how and under what circumstances it can function to bring about a therapeutic response.

After listening to the presentations concerned with immune RNA at a recent national meeting, it became clear that there was considerable confusion regarding methodology and terminology to the extent that it was difficult to compare —and sometimes difficult to interpret—the results. Thus the idea arose that it might be profitable to hold a workshop in immune RNA to address these problems. After discussions with various people, the possible value of a dialogue between the biologists working with immune RNA and molecular biologists concerned with the fine points of characterization of RNA became apparent. With this in mind, the conference evolved under the enthusiastic co-chairmanship of Drs. Yosef H. Pilch and Robert J. Crouch. It was held at the Marine Biological Laboratory, Woods Hole, Massachusetts, October 8-11, 1975. This volume includes the papers presented and session chairmen's summaries of discussions, as well as two overviews.

For the success of the conference and this volume, we are indebted to a number of people: to Robert Crouch and his colleagues for their altruistic gift of time and energy to help solve the problems of scientists in a different area; to Stewart Sell for his unflagging objectivity; to Frances Cohen for excellent editorial assistance; to Barbara Huffman and other Division secretaries for their aid and support; and especially to Dennis Fink for his understanding and help throughout this endeavor.

AN INTRODUCTION TO IMMUNE RNA[1]

Robert J. Crouch

Laboratory of Molecular Genetics
National Institute of Child Health and Human Development
National Institutes of Health
Bethesda, Maryland 20014

What is Immune RNA?*

As is the case with many biological processes, the concept of "immune" RNA is rather straightforward. Cells incubated in the presence of immune RNA can be converted to the production of antibodies of the specificity of the cell from which the immune RNA is isolated. One of the earliest examples of this phenomenon was a demonstration by Fishman (1) of the transfer of immunological specificity employing T2 bacteriophage as antigen. Macrophages were incubated with bacteriophage T2, a lysate was prepared and filtered, and the filtrate was added to a culture of lymphnode cells. Antibody directed against T2 was generated and a search for the source of this magic substance began. Soon at least a partial explanation appeared. Fishman and Adler (2) demonstrated that if RNA is isolated from macrophages incubated with T2, this RNA can be used to convert the recipient cell to production of antibodies which inhibit T2 plaque formation. RNase destroyed the converting activity but not DNase.

Is immune RNA simply the carry over of antigen?

One of the most plausible and earliest explanations for the phenomenon of immune RNA is that an antigen (such as T2) is still present in the RNA extract and it is the antigen which elicits the formation of antibody in the lymphnode cells. The early reports of "immune RNA" seemed to exclude this possibility since the T2 antigen should be readily detectable and, moreover, T2 added to the lymphnodes did not result in detectable antibody. These results were challenged on two fronts: (a) RNA might act as an adjuvant (i.e., very low levels of T2 would be sufficient to stimulate T2 antibody in the presence of RNA and (b) T2 bacteriophage might be altered in such a way

---

[1] This summary of the status of immune RNA prior to the meeting on Immune RNA in Neoplasia at Wood's Hole, Massachusetts was presented with the intent that the reader would know some of the ideas and work in this area. A complete bibliography is not presented and references are presented which may omit some of the same work published by other workers.

that special techniques would be required to detect the modified antigen or (c) both of the possibilities might occur (e.g., modified bacteriophage associated with RNA). RNA-antigen complexes have indeed been reported in amounts which seemed to account for all of the immunological activity of immune RNA preparations (3). Subsequent studies (4) employing a similar procedure revealed a fraction of immune RNA that appeared to be free of any antigen yet the properties described for immune RNA remained intact. One of the most sensitive tests to demonstrate the absence of antigen in immune RNA utilized an antigen (mono-(p-azobenzene arsonate)-N-chloracetyl-L-tryrosine) [ARS-NAT] containing arsenic (5). By atomic absorption spectra, it was possible to say that immune RNA contained less than 0.0000065% by weight of ARS-NAT, yet was able to confer the ability to make antibody to ARS-NAT when incubated with appropriate cells.

Certainly, such a low level of contamination is an indication that only RNA is involved but, if the following calculation is made, even this low level does not exclude the involvement of antigen. Since most of the RNA isolated from these cells is ribosomal or t-RNA, all but five percent of the RNA should be excluded from these calculations. Starting with 1 mg of total RNA we then exclude ribosomal and t-RNA giving us 50 µg of RNA. Estimates of the size of immune RNA, as determined by sucrose gradient sedimentation are on the order of 14-18S (6) or about 2,000 nucleotides. If we ascribe all 50 µg of RNA as being immune RNA, the following calculation can be made:

$$\frac{50 \times 10^{-6} \text{ g RNA}}{2 \times 10^3 \text{ nucleotides } .320 \text{ g/mole nucleotide}} = 7.81 \times 10^{-11} \text{ moles immune RNA}$$

6.5×10^{-8} g ARS-NAT/g RNA

or 6.5×10^{-8} mg ARS-NAT/mg RNA

$$\frac{6.5 \times 10^{-11} \text{ g ARS-NAT}}{486 \text{ g/mole ARS-NAT}} = 1.11 \times 10^{-13} \text{ moles ARS-NAT}$$

giving $$\frac{11.1 \times 10^{-14} \text{ moles ARS-NAT}}{7.81 \times 10^{-11} \text{ moles immune RNA}} = 1.04 \times 10^{-3} \frac{\text{moles ARS-NAT}}{\text{moles immune RNA}}$$

or 1 molecule of ARS-NAT per 1000 molecules of immune RNA.

There are, of course, limits to this calculation. First, we have assumed that all of the non-ribosomal and t-RNA is immune RNA which can only be an overestimate of the ratio of moles ARS-NAT/moles immune RNA. Second, the ARS-NAT value used is the maximal amount possible and would give an underestimate of moles ARS-NAT/moles immune RNA. It seems unlikely that all 50 µg of RNA is immune RNA, at least immune RNA specific for

ARS-NAT; and that one molecule of ARS-NAT per 1000 molecules of immune RNA might represent one molecule of ARS-NAT per one molecule of immune RNA specific for ARS-NAT. The purpose of this calculation is to demonstrate the difficulty of ever resolving the question of antigen contamination by this approach.

Genetic evidence supporting the concept of immune RNA

Experiments that give the strongest support for the concept of immune RNA are based on genetic differences among immunoglobulins of different animals. Slight changes in the amino acid sequence of the immunoglobulin molecule from one rabbit (A) to a second (genetically distinct) rabbit (B) have been demonstrated. Furthermore, it is possible to prepare antibodies to these different immunoglobulins (which are now being utilized as antigens) and make antisera specific for either rabbit A or rabbit B type immunoglobulins. Differences of this sort are called allotypic and show normal genetic inheritance on breeding. Allotypic differences permit the following class of experiments to be performed: immune RNA from a rabbit of allotype A can be incubated with cells from a rabbit of allotype B and the immunoglobulins produced in these cells can be challenged with antisera against type A or type B immunoglobulins. If immune RNA from type A cells is carrying the antigen into the cells of type B, the immunoglobulin produced in type B cells should be of B allotype and not A allotype. On the other hand, if immune RNA is carrying information for the production of immunoglobulins, it might be expected that immunoglobulins produced in cells (which are of B allotype) would have allotypic characteristics of the A type. The results of such experiments (7,8) clearly demonstrate the conversion of cells of B allotype by immune RNA of A allotype to the production of immunoglobulins of A allotype. A straightforward explanation of these results based on the carry over of antigen in immune RNA is difficult to conceive.

It follows from the experiments just described that immune RNA from type A cells carries information concerning the structural features of type A immunoglobulins, including both the allotypic determinants and the antigenic specificity. Several lines of evidence have suggested that mRNA for immunoglobulins is the vehicle for this information transfer. First, the "converting" activity is sensitive to RNase but not DNase or pronase. Second, drugs which block protein synthesis, but not those that block RNA synthesis, are able to prohibit conversion of a cell to produce immunoglobulins of the immune RNA type. Third, this material is isolated by techniques normally employed in the isolation of RNA. Fourth, immune RNA sediments in a sucrose gradient in the region of mRNA for immunoglobulins (6). Fifth, immune RNA binds to oligo(dT)-cellulose columns (6), a technique frequently used to isolate many types of mRNA from eukaryotic systems. Finally, some evidence is being developed which indicates that immune RNA can be translated in an *in vitro* protein synthesizing system to produce active immunoglobulins (9).

Detection of changes induced by immune RNA

This brief outline of some of the properties of immune RNA is based on experiments utilizing a variety of animals and assays for the detection of the changes induced by immune RNA. Unlike most of the assays that molecular biologists use, in which there is a very simple and direct assay of the product (e.g., incorporation of ATP into RNA by RNA polymerase), most immunological assays use properties of the immunoglobulins (e.g., antibodies against T2 bacteriophage are detected by the inhibition of phage formation of T2 - not an assay that is for the T2 antibody complex).

Probably the greatest difficulty that a molecular biologist encounters in understanding experiments concerning immune RNA lies in the techniques of the system and the estimation of the validity of such techniques. Papers presented in this volume are based on experiments in which techniques of immunoglobulin detection range from inhibition of T2 phage plaque formation to the Jerne plaque technique to a rosette assay and to the production of migration inhibitory factor (MIF). A brief description of these assays follows and reflects a "text book" interpretation to generate some basic understanding of immunological properties exploited in these assays. Inhibition of T2 plaque formation represents the more classical demonstration of antibody production in which the antigen and antibody are permitted to interact and the formation of antibody is measured by a decrease in the titer of the phage. The rosette procedure is described in an accompanying paper (10). The Jerne plaque assay involves plating antibody producing cells – in this case converted to specific antibody production by immune RNA – on a lawn of sheep red blood cells (SRBC). If the RNA is from cells or animals immunized with SRBC, addition of complement to the antigen-antibody complex results in hemolysis of the SRBC, and a cell producing antibody to SRBC is observed as a clear plaque in the lawn of SRBC. Assays employing MIF rely on the observation that the motility of macrophages is inhibited in the presence of antigen for which there is a corresponding type of macrophage (11). Migration of all macrophages is presumably inhibited if any "cells" are specifically stimulated to make MIF by the antigen. As an example, if sensitive cells are stimulated by ARS-NAT, addition of this antigen to macrophage-containing cell suspensions in a capillary tube will inhibit migration of the macrophages from the capillary tube. Each system relies on slightly different immunological properties to detect specific antibody production and has different levels of sensitivity. Theoretically, the T2 system could detect one molecule of antibody, and the Jerne plaque assay or the rosette assay may detect single cells producing antibodies. The actual level of resolution in these systems is far from theoretical and, in many cases, quantitation is relative.

Immune RNA and myeloma proteins

In the past few years, immunologists have brought the study of myeloma proteins to an extremely useful state. A series of recent papers utilizing myelomas has tended to give strong support to the concept of immune RNA as well as to provide evidence that immune RNA is identical to mRNA for immunoglobulins. Since myelomas produce a single species of immunoglobulin, immune RNA obtained from these tumors should transfer very specific information to the recipient cells. Certainly, the mRNA for immunoglobulins in these cells direct the *in vitro* synthesis of the myeloma protein. Transfer of this mRNA to a recipient cell should convert that cell to the production of immunoglobulins of the myeloma type. At this point another bit of terminology must be considered. Alterations in the immunoglobulin molecule which allow the antibody to interact with an antigen (or, more precisely, a portion thereof) are thought to occur such that the portion of the immunoglobulin interacting with the antigen is constant from one molecule to another, irrespective of the allotypic characteristics of the immunoglobulin. These determinants are said to specify the idiotype of an immunoglobulin molecule. In contrast to allotype, in which all immunoglobulins from an individual rabbit of genotype A possess the allotypic determinants A but many different antigenic responses, all immunoglobulins of idiotype I should contain a common region which is dependent on its antigenic specificity. Myeloma proteins represent a situation in which a cell is producing proteins of one idiotype. If these myeloma proteins are used as antigens to immunize syngeneic hosts, antibodies are produced which interact specifically with the region containing the idiotypic determinants. Such antibodies are said to be anti-idiotypic. Antibodies of this type are extremely useful in detecting the production of myeloma proteins and make the myeloma system amenable to manipulation of immune RNA.

RNA extracted from myeloma tissue can cause lymphocytes from animals which are free of myelomas and are not producing myeloma protein to begin the production of immunoglobulin molecules with idiotypic characteristics of the myeloma protein from which the immune RNA is extracted (12). Again it has been shown that such transfer of information can be inhibited by pretreatment of the immune RNA with RNase but not DNase or pronase. Establishment of the conversion is inhibited by cycloheximide and puromycin but not actinomycin D.

It is known that lymphocytes from animals bearing myelomas have surface immunoglobulins of the myeloma type and such animals have a greatly reduced ability to respond to antigens (i.e., they are immunosuppressed). RNA from myelomas is able to confer on animals the same state of immunosuppression observed in the diseased animal and, in addition (13), attempts to convert cells which are immune suppressed with immune RNA have been successful (14). These results suggest that immunosuppression occurs as a result of the disease and, in some manner, the lymphocytes of the diseased animal are exposed to a substance similar (or identical) to immune RNA. Particles have been observed in the plasma of animals with plasmacytomas which can convert normal lymphocytes to lymphocytes whose surface immunoglobulins are the plasmacytoma type (15). Also plasma has yielded immune RNA with the ability to convert cells to lymphocytes bearing plasmacytoma proteins (17).

Stable conversion by immune RNA

Is it possible to use immune RNA to convert cells to a new, genetically distinct immunoglobulin type? Immunosuppression seems to indicate that plasmacytomas have altered the normal immune system in a stable manner. But will the same be true if nontumor immune RNA is utilized? Phenotypic changes due to immune RNA do not depend on a cellular system sensitive to actinomycin D but protein synthesis inhibitors prevent conversion. A stable or long term change should, in some way, depend on a process that amplifies the immune RNA. Bhoopalam *et al.* (16) have performed a serial passage of immune RNA to test for increased amounts of immune RNA. RNA was extracted from a plasmacytoma and injected into a disease-free mouse. After one hour the spleen was removed from the mouse, one-half of the spleen was put into culture for 7 days or extracted immediately to yield immune RNA. *In vitro* conversion was demonstrated with the RNA isolated from the spleen immediately after removal of the spleen; but more conversion was seen with the RNA isolated from the cells that had been cultured for seven days. Also RNA prepared in a similar manner was passed sequentially through 5 animals each time injecting immune RNA in the mouse then extracting the RNA from the spleen, injecting that RNA into a second mouse, and so on. Conversion activity was lost after two or three passages when the "immediate" extract RNA was used but even after 5 passages with "7 day" immune RNA cell converting activity remained. These results were taken to indicate RNA replication in the recipient cells. Previous experiments (17) demonstrated a similar transfer but there was no genetic evidence to exclude antigen contamination.

Comparison with other systems

Attempting to understand the mechanism involved in what can almost be described as cellular transformation (here phenotypic changes and viral production will be included as examples of transformation) with RNA requires some new assumptions not previously invoked to explain other transformation. If immune RNA simply enters the cell and becomes an active mRNA without any amplification of the immune RNA either via an RNA or DNA intermediate, then we have to wonder why the cell takes up this RNA, why the cell is converted to production of proteins with this immune RNA, and why once converted the cell seems to be resistant to further change (i.e., becomes immune suppressed).

There are many examples of bacterial cells transformed with DNA (18) or transfected with bacteriophage DNA (19) or RNA (20). Polio RNA is infectious (21), DNA of Adenovirus 2 can transform cells (22), and there are other demonstrations of DNA or RNA transformation. All of these examples are easily understood. Transformation of bacteria by DNA simply involves the host mechanism of recombination. Most examples of viral nucleic acid can be explained in terms of entry of the nucleic acid into the cell. In most instances nucleic acid is injected or carried into the cell via the virus particle. Once inside the cell, the free nucleic acid is able to direct the synthesis of new viruses. A good example of this is shown by a comparison of poliovirus and vesicular stomatitis virus (VSV). Poliovirus contains the strand of RNA which can be translated into poliovirus proteins, whereas VSV contains the strand complementary to the translatable strand, along with an RNA dependent RNA polymerase to generate the translatable strand. RNA from poliovirus is infectious while VSV RNA is uninfectious.

There are enzymes capable of amplifying immune RNA, either a RNA dependent DNA polymerase or a RNA dependent RNA polymerase (23-26). When RNA is isolated from nonimmune animals or from tissues such as liver, cellular conversion to new types of immunoglobulins does not occur. Are these RNAs taken up and treated in the same way as immune RNA? Could globin mRNA be translated and replicated in this system?

There is one other phenomenon in which either DNA or RNA is used to transform a cell. In no case is viral nucleic acid as effective as the normal mechanism of infection. Lambda DNA transfects *E. coli* with an efficiency of one in 10^5 and polio RNA is about a thousand fold less efficient than poliovirus. Very few, if any, biochemical techniques are available which can tell us much about those nucleic acid molecules that are infectious. Will the efficiency of utilization of immune RNA be any different? Clearly, a technique which permits the survival of most of the immune RNA would, from a biochemical approach, be extremely useful. Micro injection techniques are available and should be a tool of great importance for solving some of the problems of immune RNA biochemistry.

REFERENCES

1. Fishman, M. (1961). Antibody formation *in vitro. J. Exp. Med. 114*:837-856.
2. Fishman, M. and F. L. Adler (1963). Antibody formation initiated *in vitro*. II. Antibody synthesis in x-irradiated recipients of diffusion chambers containing nucleic acid derived from macrophages incubated with antigen. *J. Exp. Med. 117*:595-602.
3. Gottlieb, A. A. and R. H. Schwartz (1972). Review – antigen-RNA interactions. *Cell. Immunol. 5*:341-362.
4. Fishman, M. (1973). The role of macrophage RNA in the immune response. *The Role of RNA in Reproduction and Development.* Niu and Segal, Eds. North-Holland Publishing Co.

5. Schlager, S. I., S. Dray and R. E. Paque (1974). Atomic spectroscopic evidence for the absence of a low-molecular weight (486) antigen in RNA extracts shown to transfer delayed-type hypersensitivity *in vitro*. *Cell. Immunol.* 14:104-122.

6. Giacomoni, D., V. Yakulis, S. R. Wang, A. Cooke, S. Dray and P. Heller (1974). *In vitro* conversion of normal mouse lymphocytes by plasmacytoma RNA to express idiotypic specificities on their surface characteristic of the plasmacytoma immunoglobulin. *Cell. Immunol.* 11:389-400.

7. Adler, F. L., M. Fishman and S. Dray (1966). Antibody formation initiated *in vitro*. III. Antibody formation and allotypic specificity directed by ribonucleic acid from peritoneal exudate cells. *J. Immunol.* 97:554-563.

8. Bell, C. and S. Dray (1969). Conversion of non-immune spleen cells by ribonucleic acid of lymphoid cells from an immunized rabbit to produce γM antibody of foreign light chain allotype. *J. Immunol.* 103:1196-1211.

9. Bilello, P. A., G. Koch and M. Fishman (1975). mRNA activity of immunogenic RNA. *Fed. Proc.* 34:1030.

10. Heller, P., N. Bhoopalam, Y. Chen and V. Yakulis (1976). The relationship of myeloma "RNA" to the immune response. *Immune RNA in Neoplasia*. Mary A. Fink, Ed. Academic Press, New York.

11. Weiser, R. L., Q. N. Myrvik and N. N. Pearsall (1969). *Fundamentals of Immunology*, pp. 196-197. Lea and Febiger, Philadelphia.

12. Bhoopalam, N., V. Yakulis, N. Vostea and P. Heller (1972). Surface immunoglobulin of circulating lymphocytes in mouse plasmacytoma. II. The influence of plasmacytoma RNA on surface immunoglobulins of lymphocytes. *Blood* 39:465-471.

13. Yakulis, V., V. Cabana, D. Giacomoni and P. Heller (1973). Surface immunoglobulins of circulating lymphocytes in mouse plasmacytoma. III. The effect of plasmacytoma RNA on the immune response. *Immunol. Comm.* 2:129-139.

14. Adler, F. L. and J. Fishman (1975). *In vitro* studies on information transfer in cells from allotype-suppressed rabbits. *J. Immunol.* 115:129-134.

15. Katzmann, J., D. Giacomoni, V. Yakulis and P. Heller (1975). Characterization of two plasmacytoma fractions and their RNA capable of changing lymphocyte surface immunoglobulins (cell conversion). *Cell. Immunol.* 18:98-109.

16. Bhoopalam, N., V. Yakulis, D. Giacomoni and P. Heller. Surface immunoglobulins of lymphocytes in mouse plasmacytoma. IV. Evidence for the persistence of the effect of plasmacytoma-RNA on the surface immunoglobulins of normal lymphocytes *in vivo* and *in vitro*. *Clin. Exp. Immunol.*, in press.

17. Saito, K., S. Kurashige and S. Mitsuhoshi (1969). Serial transfer of immunity through immune RNA. *Jap. J. Microbiol.* 13:122-124.

18. Avery, O. T., C. M. MacLeod and M. McCarty (1944). Studies on the chemical nature of the substance inducing transformation of pneumococcal types. *J. Exp. Med.* 79:137-158.

19. Kaiser, A. D. and R. Wu (1968). Structure and function of DNA cohesive ends. *Cold Spring Harbor Symp.* 23:329-734.

20. Pace, N. R. and S. Spiegelman (1966). The synthesis of infectious RNA with a replicase purified according to its size and density. *Proc. Nat. Acad. Sci., USA* 55:1608-1615.

21. Koch, G. (1973). Stability of polycation-induced cell competence for infection by viral RNA. *Virology* 45:841-843.

22. Graham, F. L. and A. J. von der Eb (1973). A new technique for the assay of infectivity of human adenovirus 5 DNA. *Virology* 52:456-467.

23. Jacherts, D., U. Opitz and H-G. Opitz (1972). Gene amplification in cell-free synthesis. *Z. Immun.-Forsch., Bd* 144:s. 260-272.

24. Kurashige, S. and S. Mitsuhashi (1973). The possible presence of ribonucleic acid-dependent deoxyribonucleic acid polymerase in the immune response. *Japan. J. Microbiol.* *17*:105-109.

25. Saito, K. and J. Mitsuhashi (1973). Ribonucleic acid-dependent ribonucleic acid replicase in the immune response. *Japan J. Microbiol.* *17*:117-121.

26. Downey, K. M., J. J. Byrnes, B. S. Jurmark and A. G. So (1973). Reticulocyte RNA-dependent RNA polymerase. *Proc. Nat. Acad. Sci. USA* *70*:3400-3404.

SESSION I
Molecular Biology of RNA

Chairman, Hugh Robertson

CHEMICAL STUDIES ON RNA

P. T. Gilham

*Department of Biological Sciences
Purdue University
Lafayette, Indiana 47907*

Ribonucleic acids contain a number of reactive centers and structural characteristics that can be readily exploited by the organic chemist in the study of these macromolecules. The most important of these features include (a) the weakly basic character of adenosine, cytidine, and guanosine moieties, (b) the weakly acidic behavior of uridine and guanosine residues, (c) the terminal cis-diol groups at the 3' terminals of RNA chains, (d) the 2'-hydroxyl group associated with each nucleoside residue and (e) the special reactivity of terminal mono-esterified phosphate or polyphosphate groups. Studies in this laboratory during the last few years have been directed towards employing the specific chemical behavior exhibited by these reactive centers in the determination of the primary and secondary structure of RNA species, the isolation of specific RNA molecules and RNA fragments, and the chemico-enzymatic synthesis of polyribonucleotides.

CHEMICAL MODIFICATION OF RNA BASES

The study of the fine structure of nucleic acids has depended on the availability of methods for the specific degradation of polynucleotide chains; and, while there are known nucleases that have useful specificities in their action on nucleic acids, the number of such enzymes and the variety of their cleavage specificities remain limited. One approach, by which the range of specificity in degradation has been increased, has involved the reversible chemical modification of the substrate in such a way that the types of bonds cleaved by a particular enzyme are either restricted or changed. The application of this idea became possible when it was observed that those nucleosides which have pK_a values in the range 9-10 (uridine, thymidine, guanosine, and deoxyguanosine), as well as the corresponding residues in polynucleotides and nucleic acids could be specifically and reversibly modified by reaction with reagents of the type, $R\text{-}N=C=N\text{-}(CH_2)_2\overset{+}{N}R_3$ (1,2). With the reagent, N-cyclohexyl-N'-β-(4-methylmorpholinium)ethylcarbodiimide (Cmc), the reaction takes place under conditions (20° in water at pH 8-8.5) which are of a sufficiently mild nature that the structural integrity of the polynucleotide chain is conserved.

The products have been shown to consist of guanidine derivatives in which the carbodiimide group of the Cmc cation is attached to the N^3 positions of the guanine residues and to the N^1 positions of the uracil or thymine residues in the nucleic acid. The substituent Cmc groups are stable at pH values below 8 and can be removed by a relatively mild treatment with dilute ammonia. Inosine and pseudouridine, two of the minor nucleosides that occur in tRNA, are also subject to reaction with the Cmc cation (3). Inosine forms a derivative that is analogous to that derived from guanosine, whereas pseudouridine is capable of forming a di-substituted (at N^1 and N^3) derivative. However, while Cmc-inosine can be converted back to the nucleoside under the mild alkaline conditions necessary for the removal of the Cmc groups from guanosine and uridine, the di-Cmc-pseudouridine is converted to a mono-Cmc derivative that is stable under these conditions.

All nucleases tested so far are unable to recognize Cmc-blocked nucleoside residues in polynucleotides, and accordingly, they all exhibit altered cleavage patterns in their action on Cmc-modified nucleic acids. For example, the normal action of pancreatic ribonuclease involves the cleavage of uridine and cytidine 3'-phosphoryl bonds in RNA, whereas, with Cmc-derivatized RNA, this action is restricted to the cleavage of cytidine 3'-phosphoryl bonds only (1,2,4). In addition, it has been shown that, in its action on this particular bond, the enzyme is not sensitive to a modified nucleoside on either side of the cytidine residue. A similar situation exists for pseudouridine positions in tRNA. Pseudouridine 3'-phosphoryl bonds are normally cleaved by pancreatic ribonuclease but are resistant to attack after the nucleoside has been modified by reaction with the Cmc reagent (2,3,5). Thus, the hydrolysis of Cmc-derivatized RNA with this enzyme constitutes a method for the specific cleavage of RNA at the cytidine positions in the chain.

A study of the action of two well-known exonucleases on a number of dinucleoside phosphates and their Cmc derivatives has also been carried out (5). The action of snake venom phosphodiesterase, which normally attacks polynucleotides exonucleolytically from the 3' terminus, is blocked by a Cmc-modified nucleoside at the 3' terminus of the dinucleotide. In contrast to this effect, the action of spleen phosphodiesterase, which normally degrades polynucleotides in an exonucleolytic fashion from the 5' terminus, is blocked by a Cmc-modified nucleoside adjacent to the 5' terminus. Three other endonucleases, T_1, T_2, and U_2, have also been studied with respect to their behavior towards Cmc-blocked nucleoside residues in polyribonucleotides (6). Ribonuclease T_1, which normally cleaves guanosine 3'-phosphoryl bonds in RNA, is unable to attack Cmc-guanosine residues in polynucleotides. The action of ribonuclease T_2, which attacks all nucleoside 3'-phosphoryl bonds, is restricted to the cleavage of adenosine and cytidine positions only in Cmc-modified RNA. In contrast to the other cyclizing endonucleases, ribonuclease

U_2 is sensitive to blocking groups on either side of those phosphodiester bonds that would normally be attacked by the enzyme. Since the enzyme hydrolyzes adenosine and guanosine 3′-phosphoryl bonds, the modification of uridine and guanosine residues restricts the action of this enzyme to the cleavage of -A-A- and -A-C- sequences, and this constitutes one of the most specific methods for the fragmentation of RNA. Other enzymes that display restricted cleavage patterns with Cmc-modified nucleic acids are micrococcal nuclease (7), Sl nuclease (8), deoxyribonucleases I and II (8), and SP3 deoxyribonuclease (8). The applications of some of these specific fragmentation methods in the study of RNA sequence analysis have been reviewed (9).

The Cmc reagent may also be used as a chemical probe for the study of nucleic acid secondary structure. This application arises from the fact that the positions of substitution of the Cmc group in uridine, guanosine, and their deoxyribo-counterparts correspond to the atoms that are normally involved in the H-bonded structures that are formed in nucleic acid helical complexes. Thus, during the initial phases of the derivatization reaction, the nucleotides that are involved in such helical structures should react at much slower rates than non-complexed residues. The reagent has been used in a number of laboratories to detect, by partial derivatization, those nucleosides that are located in the relatively open regions (such as loops) present in RNA structures. Examples of molecules that have been studied in this way include yeast tRNA^{Ala} (10), E. coli 5S RNA (11), E. coli tRNA^{Tyr} (12), E. coli tRNA$_f^{Met}$ (13), E. coli tRNA^{Leu} (14), and yeast tRNA^{Phe} (15).

REACTIONS AT THE TERMINAL CIS-DIOL GROUP

Two of the special properties of the terminal cis-diol group that can be used in RNA studies are its susceptibility to oxidation by periodate and its ability to form cyclic complexes with compounds containing the dihydroxyboryl group. A study of these reactions has led to the development of methods for the sequence analysis of polyribonucleotides, the isolation of RNA molecules and specific RNA fragments, and the immobilization of RNA and polyribonucleotides.

The use of periodate oxidation in RNA structural studies was based on the suggestion (16,17) that a method for the sequence analysis of polyribonucleotides might be perfected by studying the successive chemical removal of nucleotides from their 3′ terminals. The procedure involves (a) the periodate oxidation of the terminal cis-diol group of the polynucleotide chain; (b) the removal of the terminal nucleoside moiety by a base-catalyzed β-elimination reaction; and (c) the removal of the terminal phosphate group, so formed, by phosphatase, leaving the polynucleotide chain in a condition that is suitable for a second cycle of degradation. Work in this laboratory over the last few years has been directed to the study of a number of aspects of the degradation scheme with a view to developing a general method for the sequence analysis of short polyribonucleotides. Some of the problems that have been investigated include (a) the specification of reaction conditions that would allow essentially complete reactions to occur in all phases of the degradation without altering the structural integrity of the rest of the polynucleotide chain, (b) the development of methods for the separation of the polynucleotide after each degradation cycle and for the separation and identification of the released nucleoside fragment, and (c) the study of the chemical control of each degradation cycle to prevent either over- or under-reaction (18-22).

These studies have led to the application of the technique to the sequence analysis of the 3'-terminal decanucleotide from the RNA of bacteriophage f2 (18), the 3'-terminal pentadecanucleotide from phage Qβ RNA (19,23), the 3'-terminal decanucleotide from phage GA RNA (24), a dodecanucleotide derived from yeast tRNA^{Asp} (22), and the 3'-terminal nonadecanucleotide from yeast initiator tRNA (22) as well as a number of synthetic oligoribonucleotides prepared by the methods described below. The stepwise degradation technique is of particular value in the structural analysis of those cases where the use of methods that have been developed for ^{32}P-labeled polynucleotides is not feasible or where the application of sequencing methods involving partial degradation with exonucleases is likely to produce ambiguous results.

Another application of the periodate oxidation of terminal diol groups concerns the immobilization of RNA in the preparation of affinity chromatographic materials. Periodate-oxidized RNA condenses with the amino groups of aminoethylcellulose, and the complex formed can be stabilized by reduction with sodium borohydride (25). The use of immobilized nucleic acids in affinity chromatographic separations has recently been reviewed (26). The interaction of oxidized RNA or RNA fragments with aminoethylcellulose has also been exploited in a method for the isolation of molecules that contain the terminal cis-diol group. This application has been demonstrated in the isolation of the 3'-terminal polynucleotides from bacteriophage ribonucleic acids (27,28). The method consists of the enzymatic fragmentation of the RNA molecule followed by the specific oxidation of the 3'-terminal fragment and the subsequent condensation of the dialdehyde formed with aminoethylcellulose. After removal of the unbound internal fragments, the terminal polynucleotide is recovered by an amine-catalyzed β-elimination reaction in which the terminal oxidized nucleoside moiety is released from the polynucleotide chain. While this method represents an efficient procedure for the isolation of terminal sequences, the loss of the terminal nucleoside during the release step requires that its identity must be determined by a separate technique (29).

More recently, a new isolation technique has been developed. The method involves the recognition of those fragments that contain the 2',3'-diol group as before but avoids the introduction of chemical modification into the polynucleotide structure. This new method not only permits the isolation of polynucleotide fragments intact but also allows the isolation of molecules (such as tRNA) that might contain internal structures that are sensitive to periodate oxidation and β-elimination reactions. The procedure exploits the discovery that arylboric acids can form cyclic structures with the cis-diol groups of nucleosides, nucleotides, and polynucleotides, much in the same way as does the borate anion. The two cellulose derivatives, N-(m-dihydroxyborylphenyl)-carbamylmethylcellulose and N-[N'-(m-dihydroxyborylphenyl)succinamyl]-aminoethylcellulose (DBAE-cellulose), have been synthesized, and a detailed analysis of the interaction of these derivatives with nucleosides, sugars, and sugar alcohols has been carried out (30). It appears that, for complex formation, the active form of the boronic acid group is the tetrahedral boronate anion and thus, compounds containing the appropriate cis-diol group are bound tightly to the cellulose derivatives at pH values above 7.5 and can be subsequently released when exposed to buffers of pH 6. Further studies on the binding of nucleotides, dinucleoside phosphates, polynucleotides, and nucleic acids have been documented (31), and it has been shown that molecules which possess the terminal cis-diol group can be readily separated from those in which the group is either absent or blocked. The method has been applied to the isolation of the terminal polynucleotide fragments from the RNA molecules of bacteriophages, f2, Qβ, and GA (24,32), and has been used in the preparative separation of oligodeoxyribonucleotides from similar species containing a terminal ribonucleoside (33).

Of some special interest has been the potential of this chromatographic system to separate particular aminoacylated tRNA species from crude tRNA. Since aminoacylated tRNA carries the aminoacyl group at the terminal $2'(3')$ positions, it was considered that the dihydroxyboryl-substituted celluloses should be capable of effecting the separation of tRNA isoacceptors from other tRNA species in a single chromatographic step. The method, which involves the enzymatic aminoacylation of crude tRNA with a particular amino acid followed by the chromatographic separation, has recently been developed (34,35).

Other applications of DBAE-cellulose chromatography include the isolation of the 3'-terminal regions of the 4S and 6S RNA transcripts synthesized from phage λ DNA (36,37), and the terminal oligonucleotides from the phage T7 early messenger RNAs synthesized *in vivo* and *in vitro* (38,39). The chromatographic material has also been used to separate tRNA from modified tRNA species in which the terminal adenosine moiety has been replaced by a 2'- or 3'-deoxyadenosine (40). Recently, a number of eukaryotic messenger and viral ribonucleic acids have been found to contain a 5'-linked purine nucleoside attached to their 5'-terminal polyphosphate groups. Since these structures contain the unsubstituted cis-diol group, DBAE-cellulose chromatography should also be of value in isolating the 5'-terminal regions of these molecules.

REACTIONS INVOLVING THE TERMINAL PHOSPHATE GROUP

As described above, the uridine and guanosine residues in RNA are modified when exposed to the water-soluble carbodiimide in aqueous solution at pH values above 7. In solutions of pH values, 5-6, however, the reagent has the capacity to specifically activate the mono-esterified phosphate groups of nucleotides and polynucleotides such that, in the presence of alcohols, the phosphate group is converted to the corresponding phosphodiester (41). This principle has been exploited in the development of procedures for the covalent binding to cellulose of the terminals of oligo- and polynucleotides and nucleic acids (25,42). These immobilized polynucleotides and nucleic acids have been employed both in the isolation of nucleic acid fragments which have sequences that are complementary to the immobilized polynucleotide chains, and in the study of the mechanism of action of polynucleotide kinases, nucleic acid polymerases and nucleases (26). The same principle of immobilization has been used in the synthesis of celluloses containing covalently-bound homo-oligonucleotides except that, in these cases, the oligonucleotides are prepared by chemical polymerization of nucleotides and then attached to the cellulose at their 5' terminals by reaction in anhydrous solution with dicyclohexylcarbodiimide (43). These cellulose derivatives can be employed in the chromatographic fractionation of mixtures of polynucleotides on the basis of the temperature-dependent stability of their base-paired complexes formed with the immobilized oligonucleotides (43,44). For example, it was shown that oligo(dT)-cellulose can effect the separation of a series of deoxyadenosine oligonucleotides by this form of chromatography, and this particular cellulose derivative is now widely used in the isolation of RNA and RNA fragments that contain poly(A) sequences (26).

The specific activation of the mono-esterified phosphate groups of nucleic acids and polynucleotides may also be used to install a chemical "handle" at their terminals. The treatment of the polynucleotide with the Cmc reagent in the presence of a high concentration of sorbitol results in the formation of a phosphodiester linkage between its terminal phosphate group and the sorbitol molecule (31). The sorbitol moiety is capable of forming strong complexes with the dihydroxyboryl group and thus, nucleic acids or nucleic acid fragments labeled in this way may be readily isolated by chromatography on DBAE-cellulose (31,45). Another application of the activation by the Cmc reagent concerns the synthesis of polynucleotides through the head-to-tail joining of oligonucleotides that are located, in complexed form, on a suitable polynucleotide template (41). This application constitutes the action of a "chemical ligase" in that the terminal 5'-phosphate group of one oligonucleotide molecule is connected to the 3'-hydroxyl group of its neighbor in the base-paired complex. The extension of this technique to the synthesis of polynucleotides containing specific sequences is currently under investigation.

CHEMICAL BLOCKING OF NUCLEOTIDE 2'-HYDROXYL GROUPS

The specific chemical blocking of nucleotide 2'-hydroxyl groups forms an important aspect of a new method for the stepwise synthesis of polyribonucleotides of predetermined sequence. This application arose out of a study of the reactions of nucleotide polymerizing enzymes with chemically modified substrates. Various chemical modifications of the nucleotide substrates have been tested for their capacity to permit, in place of the normal polymerization reaction, the addition of a single nucleotide to an oligonucleotide acceptor molecule. The system that is presently under investigation is that of the reaction of polynucleotide phosphorylase with nucleoside 5'-diphosphates that contain blocking groups at their 2'-hydroxyl groups. Initial experiments showed that, in the presence of 2'(3')-O-(α-methoxyethyl)nucleoside 5'-diphosphates, the activity of the enzyme is restricted to the addition of a single nucleotide to the 3' terminus of an oligoribonucleotide. The blocking group can be removed with a mild acid treatment to yield on oligonucleotide product that is then available for a second single addition reaction (46). Subsequently, it has been shown that, of the mixed isomers of the blocked substrate, it is the 2' isomer that is the enzymatically active species (47), and methods for the efficient synthesis and characterization of the four 2'-O-(α-methoxyethyl)nucleoside 5'-diphosphates have been recently documented (48). Although these observations constitute a method for the stepwise synthesis of polynucleotides of specific sequence, some difficulty has been experienced with the back-reaction (phosphorolysis) that is also catalyzed by the enzyme in the presence of one of the products of the synthetic reaction: the phosphate anion. This difficulty has been overcome by including in the synthetic reaction an ancillary enzyme system for the removal of phosphate anion as it is formed (49).

The 2'-O-(α-methoxyethyl) blocking group may also be used to effect the specific joining of oligoribonucleotides through the catalysis of RNA ligase. This application is based on the observation that the *E. coli* phage T4-induced RNA ligase has the capacity to join oligoribonucleotides in the absence of a template (50). Recent studies have shown that the presence of a terminal 2'-O-(α-methoxyethyl) group in an oligonucleotide prevents the enzymatic joining of another oligonucleotide to its 3'-hydroxyl group (51). Since oligonucleotides possessing a non-phosphorylated 5'-hydroxyl group and oligonucleotides containing a terminal 5'-phosphate and a terminal 2'-O-(α-methoxyethyl) group can be readily prepared by the methods described above, these observations constitute a method for the specific joining of two oligonucleotides. For example, initial experiments have shown that the reaction of equimolar amounts of $pA-A_2$-Ame [adenosine tetranucleotide containing a terminal $2'$-O-(α-methoxyethyl) group] and $A-A_2$-A results in the formation of $A-A_6$-Ame as the only product (51). The dependency of this type of specific condensation on the chain lengths and the nucleotide compositions of the two oligonucleotide substrates is currently under investigation.

REFERENCES

1. Gilham, P. T. (1962). An addition reaction specific for uridine and guanosine nucleotides and its application to the modification of ribonuclease action. *J. Amer. Chem. Soc. 84*:687.
2. Ho, N. W. Y. and P. T. Gilham (1967). The reversible chemical modification of uracil, thymine and guanine nucleotides and the modification of the action of ribonuclease on ribonucleic acid. *Biochemistry 6*:3632-3639.
3. Ho, N. W. Y. and P. T. Gilham (1971). Reaction of pseudouridine and inosine with N-cyclohexyl-N'-β-(4-methylmorpholinium)ethylcarbodiimide. *Biochemistry 10*:3651-3657.
4. Lee, J. C., N. W. Y. Ho and P. T. Gilham (1965). Preparation of ribotrinucleotides containing terminal cytidine. *Biochim. Biophys. Acta 95*:503-504.
5. Naylor, R., N. W. Y. Ho and P. T. Gilham (1965). Selective chemical modifications of uridine and pseudouridine in polynucleotides and their effect on the specificities of ribonuclease and phosphodiesterases. *J. Amer. Chem. Soc. 87*:4209-4210.
6. Ho, N. W. Y., T. Uchida, F. Egami and P. T. Gilham (1969). A new cleavage method for the study of nucleotide sequences in RNA. *Cold Spring Harbor Symp. Quant. Biol. 34*:647-650.
7. Ho, N. W. Y. and P. T. Gilham (1974). Action of micrococcal nuclease on chemically modified deoxyribonucleic acid. *Biochemistry 13*:1082-1087.
8. Ho, N. W. Y. and P. T. Gilham, unpublished experiments.
9. Gilham, P. T. (1970). RNA sequence analysis. *Annu. Rev. Biochem. 39*:227-250.
10. Brostoff, S. W. and V. M. Ingram (1967). Chemical modification of yeast alanine tRNA with a radioactive carbodiimide. *Science 158*:666-668.
11. Lee, J. C. and V. M. Ingram (1969). Reaction of 5s RNA with a radioactive carbodiimide. *J. Mol. Biol. 41*:431-441.
12. Chang, S. E., A. R. Cashmore and D. M. Brown (1972). Selective modification of uridine and guanosine residues in tyrosine transfer ribonucleic acid. *J. Mol. Biol. 68*:455-464.
13. Chang, S. E. (1973). Selective modification of cytidine and uridine residues in *Escherichia coli* formylmethionine transfer ribonucleic acid. *J. Mol. Biol. 75*:533-547.
14. Chang, S. E. and D. Ish-Horowicz (1974). Selective modification of cytidine, uridine, guanosine, and pseudouridine residues in *Escherichia coli* leucine transfer ribonucleic acid. *J. Mol. Biol. 84*:375-388.
15. Rhodes, D. (1975). Accessible and inaccessible bases in yeast phenylalanine transfer RNA as studied by chemical modification. *J. Mol. Biol. 94*:449-460.
16. Brown, D. M., M. Fried and A. R. Todd (1953). The determination of nucleotide sequence in polyribonucleotides. *Chem. Ind. 1953*:352-353.
17. Whitfeld, P. R. and R. Markham (1953). Natural configuration of the purine nucleotides in ribonucleic acids. *Nature 171*:1151-1152.

18. Weith, H. L. and P. T. Gilham (1967). Structural analysis of polynucleotides by sequential base elimination. The sequence of the terminal decanucleotide fragment of the ribonucleic acid from bacteriophage f2. *J. Amer. Chem. Soc. 89*:5473-5474.

19. Weith, H. L. and P. T. Gilham (1969). Polynucleotide sequence analysis by sequential base elimination: 3'-terminus of phage Qβ RNA. *Science 166*:1004-1005.

20. Weith, H. L. and P. T. Gilham (1970). Automated polynucleotide sequence analysis. 160th National Meeting of the American Chemical Society, Chicago, Ill. BIOL-18.

21. Schwartz, D. E. and P. T. Gilham (1972). The sequence analysis of polyribonucleotides by stepwise chemical degradation. A method for the introduction of radioactive label into nucleoside fragments after cleavage. *J. Amer. Chem. Soc. 94*:8921-8922.

22. Keith, G. and P. T. Gilham (1974). Stepwise degradation of polynucleotides. *Biochemistry 13*:3601-3606.

23. Weith, H. L., G. T. Asteriadis and P. T. Gilham (1968). Comparison of RNA terminal sequences of phages f2 and Qβ: chemical and sedimentation equilibrium studies. *Science 160*:1459-1460.

24. Rosenberg, M., G. T. Asteriadis, H. L. Weith and P. T. Gilham (1971). New method for isolation of RNA 3'-terminal polynucleotides. Comparison of RNA terminal sequences in three serologically unrelated bacteriophages. *Fed. Proc. 30*:1101.

25. Gilham, P. T. (1971). The covalent binding of nucleotides, polynucleotides, and nucleic acids to cellulose. *Methods Enzymol. 21*:191-197.

26. Gilham, P. T. (1974). Immobilized polynucleotides and nucleic acids. *Immobilized Biochemicals and Affinity Chromatography*. R. B. Dunlap, Ed. Plenum Press, New York.

27. Lee, J. C. and P. T. Gilham (1966). A method for the determination of nucleotide sequences near the terminals of ribonucleic acids of large molecular weight. *J. Amer. Chem. Soc. 88*:5685-5686.

28. Lee, J. C., H. L. Weith and P. T. Gilham (1970). Isolation and characterization of terminal polynucleotide fragments from bacteriophage ribonucleic acids. *Biochemistry 9*:113-118.

29. Lee, J. C. and P. T. Gilham (1965). Determination of terminal sequences in viral and ribosomal ribonucleic acids. *J. Amer. Chem. Soc. 87*:4000-4001.

30. Weith, H. L., J. L. Wiebers and P. T. Gilham (1970). Synthesis of cellulose derivatives containing the dihydroxyboryl group and a study of their capacity to form specific complexes with sugars and nucleic acid components. *Biochemistry 9*:4396-4401.

31. Rosenberg, M., J. L. Wiebers and P. T. Gilham (1972). Studies on the interactions of nucleotides, polynucleotides, and nucleic acids with dihydroxyboryl-substituted celluloses. *Biochemistry 11*:3623-3628.

32. Rosenberg, M. and P. T. Gilham (1971). The isolation of 3'-terminal polynucleotides from RNA molecules. *Biochem. Biophys. Acta 246*:337-340.

33. McCutchan, T. F. and P. T. Gilham (1973). Studies on polynucleotides containing hybrid sequences. Synthesis of oligonucleotides containing thymidine, adenosine, and a single deoxyribonucleotidyl-(3'-5')-ribonucleotide linkage. *Biochemistry 12*:4840-4845.

34. Duncan, R. E. and P. T. Gilham (1975). Isolation of transfer RNA isoacceptors by chromatography on dihydroxyboryl-substituted cellulose, polyacrylamide, and glass. *Anal. Biochem. 66*:532-539.

35. McCutchan, T. F., P. T. Gilham and D. Soll (1975). An improved method for purification of tRNA by chromatography on dihydroxyboryl-substituted cellulose. *Nucleic Acids Res. 2*:853-864.

36. Rosenberg, M. (1974). Isolation and sequence determination of the 3'-terminal regions of isotopically labelled RNA molecules. *Nucleic Acids Res. 1*:653-671.

37. Rosenberg, M., B. deCrombrugghe and S. Weissman (1975). Termination of transcription in bacteriophage Lambda : Heterogeneous 3'-terminal oligo-adenylate additions and the effects of Rho factor. *J. Biol. Chem. 250*:4755-4764.

38. Kramer, R. A., M. Rosenberg and J. A. Steitz (1974). Nucleotide sequences of the 5' and 3' termini of bacteriophage T7 early messenger RNAs synthesized *in vivo*: Evidence for sequence specificity in RNA processing. *J. Mol. Biol. 89*:767-776.

39. Rosenberg, M., R. A. Kramer and J. A. Steitz (1974). T7 early messenger RNAs are the direct products of ribonuclease III cleavage. *J. Mol. Biol. 89*:777-782.

40. Sprinzl, M., K. Scheit, H. Sternbach, F. von der Haar and F. Cramer (1973). *In vitro* incorporation of 2'-deoxyadenosine and 3'-deoxyadenosine into yeast tRNA^{Phe} using tRNA nucleotidyl transferase, and properties of tRNA^{Phe}-C-C-2'dA and tRNA^{Phe}-C-C-3'dA. *Biochem. Biophys. Res. Commun. 51*:881-887.

41. Naylor, R. and P. T. Gilham (1966). Studies on some interactions and reactions of oligonucleotides in aqueous solution. *Biochemistry 5*:2722-2728.

42. Gilham. P. T. (1968). The synthesis of celluloses containing covalently bound nucleotides, polynucleotides, and nucleic acids. *Biochemistry 7*:2809-2813.

43. Gilham, P. T. (1964). The synthesis of polynucleotide-celluloses and their use in the fractionation of polynucleotides. *J. Amer. Chem. Soc. 86*:4982-4985.

44. Gilham, P. T. and W. E. Robinson (1964). The use of polynucleotide-celluloses in sequence studies of nucleic acids. *J. Amer. Chem. Soc. 86*:4985-4989.

45. Ho, N. W. Y. (1975). Esterification of terminal phosphate groups of polynucleotides with sorbitol and its use in the isolation of terminal polynucleotide fragments. *Fed. Proc. 34*:607.

46. Mackey, J. K. and P. T. Gilham (1971). New approach to the synthesis of polyribonucleotides of defined sequence. *Nature 233*:551-553.

47. Bennett, G. N., J. K. Mackey, J. L. Wiebers and P. T. Gilham (1973). 2'-O-(a-Methoxyethyl)nucleoside 5'-diphosphates as "single addition" substrates in the synthesis of specific oligoribonucleotides with polynucleotide phosphorylase. *Biochemistry 12*:3956-3962.

48. Bennett, G. N. and P. T. Gilham (1975). "Single addition" substrates for the synthesis of specific oligoribonucleotides with polynucleotide phosphorylase. Synthesis of 2'-O-(a-methoxyethyl)nucleoside 5'-diphosphates. *Biochemistry 14*:3152-3158.

49. Sninsky, J. J., G. N. Bennett and P. T. Gilham (1974). "Single addition" and "transnucleotidation" reactions catalyzed by polynucleotide phosphorylase. Effect of enzymatic removal of inorganic phosphate during reaction. *Nucleic Acids Res. 1*:1665-1674.

50. Walker, G. C., O. C. Uhlenbeck, E. Bedows and R. I. Gumport (1975). T4-Induced RNA ligase joins single-stranded oligoribonucleotides. *Proc. Nat. Acad. Sci. USA 72*:122-126.

51. Sninsky, J. J. (1975). 2'-O-(a-Methoxyethyl)-oligoribonucleotides as "blocked condensation" substrates in the synthesis of specific oligoribonucleotides with RNA ligase. 11th Midwest Regional Meeting of the American Chemical Society, Carbondale, Ill., Abstract 209.

THE FRACTIONATION AND CHARACTERIZATION OF THE LOW MOLECULAR WEIGHT RNA OF RNA TUMOR VIRUSES

Gordon G. Peters

Department of Physiological Chemistry
University of Wisconsin
Madison, Wisconsin 53706

In general terms, the molecular biologist tries to explain how the structures and chemical properties of macromolecules allow them to interact with one another and perform particular biological functions. This ultimate goal is usually far removed from the initial observation that the component responsible for some measurable biological activity can be classified as being, for example, an RNA molecule. In order to approach a correlation between structure and function, it is necessary to have some means of fractionating the components of the system and of isolating the active molecule in a pure and functional state. In turn, fractionation depends on being able to detect the presence of the molecule of interest either by virtue of its biological activity or by some other means of characterization.

Although nucleic acids are chemically rather uniform, RNA molecules can vary enormously in terms of chain length, nucleotide sequence, extent of methylation, and the presence of secondary or tertiary structure. It is becoming clear that the three dimensional shapes of some RNAs can be remarkably complex and subtle. There is also a great deal of functional variability in RNA reflected by differences in turnover rates, sub-cellular location, and overall abundance — all important considerations when embarking on fractionation schemes. The purpose of this presentation is not a complete survey of all the methods available for RNA fractionation each with its own merits and applications, but rather to discuss the approach we have taken (most of the work presented here was carried out by Drs. Robert Sawyer, Fumio Harada, Gordon Peters and James Dahlberg in Professor Dahlberg's laboratory at the University of Wisconsin) to study the low molecular weight RNAs of RNA tumor viruses. Hopefully, this may serve as an illustration of fractionation and characterization techniques which may prove applicable to "immune RNA."

RNA tumor viruses contain single-stranded RNA as their genetic material, but replicate via a double-stranded DNA intermediate which becomes integrated into the chromosomal DNA of the host cell as the "provirus" (1,2). The initial copying of the genetic information on the viral RNA into DNA is performed by a virion associated RNA-dependent DNA polymerase commonly called the reverse transcriptase. In common with other DNA polymerases, this enzyme will not initiate synthesis *de novo* but requires a nucleic acid primer molecule on which to begin polymerization (3). In the case of Rous sarcoma virus which infects chicken cells and Moloney murine leukemia virus, the primers used during reverse transcription, at least *in vitro*, have been shown to be small RNA molecules (4,5,6). The isolation and characterization of these RNAs is the subject of this presentation.

Rous sarcoma virus (RSV) and murine leukemia virus (MLV) are convenient laboratory viruses in that they can grow and transform cells in tissue culture. Infected cells release viral progeny into the culture fluid from which they can be readily harvested. Tissue culture also permits radioactive labeling of the virus under reasonably controlled conditions, normally with ^{32}P-labelled phosphate, since this isotope can be detected by autoradiography and counted by Cerenkov radiation without having to modify or destroy the sample (7). However, techniques are now available for efficient and quantitative detection of ^{3}H on x-ray film by fluorography, by impregnating a gel or chromatogram with scintillation fluor (8). A further alternative is to label purified RNA with ^{125}I; iodination occurs at the cytosine residues (9).

The RNA obtained by phenol extraction of RSV virions sediments in sucrose gradients as two major size classes (Figure 1): the high molecular weight material sedimenting as 70S and some small RNA in the 4 to 7S size range. This latter RNA is operationally defined as "free" small RNA since it is released from the virus by protein denaturants. This distinguishes it from a second class of small RNA which is associated with the 70S material. The 70S RNA is in fact a complex which can be dissociated by heating into 2 to 4 subunits of 35S RNA and the "associated" small RNAs. The 35S RNA is single-stranded, approximately 10,000 nucleotides long, with poly A at the 3'-end and m^{7}Gppp at the 5'-end, analagous to most eukaryotic mRNAs. Current evidence indicates that the 35S RNA subunits are identical, have a unique sequence, and represent the genomic material of the virus.

The free and associated small RNAs are not single species but rather mixtures of several molecules. Initial indications were that at least some of these resembled transfer RNA (10), but further study required fractionation of the individual small RNAs. This was achieved by electrophoresis in two dimensions in polyacrylamide gels (11,12). Figure 2 shows the results of electrophoresis of total, free and associated small RNA from RSV in the first dimension which is a 10% acrylamide slab gel. Here fractionation appears to be on the basis of chain length in that there is an inverse linear relationship between

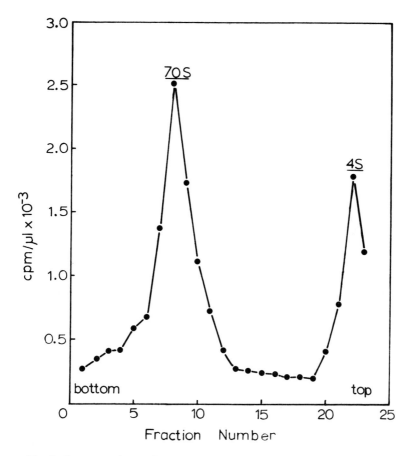

Fig. 1. *Sucrose gradient velocity sedimentation of RSV RNA.*
Total RNA was prepared from purified ^{32}P-labeled virion of the Schmidt-Ruppin-D strain of Rous sarcoma virus, by pronase treatment, phenol-sodium dodecyl sulfate extraction, and ethanol precipitation (12). The RNA was dissolved in 0.1 ml of TSE buffer (0.02M Tris·HCl, pH 7.8; 0.01M NaCl; 0.001M EDTA) and loaded onto a 5 ml linear gradient of 5 to 20% sucrose in TSE containing 0.5% Sarkosyl. Centrifugation was at 200,000 × g for 1h at 4°C in the Beckman Sw50.1 rotor. 0.2 ml fractions were collected and aliquots counted for ^{32}P.

the mobility of the RNA in the gel and the logarithm of its molecular weight (13,14). Clearly this first dimension can resolve 4S RNAs from 5S RNA and shows that the small RNA of RSV includes probably several 4S molecules, some 5S, 6S and 7S RNA. RNA larger than about 8-10S cannot enter the 10% gel and remains trapped at the origin. Acrylamide gel electrophoresis can be used to separate much larger RNA molecules, including 35S, if the acrylamide concentration is reduced (14,15).

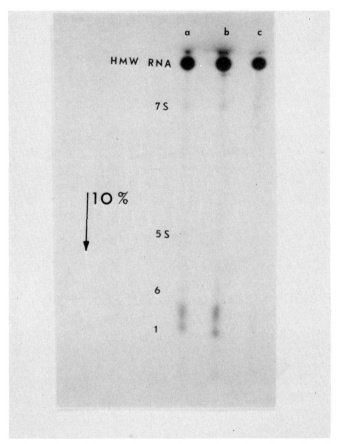

Fig. 2. Separation of RSV small RNAs by one-dimension electrophoresis in a 10% polyacrylamide gel. (a) Free small RNAs - total viral RNA prepared as described in Figure 1 was applied to the gel in molten 1% agarose at 40°C. (b) Total small RNAs - viral RNA was heated to 95°C for 2 min before loading on the gel in molten agarose. (c) Associated small RNAs - the 70S RNA was isolated as in Figure 1 and heated denatured at 95°C for 2 min before loading. Electrophoresis was at 400V for 3 h at 15°C in 0.045M Tris base; 1.4 mM EDTA; 0.045M boric acid; pH 8.3. (12). The figure is an autoradiogram of the ^{32}P-labeledd RNAs in the gel. The location of high molecular weight 7S and 5S RNA are indicated as well as two 4S RNA species numbered 1 and 6.

This single dimension electrophoresis only partially resolves the mixture of 4S RNAs. To fractionate these further, the gel strip from the 10% acrylamide slab is cut out, turned 90°, and embedded across the top of a 20% slab gel during polymerization. The electrophoretic mobility of even the 4S RNA in this high concentration gel is very slow, and resolution appears to be influenced by molecular shape as well as size. Figure 3 shows the typical results for the 4S RNA of RSV which resolves into 10-15 species or at least "spots" on the autoradiograph of the gel. Since discrete "spots" are observed, it is unlikely that the 4S RNA is derived simply by degradation of the high molecular weight RNA during isolation. However, a single spot on the autoradiograph is not sufficient to claim that these represent single molecular species. Two or more different molecules with identical gel mobilities could be superimposed. This point can be readily checked by fingerprint analysis of the RNA.

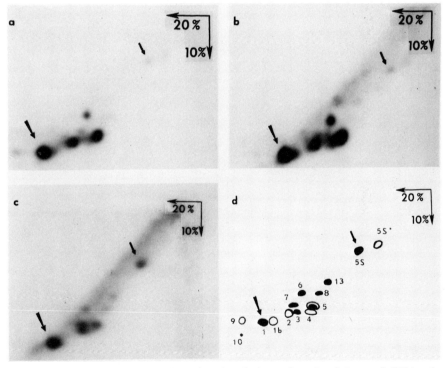

Fig. 3. Two-dimensional polyacrylamide gel electrophoresis of the small RNAs of Rous sarcoma virus.

(a) Free, (b) total, (c) associated small RNAs, and (d) schematic drawing indicating location and numbering of individual spots in the two dimensional gel. Strips of the first dimension gel indicated in Figure 2 were used as the origins for the second dimension gels. These were 20% acrylamide, electrophoresed at 400V for 16h at 15°C in the buffer described in Figure 2. The arrows indicate 5S RNA and spot-1, a major 4S RNA species. Closed circles in the schematic drawing indicate species found both free and associated with the 70S RNA while open circles show species found only in the free fraction.

The segment of gel corresponding to a particular radioactive spot can be cut out, crushed and the RNA eluted (12). Digestion of the RNA with ribonuclease T_1 generates a collection of oligonucleotides all with -Gp at the 3'-end. The number and base sequence of these oligonucleotides depends directly on the primary sequence of the RNA under study. The oligonucleotides can be analyzed by two dimensional electrophoresis using the techniques described by Sanger and his colleagues (16). Normally the digest is fractionated by electrophoresis at pH 3.5 on a strip of cellulose acetate, followed by a second dimension electrophoresis in 7% formic acid on DEAE-cellulose paper, or alternatively chromatography on thin layers of polyethylene imine cellulose. The mobility of each oligonucleotide depends on its base composition and sequence. For instance, the presence of cytosine residues slows the mobility at pH 3.5, while uracil residues retard during electrophoresis in 7% formate. Families of nucleotides – such as CG, C_2G, C_3G and so on – therefore have quite predictable mobilities; and the complete set of possible oligonucleotides all with only one guanine at the 3'-end fall on a readily interpretible pattern (17).

The pattern of oligonucleotide spots obtained by fingerprinting is characteristic of the nucleotide sequence of the particular RNA. In addition, the number of spots serves as an indication of purity. For example, a 4S RNA about 60-80 nucleotides long like tRNA would be expected to contain roughly 15-20 guanine residues and therefore generate an equivalent number of T_1 oligonucleotides. If substantially more products are detected in the digest, then the RNA is very likely a mixture of more than one species. Many of the 4S RNA spots from RSV gave pure fingerprints, an example being the RNA numbered as "spot-1" in Figure 3. Figure 4 gives a schematic representation of the T_1 and pancreatic ribonuclease fingerprints of spot-1 RNA.

Fingerprint analysis is a rapid and simple means of structural characterization and allows one to decide whether RNAs isolated from two different sources are identical at least at the level of primary sequence. Figure 5 shows a comparison of the 4S RNA from RSV and uninfected chick cells fractionated by two dimensional gels. A molecule with the same mobility as spot-1 appears to be present in normal chicken cells and its identity was confirmed by fingerprint analysis (18). Obviously the pattern of RSV 4S RNA is much simpler than that of cellular 4S RNA, but the identity of some of the molecules suggests that the viral RNAs may represent a subset of the host cell's transfer RNA. However, fingerprint analysis alone gives no indication of biological function, only of structure.

Spot-1 RNA is clearly the most abundant 4S molecule in RSV occurring in between 0.5 and 1 copies per 35S subunit in the associated fraction and 2 to 4 copies free (12). Many of the other small RNAs are present in less than one copy per virion. However, it is clear from Figure 5 that spot-1 is only a

FRACTIONATION AND CHARACTERIZATION OF LOW MOLECULAR WEIGHT 19

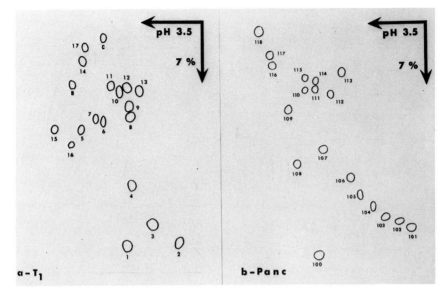

Fig. 4. T_1 and pancreatic ribonuclease fingerprints of spot-1 RNA.
^{32}P-labeled spot-1 RNA was eluted from the gel, concentrated by ethanol precipitation, and digested with either T_1 or pancreatic RNase for fingerprinting (12,16). The first dimension, right to left in the figure, was electrophoresis on cellulose acetate at pH 3.5 in pyridine acetate in 6M urea. The second dimension, top to bottom, was electrophoresis on DEAE-cellulose paper in 7% formic acid. The figure is a schematic illustration of the autoradiographs obtained and the numbering system used for identification of the individual oligonucleotides.

minor component of cellular 4S RNA. This makes it appear unlikely that the virus acquires spot-1 simply by *random* packaging of the host cell's RNA. Further studies on the structure and biological activity of spot-1 confirmed that it is indeed a transfer RNA. These included determining the complete nucleotide sequence. This showed the presence of several unusual and modified bases and could be arranged into the clover-leaf shape characteristic of a tRNA (19, and Figure 6). The anti-codon is -Cm-C-A implying that it should be chargeable with the amino acid tryptophan whose codon is U-G-G. This was verified by using radioactive spot-1 RNA as a tracer to isolate an equivalent molecule from chicken liver tRNA by chromatography on a series of ion exchange columns (19). Figure 7 shows a typical elution profile from DEAE-Sephadex. When amino-acid charging assays were performed on the fractions from such a column, it was observed that the ^{32}P-marker co-eluted with the peak of acceptance for tryptophan.

Recently a very simple and useful test for a tRNA was developed in our laboratory (20). When tRNA^{Trp} is treated with the single strand specific nuclease S_1 from *Aspergillus oryziae* at 37°C, the RNA is completely

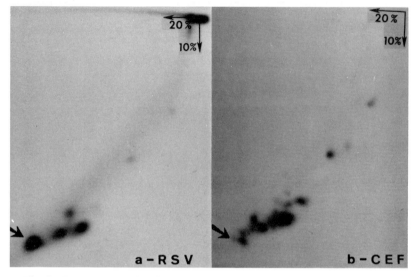

Fig. 5. *Two dimensional polyacrylamide gel fractionation of chicken embryo fibroblast small RNAs.*

(a) RSV small RNAs as in Figure 3; (b) 4S RNA from uninfected chick cells. ^{32}P-RNA was prepared from chicken embryo fibroblast cells by phenolsodium dodecyl sulfate extraction and ethanol precipitation (18). The small RNAs were fractionated by gel electrophoresis as described in the legend to Figure 3. The arrows indicate viral spot-1 and a cellular 4S RNA with the same gel mobility and RNase T_1 fingerprint as spot-1.

degraded; but at 20°C the nuclease cuts the molecule at the anti-codon loop and at the exposed single-stranded -C-C-A end as indicated in Figure 6. Analysis of the products on a two dimensional gel run for half the normal time gives the results shown in Figure 7. The spot marked (a) represents intact tRNA^{Trp} but with the -C-C-A end removed; (b) and (c) are both the 5'-"half" of the tRNA, differing only in the position at which the nuclease has cut, while (d) is the 3'-fragment (20). This means that it is possible to identify which oligonucleotide on a fingerprint corresponds to the anticodon by doing fingerprints with and without pretreatment with S_1 nuclease. The advantage is that only one oligonucleotide need be fully sequenced to deduce the anticodon of the tRNA.

What then is the biological role, if any, of the normal cellular tRNA^{Trp} in the life cycle of RSV? The natural template for the viral RNA-dependent DNA polymerase is the 70S RNA complex, so that it is conceivable that one or more of the associated small RNAs may serve as primer for DNA synthesis. The ability of 70S RNA to serve as template for reverse transcription is abolished by heating the complex (21). Heating to 65°C dissociates the com-

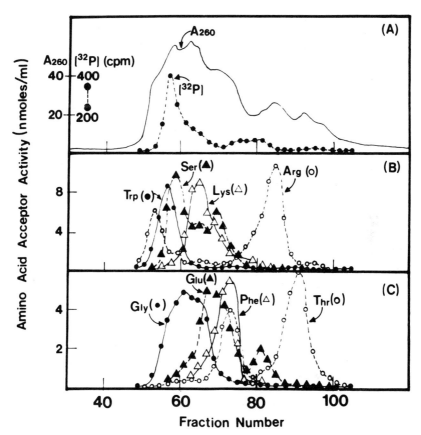

Fig. 6. *Fractionation of chicken liver tRNA by DEAE-Sephadex A-50 column chromatography at pH 4.0.*

450 mg of total chicken liver tRNA were mixed with 8,000 cpm of purified ^{32}P-labeled spot-1 RNA, applied to a column (1 × 100 cm) of DEAE-Sephadex A-50, and eluted with a 1,000 ml linear gradient of 0.5M to 0.75M NaCl in 0.01M $MgCl_2$; 0.02M sodium acetate, pH 4.0 (19). Fractions of 4.0 ml were collected at a flow rate of 7 ml per hour and aliquots rested for amino acid acceptor activity (19).

(A) Optical density and ^{32}P elution profiles; (B), (C) aminoacylation with various amino acids.

plex into subunits and releases the majority of the associated small RNAs but does not abolish template activity. If the high molecular weight RNA recovered at 65°C is heated further to 80°C, template activity is lost and a single 4S RNA species is released having both the gel mobility and fingerprint of tRNA^{Trp} (22). This was suggestive evidence that tRNA^{Trp} may be the primer. Confirmation that this was the case, at least *in vitro*, was obtained by carrying out limited DNA synthesis on the 70S RNA. Virions disrupted with a

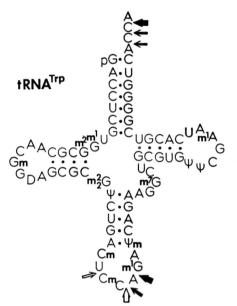

Fig. 7. *Nucleotide sequence of tRNA^{Trp}.*

The complete nucleotide sequence of spot-1 RNA was determined as described (19). The arrows indicate position of cleavage by nuclease S_1 (20).

non-ionic detergent are capable of incorporating exogenously supplied deoxynucleotide triphosphates into DNA using the endogenous 70S RNA as template. By supplying such a system with the appropriate dNTP (in this case dATP) labeled in the α-phosphate with ^{32}P, it was possible to radioactively tag any primer molecules at their 3'-ends. With RSV, a single RNA molecule with the same mobility in gel electrophoresis as spot-1 acquires a ^{32}P-dAMP label in such an endogenous DNA synthesis reaction (22). However, as discussed earlier, similar electrophoretic mobility cannot be taken as proof of identity; and since the label was only attached to the 3'-end, no fingerprint analysis was possible. When ^{3}H-dATP was used as the tag using uniformly labeled ^{32}P-virions to direct the DNA synthesis, the 70S associated tRNA^{Trp} contained ^{3}H attached to the 3'-terminal oligonucleotide produced by T_1 digestion (22).

The situation in RSV is not unique. Recently we have been able to show that the Moloney strain of MLV has an analogous but different set of host cell transfer RNA in the virion. One of these is more tightly associated with the 35S RNA and can be tagged with dAMP in a limited DNA synthesis reaction (6). It has the nucleotide sequence and amino acid acceptor activity consistent with it being a normal cellular tRNA for proline (23). Neither the tRNA^{Trp} or tRNA^{Pro} have any unique properties that make them appear different from other transfer RNAs except that they both have the sequence -G-ψ-ψ-C in place of the normal -G-T-ψ-C in the so-called pseudouridine loop.

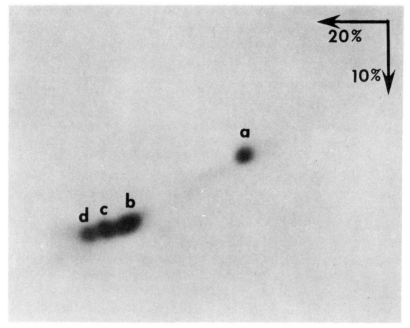

Fig. 8. *Two dimensional polyacrylamide gel electrophoresis of the products of S_1 nuclease digestion of ^{32}P-tRNA^{Trp} at $20°C$.*

^{32}P-tRNA^{Trp} *from chicken cells was digested with S_1 nuclease for 1h at $20°C$ (20) and electrophoresed in two dimensions on polyacrylamide gels as described in Figure 3, except that the first dimension was at 200V for 3h and the second at 200V for 16h. The products a,b,c and d are identified in the text.*

Work is currently in progress to determine the significance of this sequence, if any; the interaction between primer and 35S and between primer and reverse transcriptase; and the mechanism whereby these RNA tumor viruses select the subset of host tRNAs to be included in the virion particle.

The aim of this paper has been to introduce a number of techniques for the fractionation and characterization of RNA and to illustrate their use with particular reference to tumor virus RNA. Reference has been made to articles of general interest and those which cover pertinent practical details, rather than attempting a comprehensive survey of the field. Although the discussion was limited to the analysis of low molecular weight RNAs, most of the techniques described are generally applicable — possibly with slight modification — to most size and functional classes of RNA, hopefully including "immune RNA."

REFERENCES

1. Temin, H. M. (1964). Nature of the provirus of Rous sarcoma. *Nat. Cancer Inst. Monogr. 17*:557-570.
2. Temin, H. M. (1971). Mechanism of cell transformation by RNA tumor viruses. *Ann. Rev. Microbiol. 25*:609-628.
3. Temin, H. M. and D. Baltimore (1972). RNA-directed DNA synthesis and RNA tumor viruses. *Advances in Virus Research 17*:129-186.
4. Bishop, J. M., C. Tsan-Deng, A. J. Faras, H. M. Goodman, W. E. Levinson, J. M. Taylor and H. E. Varmus (1973). Transcription of the Rous sarcoma virus genome by RNA-directed DNA polymerase. *Virus Research*. C. F. Fox, Ed. Academic Press, New York.
5. Dahlberg, J. E., F. Harada and R. C. Sawyer (1974). Structure and properties of an RNA primer for initiation of Rous sarcoma virus DNA synthesis in vitro. *Cold Spring Harbor Symp. Quant. Biol. 39*:925-932.
6. Peters, G., F. Harada, J. E. Dahlberg, W. A. Haseltine, A. Panet and D. Baltimore, manuscript in preparation.
7. Clausen, T. (1968). Measurement of ^{32}P activity in a liquid scintillation counter without use of scintillator. *Anal. Biochem. 22*:70-73.
8. Laskey, R. A. and A. D. Mills (1975). Quantitative film detection of ^3H and ^{14}C in polyacrylamide gels by fluorography. *Eur. J. Biochem. 56*:335-341.
9. Prensky, W. The radioiodination of RNA and DNA to high specific activities. *Methods in Cell Biology XIII*, in press. E. M. Prescott, Ed. Academic Press, New York.
10. Erikson, E. and R. L. Erikson (1970). Isolation of amino acid acceptor RNA from purified avian myeloblastosis virus. *J. Mol. Biol. 52*:387-390.
11. Ikemura, T. and J. E. Dahlberg (1973). Small RNAs of *E. coli*. I. Characterization by polyacrylamide gel electrophoresis and fingerprint analysis. *J. Biol. Chem. 248*:5024-5032.
12. Sawyer, R. C. and J. E. Dahlberg (1973). Small RNAs of Rous sarcoma virus: characterization by two-dimensional polyacrylamide gel electrophoresis and fingerprint analysis. *J. Virol. 12*:1226-1237.
13. Loening, U. E. (1969). The determination of the molecular weight of ribonucleic acid by polyacrylamide-gel electrophoresis. *Biochem. J. 113*:131-138.
14. Peacock, A. C. and C. W. Dingman (1968). Molecular weight estimation and separation of ribonucleic acid by electrophoresis in agaroseacrylamide composite gels. *Biochemistry 7*:668-674.
15. Duesberg, P. H. and P. K. Vogt (1970). Differences between the ribonucleic acids of transforming and nontransforming avian tumor viruses. *Proc. Nat. Acad. Sci. 67*:1673-1680.
16. Sanger, F., G. G. Brownlee and B. G. Barrell (1965). A two dimensional fractionation procedure for radioactive nucleotides. *J. Mol. Biol. 13*:373-398.
17. Barrell, B. G. (1971). Fractionation and sequence analysis of radioactive nucleotides. *Procedures in Nucleic Acid Research*, Vol. 2. S. L. Cantoni and D. R. Davies, Eds. Harper and Row, New York.
18. Sawyer, R. C., F. Harada and J. E. Dahlberg (1974). Virion-associated RNA primer for Rous sarcoma virus DNA synthesis: Isolation from uninfected cells. *J. Virol. 13*:1302-1311.
19. Harada, F., R. C. Sawyer and J. E. Dahlberg (1975). A primer ribonucleic acid for initiation of in vitro Rous sarcoma virus deoxyribonucleic acid synthesis. *J. Biol. Chem. 250*:3487-3497.

20. Harada, F. and J. E. Dahlberg (1975). Specific cleavage of tRNA by nuclease S_1. *Nucleic Acids Research* 2:865-871.

21. Canaani, E. and P. Duesberg (1972). Role of subunits of 60 to 70S avian tumor virus ribonucleic acid in its template activity for the viral deoxyribonucleic acid polymerase. *J. Virol.* 10:23-31.

22. Dahlberg, J. E., R. C. Sawyer, J. M. Taylor, A. J. Faras, W. E. Levinson, H. M. Goodman and J. M. Bishop (1974). Transcription of DNA from the 70S RNA of Rous sarcoma virus. I. Identification of a specific 4S RNA which serves as primer. *J. Virol.* 13:1126-1133.

23. Harada, F., G. Peters and J. E. Dahlberg, unpublished results.

DETECTION OF GLOBIN MESSENGER RNA SEQUENCES BY MOLECULAR HYBRIDIZATION*

Jeffrey Ross

McArdle Laboratory for Cancer Research
University of Wisconsin
Madison, Wisconsin 53706

INTRODUCTION

The recognition of a unique RNA molecule obviously depends on special properties of that molecule that distinguish it from others. For ribosomal or tRNA's that are abundant, size or amino acid acceptor activity are useful special properties. For a mRNA molecule that represents a minor fraction of the total cell RNA and of the total message population, other properties such as translation *in vitro* (1,2), unique nucleotide sequence (3), or molecular hybridization to complementary nucleic acids can be exploited.

The aim of this paper is to describe two techniques for mRNA detection and quantitation using molecular hybridization. These techniques are applied here to mRNA molecules that can be isolated in relatively pure form. Purified mRNA is utilized as a template to synthesize DNA complementary to it with RNA tumor virus RNA-dependent DNA polymerase. This complementary DNA (cDNA) hybridizes specifically to its mRNA template and thus serves as a specific probe for mRNA detection and quantitation. The use of mouse globin cDNA probes for detecting unlabeled as well as radioactive globin mRNA will be illustrated here.

MATERIALS AND METHODS

Methods for the purification of mouse globin mRNA, synthesis of ^{3}H-cDNA, isolation of cellular RNA — either total or cytoplasmic — and hybridization assays using ^{3}H-cDNA were performed as previously described (4). Synthesis of large quantities of unlabeled globin cDNA and hybridization assays for radioactive globin mRNA were as described (5).

Mouse erythroleukemic cells (6) were cultured in Ham F12 growth medium containing 10% unheated fetal calf serum (GIBCO). Fourteen day old

---

*This research was supported by Grant CA 07175 from the National Institutes of Health.

mouse fetal livers were dissected from timed pregnant mice (CD1) obtained from Charles River Mouse Farms (North Wilmington, Ma.), and the cells were cultured as described (5). Radioactive cellular RNA was prepared from cells incubated with uridine-5-^{3}H (New England Nuclear, 30-40 Ci/mmole).

To prepare partially purified ^{3}H-globin mRNA, 14 day old mouse fetal liver cells (1.4×10^8) were cultured for 30 min in 10 ml of growth medium. At that time ^{3}H-uridine (260 μCi/ml) was added. After 3 hr in ^{3}H-uridine, additional ^{3}H-uridine was added to bring the final concentration to 460 μCi/ml. After 4 more hr cells were harvested, and RNA was extracted and fractionated on sucrose gradients (6). The 3 peak tubes at the 9S region were pooled and used as a source of ^{3}H-globin mRNA.

RESULTS
Hybridization Techniques

Methods for preparation of radioactive DNA complementary to globin mRNA's (cDNA) have been described (7-9), and cDNA's to other messages such as immunoglobulin light chain (10) and ovalbumin (11) have been synthesized in a similar fashion. In each case it has been established by a number of criteria that hybridization of cDNA to its mRNA is highly specific. For example, rabbit globin cDNA will not form hybrids with duck globin message (9). Moreover, it is possible to prepare ^{3}H-cDNA of high specific radioactivity (5-10 $\times$ 10^6 cpm/μg), so that it can serve as an extremely sensitive reagent. In our hands, ^{3}H-globin cDNA can detect 10-20 pg of globin message (unpublished data). This is much greater sensitivity than that achieved by conventional *in vitro* protein synthesizing systems (2,12), although slightly less than that of the oocyte injection system (1). In summary, cDNA probes can be highly specific and extremely sensitive reagents for detection of specific mRNAs.

An additional advantage of cDNA hybridization probes is the rapidity and convenience of the assays. The most frequently used technique for hybrid detection (Figure 1) exploits nucleases that recognize only single-stranded molecules. RNA is incubated with ^{3}H-cDNA to form DNA-RNA hybrids (step I). Molecules of ^{3}H-cDNA that have not hybridized will be hydrolyzed by a single-strand specific nuclease such as S_1 (13) (step II) into acid-soluble nucleotides and oligonucleotides. Depending on the amount of globin message in an RNA preparation (see below), annealing reactions and S_1 nuclease analyses can be performed in a few hours. Other methods for hybrid detection rely on separation of annealed from non-annealed ^{3}H-cDNA by hydroxyapatite chromatography (14) or by equilibrium density centrifugation (8).

These hybridization techniques have been utilized in a number of laboratories to determine the amount of globin mRNA in erythroid cells at various stages in development (4,6,15,16). However, this assay detects accumulated

HYBRIDIZATION ASSAY FOR RADIOACTIVE
GLOBIN mRNA USING UNLABELED GLOBIN cDNA

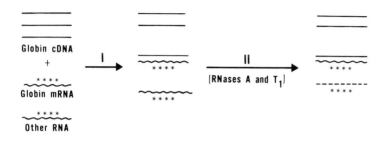

mRNA under steady-state conditions, and thus provides little or no information concerning the kinetics of mRNA metabolism. In order to study RNA kinetics, the hybridization assay must be such as to recognize only newly synthesized molecules under precise cell labeling conditions.

For this purpose a second assay has been devised to detect radioactive globin mRNA (Figure 2). Here, excess unlabeled globin cDNA is incubated with radioactive cellular RNA (step I). The reaction mixture is then treated with ribonucleases to hydrolyze non-annealed RNA sequences to acid-soluble nucleotides and oligonucleotides (step II). Two major features of this assay distinguish it from the ^{3}H-cDNA technique shown in Figure 1: (a) Excess globin cDNA must be used, because most preparations of radioactive erythroid cell RNA will contain unlabeled as well as labeled globin message, both of which will anneal with the cDNA. Therefore, to ensure that *all* of the ^{3}H-globin message has hybridized, the amount of cDNA in the hybridization reaction must be greater than the *total* amount of globin mRNA. (b) The assay for radioactive globin mRNA is much less sensitive than that for unlabeled message. It is possible to detect 20 pg of unlabeled globin message in 20 µg of total cell RNA (.0001%), because of the high specific radioactivity of the ^{3}H-cDNA. However, since ribonucleases usually fail to hydrolyze *all* of the non-annealed ^{3}H-RNA, there is a "background" of ^{3}H-RNA in the excess cDNA experiments that limits the detection sensitivity (e.g., see Table II). In this case "background" represents the percentage of the total input ^{3}H-RNA

HYBRIDIZATION ASSAY FOR UNLABELED
GLOBIN mRNA USING ^{3}H-GLOBIN cDNA

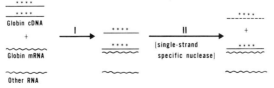

dpm that are RNase-resistant in the absence of added cDNA. If the background is 0.1% it is not possible to be confident of percentages of hybridization (in the presence of cDNA) much less than .05% above background, which is 500-fold lower than the sensitivity with ^{3}H-cDNA.

Detection and Quantitation of Globin mRNA Sequences in Erythroid Cells

The rate of nucleic acid hybridization is proportional to the concentrations of the reactants; the extent of hybridization of a reactant at a given concentration is proportional to the duration of the annealing reaction. These parameters, concentration and time, have been used to formulate a convenient system for quantitating hybridization reactions (17). The product of the concentrations of the reactants × time of annealing ($C_o t$) is compared for different RNA preparations. The more concentrated a reactant (e.g., globin mRNA), the faster the annealing rate. For example, consider two RNA preparations, one containing 100%, the other 1%, globin message. If the same total amount of RNA of each sample is used in the hybridization reaction, the more concentrated sample will hybridize approximately 100-fold faster than the less concentrated sample.

Many laboratories have utilized "$C_o t$ analysis" to quantitate the amount of globin mRNA present in erythroleukemic tissue culture cells undergoing a form of erythrodifferentiation (4,15,16). Another example is shown in Figure 3. To determine the average amount of globin message in 14 day mouse fetal liver cells, most of which are erythroid (18), total cell RNA or purified globin mRNA was incubated with ^{3}H-cDNA and the percentage of cDNA hybridized determined by S_1 nuclease digestion at different $C_o t$ values (see Figure 1). $C_o t$ values were determined as described (17). Table 1 shows how the globin message content per cell (average) can be calculated from these data.

The $C_o t$ ½ is the $C_o t$ value at which the reaction is half-complete; in this case $C_o t$ ½ represents 50% hybridization. The fraction of globin mRNA is determined by dividing the $C_o t$ ½ of purified message by that of the RNA sample. Thus, if a mRNA standard whose purity is close to 100% is available, it is possible to quantitate the amount of that message in any RNA preparation.

An example of hybridization analysis for radioactive globin mRNA is shown in Table II. Total cytoplasmic RNA was isolated from cells cultured in ^{3}H-uridine (200 µCi/ml) for 2 hours. When this RNA is incubated with all reactants except cDNA and then treated with RNases A and T_1, .08 - .10% remains acid-precipitable (column 3). As discussed above, this "background" limits are sensitivity of this assay. Fortunately, primary cultures of mouse fetal liver cells synthesize large quantities (greater than 13-fold over background) of ^{3}H-RNA that becomes RNase-resistant in the presence of cDNA (column 4). As a control, a similar hybridization experiment was performed

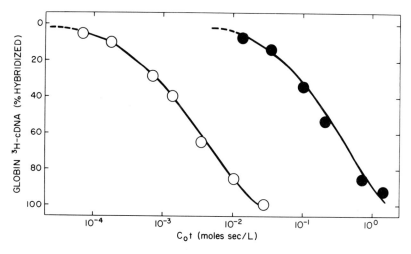

Fig. 3. *Hybridization of Mouse Fetal Liver RNA to ^{3}H-Globin cDNA.*

Mouse fetal liver cells from 14 day old embryos were cultured at a concentration of 8×10^6/ml for 2 hr without erythropoietin. Cells were harvested and total cellular RNA was extracted. This unlabeled RNA was then incubated with ^{3}H-globin cDNA (650 cpm) at 65°C for 2 hr under conditions described previously (4), except that formamide was omitted and reactions contained 0.2% (w/v) sodium dodecyl sulfate. The percentage of the cDNA that hybridized was determined by the S_1 nuclease assay, 1% of the cDNA being nuclease resistant in the absence of RNA. Appropriate controls were included to demonstrate that nuclease resistance was due to hybridization rather than to enzyme inhibition (4).

○──○ *Purified mouse globin mRNA*
●──● *Mouse fetal liver total cell RNA*

with total cytoplasmic RNA from T3-C12 cells, an erythroid precursor cell line (19,20) that contains little or no globin message (4,6). As shown in Table II, there is no detectable ^{3}H-RNA from T3-C12 cell cytoplasm that hybridizes to globin cDNA. This result indicates that hybridization occurs only with RNA from cells actively making hemoglobin.

Based on specificity studies with the ^{3}H-cDNA probe (7-9,14) and considering the purity of the globin mRNA used to make the cDNA (7), it

TABLE 1.
Globin mRNA Content of 14 Day Old Mouse Fetal Liver Cells

RNA	C_0t ½	Globin mRNA (%)	Total RNA Per Cell (pg)	Globin mRNA Per Cell (pg)
Purified Globin mRNA	2.2×10^{-3}	100	–	–
Mouse Fetal Liver	2.3×10^{-1}	1	1.8	1.8×10^{-2}

TABLE 2.
Hybridization Analysis of Radioactive RNA

Cells	Input DPM ($\times 10^{-3}$)	RNase Resistant DPM		Percent Hybridization
		−cDNA	+cDNA	
Mouse Fetal Liver (14 day)	59	48 (.08)	630 (1.07)	.99
Erythroleukemia (T3-C12)	520	520 (.10)	500 (.09)	0

seemed likely that the cDNA-dependent RNase-resistant material observed in Table II was globin message. Nevertheless, a hybridization competition experiment was performed as a control to provide further assurance to this point. Excess *unlabeled* globin mRNA was added to annealing reactions containing 14.4 ng of cDNA and a total of 2 ng of partially purified, 9S ^{3}H-globin mRNA. Figure 4 shows that hybridization of ^{3}H-RNA was blocked by 95% with a 20-fold excess of unlabeled globin message. In contrast, 30 µg of

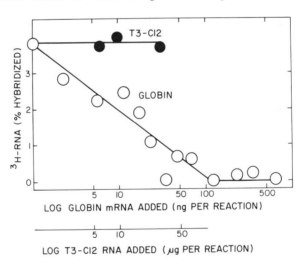

Fig. 4. Hybridization of ^{3}H-RNA from Mouse Fetal Liver Cells: Competition by Unlabeled RNA.

Approximately 2 ng (7400 dpm) of partially purified ^{3}H-globin mRNA (see Materials and Methods) was incubated with 14.4 ng of cDNA in the presence or absence of unlabeled T3-C12 cytoplasmic RNA or 9S mouse globin mRNA. All RNA was added to the reaction tubes prior to addition of cDNA. The background for this experiment was 13 dpm (.18%).

    ○———○ Mouse globin mRNA
    ●———● T3-C12 RNA

T3-C12 RNA failed to inhibit hybridization. Therefore, the hybridization reaction is globin-specific.

Another control experiment was performed to determine how much cDNA was required in order to achieve sufficient excess over globin message. As noted above, an absolute requirement for comparative quantitation of ^{3}H-globin message levels is that cDNA be in excess over the total globin mRNA. For this experiment increasing amounts of cDNA were added to 5 ng of partially purified, radioactive globin mRNA. As shown in Figure 5, the level of hybridization reached a plateau of 4.2% with 30 ng of cDNA and failed to increase further in reactions containing as much as 60 ng. Therefore, a 6-fold cDNA excess completely saturated all globin message in this experiment (see Discussion). Saturation kinetics with a 6-fold cDNA excess were also observed with total cytoplasmic fetal liver ^{3}H-RNA (unpublished data); this result indicated that the presence of other cytoplasmic RNA molecules in the annealing reaction did not affect the relative amount of cDNA required to achieve excess.

DISCUSSION

The hybridization technique with radioactive cDNA has been exploited with purified mRNAs to answer a variety of questions concerning the number of mRNA molecules and of specific genes in a number of cells under a variety of conditions (see above). In each instance the mRNA was polyadenylated. However, the technique will probably not be limited to adenylated RNA species. For example, nonadenylated histone mRNA to which poly (A) has been added enzymatically can function as a template for cDNA synthesis (21). Moreover, it should be pointed out that cDNA can be prepared from the total polyadenylated RNA of a particular cell. This cDNA can then be used as a hybridization probe to provide information about the complexity and frequency classes of polyadenylated mRNA in the cell (22).

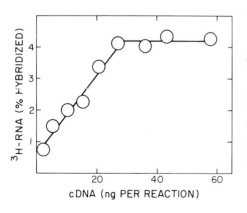

Fig. 5. *Hybridization of ^{3}H-RNA from Mouse Fetal Liver Cells: cDNA Concentration Curve.*

Partially purified ^{3}H-globin mRNA (Materials and Methods) containing 23,000 dpm and a total of 5 ng of globin message was incubated for 1.5 hr with the indicated amount of cDNA. The background for this experiment was 35 dpm (.15%).

Complementary RNA probes have also been used to detect and quantitate RNA levels. ^{3}H-RNA complementary to duck globin message has been prepared with *M. lysodeikticus* RNA polymerase (23). This RNA hybridizes to duck globin message. Recently a technique utilizing Qβ replicase to transcribe a variety of RNA's with apparent fidelity has been introduced (24). These techniques offer the advantage that the RNA being transcribed need not be adenylated. However, cRNA probes are less satisfactory than cDNA probes for hybridization assays because cDNA can be more easily separated from its RNA primer, DNA is more stable than RNA, and RNA-RNA hybridization reactions are less well understood than DNA-RNA reactions (25).

The experiments reported here indicate that globin cDNA can be used in excess to detect radioactively labeled globin mRNA from mouse fetal liver cell cytoplasm, even in the presence of a relatively large amount of endogenous unlabeled globin message. Under the conditions described here, the assay is apparently specific for globin mRNA sequences, and there is no hybridization to ^{3}H-RNA from immature erythroid cells that do not synthesize hemoglobin. This assay has also been utilized to detect cytoplasmic ^{3}H-globin mRNA synthesized by erythroleukemic tissue culture cells treated with solvents that induce the cells to "differentiate" (6,20). Thus, approximately .14% of the total cytoplasmic ^{3}H-RNA from cells treated with the inducer N-methylpyrrolidinone (26) is globin message (5). The ability to detect a newly synthesized, radioactive eukaryotic mRNA should permit us to ask a variety of questions concerning mRNA metabolism that could not be adequately approached with the assay for unlabeled message.

The major disadvantage of this cDNA excess assay, its low sensitivity as compared to the assay for unlabeled message, has been discussed. An additional factor, demonstrated in Figure 5, should also be emphasized. The data in Figure 5 demonstrate that a 6-fold excess of cDNA is required to saturate a preparation of ^{3}H-globin mRNA. While there may be several factors contributing to this result, perhaps the most important is related to the fact that not all cDNA molecules are as large or nearly as large as the message (8). Hybrids formed between mRNA and relatively small DNA molecules would thus contain single-stranded RNA regions that would be RNase sensitive. Addition of larger quantities of cDNA would provide a larger number of long molecules capable of protecting a greater fraction of each mRNA molecule from nuclease digestion.

REFERENCES

1. Moar, V. A., J. B. Gurdon, C. D. Lane and G. Marbaix (1971). Translational capacity of living frog eggs and oocytes, as judged by messenger RNA injection. *J. Mol. Biol.* 61:93-103.

2. Roberts, B. E. and B. M. Paterson (1973). Efficient translation of tobacco mosaic virus RNA and rabbit globin 9S RNA in a cell-free system from commercial wheat germ. *Proc. Nat. Acad. Sci. USA 70*:2330-2334.

3. Suzuki, Y. and D. D. Brown (1972). Isolation and identification of the messenger RNA for silk fibroin from *Bombyx mori. J. Mol. Biol. 63*:409-429.

4. Ross, J., J. Gielen, S. Packman, Y. Ikawa and P. Leder (1974). Globin gene expression in cultured erythroleukemic cells. *J. Mol. Biol. 87*:697-714.

5. Ross, J. and L. P. Beach. Newly synthesized globin mRNA in mouse fetal liver and Friend leukemia cells. Submitted for publication.

6. Ross, J., Y. Ikawa and P. Leder (1972). Globin messenger RNA induction during erythroid differentiation of cultured leukemia cells. *Proc. Nat. Acad. Sci. USA 69*:3620-3623.

7. Kacian, D. L., S. Spiegelman, A. Bank, M. Terada, S. Metafora, L. Dow and P. Marks (1972). In vitro synthesis of DNA components of human genes for globins. *Nature New Biol. 235*:167-169.

8. Ross, J., H. Aviv, E. Scolnick and P. Leder (1972). In vitro synthesis of DNA complementary to purified rabbit globin mRNA. *Proc. Nat. Acad. Sci. USA 69*:264-268.

9. Verma, I. M., G. F. Temple, H. Fan and D. Baltimore (1972). In vitro synthesis of DNA complementary to rabbit globin mRNA. *Nature New Biol. 235*:163-167.

10. Aviv, H., S. Packman, D. Swan, J. Ross and P. Leder (1973). In vitro synthesis of DNA complementary to mRNA derived from a light chain-producing myeloma tumor. *Nature 241*:174-176.

11. McKnight, G. S. and R. T. Schimke, (1974). Ovalbumin messenger RNA: Evidence that the initial product of transcription is the same size as polysomal ovalbumin messenger. *Proc. Nat. Acad. Sci. USA 71*:4327-4331.

12. Mathews, M. B. (1973). Mammalian messenger RNA. *Essays in Biochem. 9*:59-102.

13. Ando, T. (1966). A nuclease specific for heat-denatured DNA isolated from a product of *Aspergillus oryzae. Biochim. Biophys. Acta 114*:158-168.

14. Ross, J., H. Aviv and P. Leder (1973). Characterization of synthetic rabbit globin DNA. *Arch. Biochem. Biophys. 158*:494-502.

15. Preisler, H. D., D. Housman, W. Scher and C. Friend (1973). Effects of 5-bromo-2'-deoxyuridine on production of globin messenger RNA in dimethyl sulfoxide-stimulated Friend leukemia cells. *Proc. Nat. Acad. Sci. USA 70*:2956-2959.

16. Gilmour, R. S., P. R. Harrison, J. D. Windass, N. A. Affara and J. Paul (1974). Globin messenger RNA synthesis and processing during hemoglobin induction in Friend cells. I. Evidence for transcriptional control in clone M2. *Cell Differentiation 3*:9-22.

17. Britten, R. J. and D. E. Kohne (1968). Repeated sequences in DNA. *Science 161*:529-540.

18. Marks, P. A. and Rifkind, R. A. (1972). Protein synthesis: Its control in erythropoiesis. *Science 175*:955-961.

19. Furusawa, M., Y. Ikawa and H. Sugano (1971). Development of erythrocyte membrane-specific antigen(s) in clonal cultured cells of Friend virus-induced tumor. *Proc. Jap. Acad. 74*:220-224.

20. Friend, C., W. Scher, J. G. Holland and T. Sato (1971). Hemoglobin synthesis in murine virus-induced leukemic cells *in vitro*. Stimulation of erythroid differentiation by dimethyl sulfoxide. *Proc. Nat. Acad. Sci. USA 68*:378-382.

21. Thrall, L., W. D. Park, H. W. Rashba, J. L. Stein, R. J. Mans and G. S. Stein (1974). In vitro synthesis of DNA complementary to polyadenylated histone messenger RNA. *Biochem. Biophys. Res. Communs. 61*:1443-1449.

22. Bishop, J. O., J. G. Morton, M. Rosbash and M. Richardson (1974). Three abundance classes in HeLa cell messenger RNA. *Nature* 250:199-204.

23. Melli, M. and R. Pemberton (1972). New method of studying the precursor-product relationship between high molecular weight RNA and messenger RNA. *Nature New Biol.* 236:172-174.

24. Obinata, M., D. S. Nasser and B. J. McCarthy (1975). Synthesis of probes for RNA using Qβ-replicase. *Biochem. Biophys. Res. Communs.* 64:640-647.

25. Macnaughton, M., K. B. Freeman and J. O. Bishop (1974). A precursor of hemoglobin mRNA in nuclei of immature duck red blood cells. *Cell* 1:117-124.

26. Tanaka, M., J. Levy, M. Terada, R. Breslow, R. A. Rifkind, and P. A. Marks (1975). Induction of erythroid differentiation in murine virus infected erythroleukemia cells by highly polar compounds. *Proc. Nat. Acad. Sci. USA* 72:1003-1006.

RIBONUCLEASES AND FACTORS INFLUENCING THEIR ACTIVITIES

Robert J. Crouch

Laboratory of Molecular Genetics
National Institute of Child Health
and Human Development
National Institutes of Health
Bethesda, Maryland 20014

INTRODUCTION

Recent studies on ribonucleases are discussed with respect to their cellular functions and variations in activities noted under different conditions *in vitro*. The relationship of these activities to stability of RNA in isolation procedures or presentation cells is discussed.

A major concern of many investigators is the isolation and preservation of as much biologically active material as possible, a concern shared by most investigators since the inception of research on RNA. The ubiquity and tenacity of RNases has led most investigators to be almost paranoid in their approach to dealing with the problems presented by these enzymes. Gloves should be worn during the handling of materials which will come in contact with RNA to prevent contamination with "finger nuclease" (1). Solutions and glassware are often sterilized and treatment of glassware after contact with RNase by $H_2 SO_4 - Na_2 Cr_2 O_7$ solutions is common practice to remove trace amounts of RNase. One definition of a molecular biologist is a scientist who is able to explain all of his results on the basis of RNase. Problems of RNase degradation of RNA during RNA isolation have focused the attention of most investigators on the negative aspects of ribonucleases. In the discussion to follow, some of the more recent studies on ribonucleases will be presented that should give a different perspective but at the same time point out some of the difficulties in dealing with RNase from the viewpoint of a person desiring to extract intact RNA molecules. Since an excellent review of ribonucleases has been written by Barnard (2), no attempt has been made to survey ribonucleases and RNase inhibitors.

What Happens To RNA In A Cell?

If we examine the manner in which cells handle the problem of ribonucleases and RNA stability we might learn something of a "natural" protection mechanism that could be useful in other circumstances.

Prokaryotic Cells — Bacterial cells represent a situation in which there is little compartmentalization, resulting in concomitant transcription and translation of RNA, a phenomenon peculiar to prokaryotes. Both the frequency of gene transcription and the rate of turnover of the messenger RNA of bacterial cells is especially high. However, transfer RNA and ribosomal RNA, even though they are present in the same cell sap, are synthesized and maintained in a stable form for many generations. The primary transcripts containing tRNA sequences (3) and those corresponding to ribosomal RNA sequences (4,5,6) have additional sequences that are removed as the primary transcript is processed to the final, stable product. In bacteria there are examples of RNA with a high rate of turnover and RNA which has a long life time, though portions of the original transcript containing the stable sequences have a short half-life. Differences in the preservation of some species of RNA and degradation of others can most easily be explained on the basis of protection of the preserved RNAs. The specific association of ribosomal RNA with ribosomal proteins might, in fact, be a method of conservation of ribosomal RNA. It should be remembered that the function of ribosomes is in protein synthesis and many, if not all, of the ribosomal proteins relate to the protein synthetic function, not simply to protect ribosomal RNA from degradation. Also, ribosomal RNA, at least some portions, is susceptable to ribonuclease even in the form of ribosomes. The stability of t-RNA is more difficult to explain. Maybe because of the small size and/or secondary and tertiary structure of t-RNA it is resistant to the RNases it encounters in the cell.

A glaring exception to the rapid turnover of mRNA in bacterial cells is the early mRNA of *E. coli* infected with T7 bacteriophage (7). This RNA is synthesized as a single transcriptional unit, including mRNAs from five different genes (4). Subsequently, the mRNAs from this transcript are separated by RNase III of *E. coli* (4,8) generating five RNA molecules each coding for a different protein. As of this time there is no explanation for the stability of early T7 mRNA. It does not seem that an inhibitor of the normal degradative enzymes is formed since the half-life of tryptophan mRNA is the same in T7 infected cells as it is in uninfected *E. coli* (9).

Eukaryotic Cells — A major characteristic of eukaryotic cells is a compartmentalization of cellular functions. During most of the cell cycle the nuclear membrane separates the protein synthetic apparatus from the transcriptional processes. Further compartmentalization of the nucleus segregates the synthesis of ribosomal RNA, which occurs in the nucleolus, from other types of transcription. Although there has been a great deal of circumstantial evidence that compartmentalization can be a method of preventing contact between molecules in the same cell, not all material within an orgarelle is confined to that cellular location since even the nuclear membrane is permeable to many cellular substances.

Transcription of DNA in eukaryotic cells is much more complicated than prokaryotes from several viewpoints. First, there are at least three different RNA polymerases in eukaryotes (10). Second, the initial transcripts may contain more RNA than the RNA which functions in the cytoplasm (11). Third, subsequent to synthesis the RNAs are polyadenylated (12), in many cases modified at the 5' terminus (13,14) and are transported to the cytoplasm. Fourth, not all of the RNA of a specific type that is formed in the nucleus reaches the cytoplasm (11). The cellular function of these "decorations" and processing is as yet unclear, although some experiments suggest that polyadenylation is necessary for mRNA stability (15) but there is at least one example of a stable nonpolyadenylated mRNA. Modification of the 5' terminus seems to make some RNAs more readily translatable (17).

Once RNA has been transported to the cytoplasm essentially three fates await it. Translation is clearly the most direct and obvious but degradation also occurs with half-lives of message being much longer than in bacterial cells. Occasionally, RNA is found in particles, either as A-type (18) and C-type particles (19) or in informasomes (20). Both informasomes and A-type or C-type particles are considered to be a repository of RNA which can be called upon during changes in cellular metabolism (such as during development or in tumor activation). Whatever the function of these particulate RNA species, they are good examples of a stabilization of RNA by the cell.

RNases — Cells have an advantage over scientists interested in utilizing intact RNA molecules in many ways since the cell is rarely limited in production of new RNA. Preservation of RNA during isolation and inhibition of RNases are, of course, major concerns in biochemical or biological studies with RNA. Consideration of the properties of RNases might give some insight into methods of RNA isolation and utilization that would circumvent the problems of RNase degradation.

Specificity of RNases — RNases can be categorized in a number of ways: according to the structure of the RNA (i.e. specificity for single stranded RNA, for double stranded RNA, or RNA in DNA-RNA hybrids); according to a base specificity (such as pancreatic ribonuclease cleavage adjacent to pyrimidines, T1 RNase cleavage adjacent to guanylate residues, or U_2 RNase preference for cleavages next to adenylate positions); according to the position of the phosphate residue after cleavage (i.e., 5' phosphoryl or 2' (3') phosphoryl); or according to the exo- or endo- nucleolytic character of the enzyme. Of course, any given RNase will exhibit properties that will permit its classification by any of these systems. Some of the more sophisticated RNases may combine both a recognition of secondary structure and base specificity so that they can not easily be classified either by base specificity or by structure of the RNA, but would fall into a category in which degradation is observed only with a special substrate or set of substrates.

RNase III of E. coli — As an example of the specificites described above, I want to point out some of the features of *E. coli* RNase III. In the past most RNases have been examined for their abilities to form acid or ethanol soluble products, a process which requires multiple cleavages to take polynucleotides and generate acid soluble products (if the RNase is endonucleolytic is character). Robertson *et al.* (21) used such an assay for the purification of RNase III. Unlike most other RNases, RNase III is able to solubilize RNA only if the RNA exists in a duplex form (i.e. both strands of the double helix are RNA). If the products of degradation of duplex RNA by RNase III are examined, it is possible to place RNase III in the specificity catagories mentioned above. By the criterion of solubilization, RNase III is specific for degrading RNA of RNA-RNA duplexes (22,23), it produces oligonucleotides 15-20 nucleotides in length (22,23,24) terminated in 5' phosphoryl moieties (22,23) and demonstrates very little, if any, base specificity (22,23). These properties are what one observes if solubilization is the assay employed. However, certain large RNA molecules are made smaller (but not soluble) by RNase III.

Dunn and Studier purified an enzyme for *E. coli* that was able to process T7 early mRNA to the appropriate size messages *in vitro* (4). Subsequently this enzyme was identified as RNase III (23) and the cleavages observed *in vitro* were seen to be identical with those occurring *in vivo* (8). Prior to the description of RNase III as a processing enzyme, a mutant of *E. coli* defective in RNase III activity had been isolated (25). Examination of the mutant showed an inability to process T7 early mRNA *in vivo* (4) and, in addition, a precursor to ribosomal RNA was seen to accumulate in this mutant (4,5). RNase III can cleave the precursor to intermediates of RNA that are found in the cell (4,5,6). Both the cleavage of T7 early mRNA and the processing of ribosomal RNA lead to the production of RNAs smaller than the original but do not produce any significant solubilization of RNA. Processing of T7 mRNA occurs at sites which can be considered as duplex RNA by virtue of hairpin loops found at the points of cleavage (26,27), a finding consistant with the ability of RNase III to solubilize duplex RNA. Processing of rRNA, cleavage of T7 early mRNA precursor and solubilization of duplex RNA all exhibit a requirement of monovalent cation concentrations between 50 and 200mM (21). However, at least one cleavage, attributable to RNase III (27) occurs optimally in the absence of monovalent cation (28) and, more recently, Dunn (30) has observed further degradation of T7 mRNA if monovalent cation concentrations are decreased.

RNA III also seems to recognize regions within eukaryotic RNAs. Ribosomal precursor from eukaryotic cells can be cleaved to produce RNAs with a size (31,32,33) similar to intermediates observed *in vivo*. Although most of the messenger RNAs examined to date are resistant to RNase III under normal enzyme concentrations [(i.e., concentrations sufficient to degrade all of the duplex RNA present in the same reaction mixture (33,35)], addition of more enzyme results in fragmentation of several mRNAs as well as 18 and 28S ribosomal RNA of KB cell (35) and from a cell line derived from *Notophthlamus viridescens* (33).

Although these properties of RNase III are extremely interesting in light of regulatory mechanisms based on processing of the primary gene transcripts, the significance of the multifaceted activities, which RNase III is capable of exhibiting, resides in the problems generated in determining if the enzyme has been inhibited. Since RNase III is inhibited for duplex RNA degradation by lowering the ionic strength of the reaction mixture, one would assume that the enzyme would not be functional; yet when other substrates are tested, degradation of RNA can be observed. Thus, with a single enzyme (RNase III) it seems unlikely that a simple approach of inhibition with a drug or chemical inhibitor will always prevent some degradation of RNA and, moreover, with the many different types of RNases present in most organisms such an approach will almost certainly fail. An alternative to inhibition of all RNAses in an animal or cell to be presented with RNA would be to present the RNA in some form resistant to RNases (i.e., in the forms seen to preserve RNA in a cell) or, more optimistically, let the animal or cell "protect" the RNA in its own way.

REFERENCES

1. Holley, R. W., J. Apgar and S. H. Merrill (1961). Evidence for the liberation of a nuclease from human fingers. *J. Biol. Chem. 236*:42-43.

2. Barnard, E. C. (1969). Ribonucleases. *Ann. Rev. Biochem. 38*:706-732.

3. Altman, S. (1971). Isolation of tyrosine tRNA precursor molecules. *Nature New Biol. 229*:19-21.

4. Dunn, J. J. and F. W. Studier (1973). T7 early RNAs and *Escherichia coli* ribosomal RNAs are cut from large precursor RNAs *in vivo* by ribonuclease III. *Proc. Nat. Acad. Sci. USA 70*:3296-3300.

5. Nikolaev, N., L. Silengo and D. Schlessinger (1973). Synthesis of a large precursor to ribosomal RNA in a mutant of *Escherichia coli. Proc. Nat. Acad. Sci. USA 70*:3361-3365.

6. Ginsburg, D. and J. A. Steitz (1975). The 30S ribosomal precursor RNA from *Escherichia coli. J. Biol. Chem. 250*:5647-5654.

7. Siegel, R. B. and W. C. Summers (1970). The process of infection with coliphage T7. III. Control of phage-specific RNA synthesis *in vivo* by an early phage gene. *J. Mol. Biol. 49*:115-123.

8. Rosenberg, M., Kramer, R. A. and J. A. Steitz (1974). T7 early messenger RNAs are the direct products of ribonuclease III cleavage. *J. Mol. Biol. 84*:777-782.

9. Marrs, B. L. and C. Yanofsky (1971). Host and bacteriophage specific messenger RNA degradation in T7-infected *Escherichia coli. Nature New Biol. 234*:168-170.

10. Roeder, R. G., L. B. Schwartz and V. E. F. Sklar. Structure, function and regulation of eukaryotic RNA polymerases. *Hormones and Molecular Biology*: the 34th symposium for the society for developmental biology. John Papaconstantinou, Ed. Academic Press, New York, in press.

11. Darnell, J. D. (1968). Ribonucleic acids from animal cells. *Bacteriol. Rev.* *32*:262-290.

12. Edmonds, M., M. H. Vaughan and H. Nakazoto (1971). Polyadenylic acid sequences in heterogeneous nuclear RNA and rapidly-labeled polyribosomal RNA of HeLa cells: Possible evidence for a precursor relationship. *Proc. Nat. Acad. Sci. USA* *68*:1336-1340.

13. Wei, C. M. and B. Moss (1974). Methylation of newly synthesized viral messenger RNA by an enzyme in vaccinia virus. *Proc. Nat. Acad. Sci. USA 71*:3014-3018.

14. Shatkin, A. J. (1971). Methylated messenger RNA synthesis *in vitro* by purified reovirus. *Proc. Nat. Acad. Sci. USA 71*:3204-3207.

15. Huez, G., G. Marbaix, E. Hubert, M. LeClercq, U. Nudel, H. Soreq, R. Salomon, B. Lebleu, M. Revel and U. Z. Littauer (1974). Role of the polyadenylate segment in the translation of globin messenger RNA in *Xenopus* oocytes. *Proc. Nat. Acad. Sci. USA 71*:3143-3146.

16. Adesnik, M. and J. E. Darnell (1972). Biogenesis and characterization of histone messenger RNA in HeLa cells. *J. Mol. Biol. 67*:397-406.

17. Both, G. W., A. K. Banerjee and A. J. Shatkin (1975). Methylation-dependent translation of viral messenger RNAs *in vitro*. *Proc. Nat. Acad. Sci. USA 72*:1189-1193.

18. Hinglais-Guillaud, N., M. Riviere and W. Bernhard (1959). Presence de particules d'aspect viral dans un epithelioma uterin du rat (tumeur de Guerin). *C. R. Acad. Sci. 249*:1589-1591.

19. Dalton, A. J., M. Potter and R. M. Merwin (1961). Some ultrastructural characteristics of a series of primary and transplanted plasma-cell tumors of the mouse. *J. Nat. Cancer Inst. 26*:1221-1267.

20. Spirin, A. S. (1969). Informasomes. *European J. Biochem. 10*:20-35.

21. Robertson, H. D., R. E. Webster and N. D. Zinder (1968). Purification and properties of ribonuclease III from *Escherichia coli*. *J. Biol. Chem. 243*:82-91.

22. Crouch, R. J. (1974). Ribonuclease III does not degrade deoxribonucleic acid – ribonucleic acid hybrids. *J. Biol. Chem. 249*:1314-1316.

23. Robertson, H. D. and J. J. Dunn (1975). Ribonucleic acid processing activity of *Escherichia coli* ribonuclease III. *J. Biol. Chem. 250*:3050-3056.

24. Schweitz, H. and J. P. Ebel (1971). A study of the mechanism of action of *E. coli* ribonuclease III. *Biochimie* (Paris) *53*:585-593.

25. Kindler, P., T. V. Keil and P. H. Hofschneider (1972). Isolation and characterization of a ribonuclease III deficient mutant of *Escherichia coli*. *Molec. Gen. Genet. 126*:53-69.

26. Robertson, H. D. and J. J. Dunn, personal communication.

27. Rosenberg, M. and R. Kramer, personal communication.

28. Paddock, G. V., K. Fukada, J. Abelson and H. D. Robertson. *Nuc. Acid Research*, in press.

29. Paddock, G. and J. Abelson (1973). Sequence of T4, T2, and T6 bacteriophage species I RNA and specific cleavage by an *E. coli* endonuclease. *Nature New Biol. 246*:2-5.

30. Dunn, J. J., personal communication.

31. Gotoh, S., N. Nikolaev, E. Battaner, C. H. Birge and D. Schlessinger (1974). *Escherichia coli* RNase III cleaves HeLa nuclear RNA. *Biochem. Biophys. Res. Comm. 59*:972-978.

32. Yang, N. S. and L. P. Gage, personal communication.

33. Crouch, R. J., T. Honjo and D. Reese, unpublished observations.

34. Robertson, H. D. and E. Dickson (1974). RNA processing and the control of gene expression. *Brookhaven Symposium 26*:240-266.

35. Westphal, H. and R. J. Crouch (1975). Cleavage of adenovirus messenger RNA and of 28S and 18S ribosomal RNA by RNase III. *Proc. Natl. Acad. Sci. USA* 72:3077-3081.

DISCUSSION

Hugh D. Robertson, Chairman

This session featured the chemistry and biochemistry of RNA, emphasizing the nature and scope of many recent advances in this area. Most of the methods and techniques described were introduced in conjunction with studies on the involvement of RNA in one of its widely accepted conventional roles: either as transfer RNA, ribosomal RNA, or (most commonly) as messenger RNA. However, there were several approaches described where recent techniques could be applied to the question of additional roles for RNA in cells, e.g., as a regulatory substance. In the light of subsequent sessions on properties and problems of immune RNA, this dual emphasis is particularly important. It would be possible to explain many properties of immune RNA by invoking the transmission of a specific messenger RNA from one cell to another, followed by translation in the recipient cell in the normal manner to produce the additional protein(s) required for the change in the immune properties of the recipient cell. However, several facets of the immune RNA problem cannot be readily explained by this straightforward messenger model. Specifically, suggestions that the acquired immune specificity can be transmitted on cell division — i.e., that it is not merely a phenotypic change of recipient cells — would require a much more elaborate molecular explanation and might need to invoke extra, untraditional functions for the RNA involved.

Dr. P. Gilham described progress on specific chemical modifications of RNA which allow the study and utilization of some of its unique properties. Specifically, several reactions with the water-soluble carbo-diimide reagent which attaches specifically to U and G residues at pHs above 7.0 were described. This reagent blocks nuclease action adjacent to modified residues. This compound blocks the 5' end of RNA chain in a reaction carried out at pH 6.0, facilitating isolation of oligonucleotides thus modified. Finally, studies on 3' termini using borate-substituted DEAE-cellulose chromatography revealed that any oligonucleotide containing the *cis* diol configuration characteristic of ribose can be selectively isolated. In addition, chemical modification of residues to obtain greater specificity was exemplified by a reaction in which the 2' OH group of ribose acquired a methoxyethyl group. Using RNA modified in this way, Dr. Gilham has obtained blockage of polynucleotide phosphorylase beyond its initial cleavage step.

Dr. G. Peters reviewed progress in application of two-dimensional polyacrylamide gel analysis to purifying specific RNAs from complex mixtures as well as the use of the RNA fingerprinting techniques worked out by Sanger and his associates as the first step of RNA sequence determination. Both of these groups of methods were exemplified using the case of the RNA primer molecule found in association with the genome of RNA tumor viruses such as Rous sarcoma virus and murine leukemia virus. Dr. Peters explained how it was possible to demonstrate association of the primer, which is a transfer RNA originally present in uninfected cells, with the viral RNA as one of a number of RNA species found in viral particles. Isolation and sequence analysis show that the primer for the first step of DNA synthesis by the reverse transcriptase present in the viral particles forms a covalent linkage with the nascent DNA stand complementary to viral RNA. The idea that RNA can serve as a primer for the synthesis of additional nucleic acid *in vivo* is a relatively new one and its potential importance for cellular regulation remains to be assessed.

Dr. J. Ross summarized current progress on the application of nucleic acid hybridization techniques to the study of the transcription of a particular gene. The example chosen — globin genes transcribed in erythroleukemic cells produced by Friend leukemia virus infection — was studied using as a probe complementary DNA synthesized with reverse transcriptase and oligo-deoxthymidine primer using globin mRNA as template. Dr. Ross described detection of globin mRNA sequences following induction of globin synthesis in these cells. These techniques are applicable when it is impossible to separate the RNA molecule of interest from a heterogeneous mixture of RNA molecules.

Dr. R. Roeder summarized structural characteristics of eukaryotic RNA polymerases, emphasizing the complexity of their polypeptide composition and the potential relatedness of the three major types. He also described experiments with isolated nuclei in which, for the first time, the synthesis of a specific RNA species (in this case an adenovirus-coded RNA) could be monitored under controlled conditions. Finally, he described conditions where cells change significantly the level of RNA transcription without a commensurate change in the amount of the various polymerizing enzymes, suggesting the existence of additional factors regulating the activities of these enzymes.*

Dr. R. Crouch reviewed cellular ribonucleases, using an enzyme from *E. coli* — Ribonuclease III — as a prototype example. This enzyme is involved in the specific processing of cellular ribosomal RNA precursors as well as in the specific cleavage of bacteriophage T7 messenger RNA precursors. The existence of enzymes of this sort allows speculation upon the need for RNA precursors as well as potential roles for the extra RNA segments which are transcribed but which are then metabolized with a much shorter half-life than the stable portion of the RNA transcript. Dr. Crouch also described current interest in the possible existence of similar enzymes in mammalian cells and described exploratory studies on cleavage of eukaryotic viral and cellular RNA's with *E. coli* RNase III. While some cleavages were obtained at high enzyme to substrate ratio, Dr. Crouch cautioned against interpreting these cleavages (which serve largely to inactivate mRNA molecules for *in vitro* translation) as reflecting authentic *in vivo* processing events. Finally, a series of reactions with RNase III was described in which the reaction conditions were varied in such a way that the amount and specificity of cleavage events varied drastically.

Dr. A. Grassmann reviewed his experiments with micro-injection of RNA into mammalian tissue culture cells. Volumes in the range of 10 picoliters containing about 5 picograms of various RNAs are injected into the cytoplasm and the effects studied. For example, RNA synthesized by *E. coli* RNA polymerase using as template the double-stranded DNA of the tumor virus polyoma was found to stimulate production of the tumor antigen (T-antigen) characteristic of infections with polyoma virus as assayed by fluorescent antibody staining. This finding seems quite important for those interested in evaluating the "messenger" model of immune RNA action mentioned above. However, several findings reported by Dr. Grassmann in connection with these studies do not lend themselves to such straightforward interpretation. Not only is there a time lag of ten hours or more before detection of T-antigen in injected cells, but also nearly all of the injected cells are stimulated to begin DNA synthesis (as measured by incorporation of DNA precursors) at some time between 0-25 hours after RNA injection. The relationships among these various findings must be elucidated before a full understanding of the effects of polyoma RNA injection can be obtained.

Subsequently, Dr. H. Robertson described the use of chemical iodination as a handle on the study of RNA molecules which could not be readily obtained in radioactive form from cells. The example used was characterization of the genomic RNA of plant viroids (1). Viroids each consist of a single species of free RNA 250-350 nucleotides long (2) which infect and replicate in a variety of plants causing severe disease symptoms in only a few (3). Since preparations of small amounts of this RNA require large numbers of plants, growth of sufficient radioactive RNA for use in isolation, sequencing and nucleic acid hybridization has so far proven impractical. However, chemical iodination with ^{125}I allows the application of methods which will allow all three types of study (4,5). Techniques such as these are essential for the study of RNA molecules present at low levels, as may be the case with immune RNA.

Dr. Robertson also suggested that the "extra" RNA transcribed in mammalian cell nuclei, which undergoes a rapid metabolism in the nucleus without ever reaching the cytoplasm, could be directly involved in gene regulation of RNA. Referring to a proposal by himself and E. Dickson (6), he suggested that such specific regulatory RNAs produced by specific processing in the nucleus could act by priming transcription of previously inactive genetic units located adjacent to DNA sequences complementary to the regulatory RNAs.

In conclusion, this session demonstrated that techniques for the isolation and characterization of small amounts of specific RNA molecules are improving rapidly. Chemical modification along with improved polyacrylamide gel analysis point to the early isolation of many more RNA species, while both fingerprinting and hybridization techniques (used separately or in combination) should give rigorous criteria for identifying particular RNA's during their involvement in various cellular steps. Studies on RNA polymerases and specific RNA processing enzymes suggest an increased ability to synthesize and characterize particular RNA species *in vitro*, while the micro-injection method holds out promise for studying the effects of very low levels of specific RNAs. Application of these techniques should reveal whether or not immune RNA phenomena are based simply on the transmission of highly efficient, stable messenger RNAs from one cell to another. This explanation would be less satisfying to molecular biologists than the alternative choice that immune RNA causes permanent shifts in regulation of RNA synthesis in recipient cells and their descendants. If true, this latter possibility could lead to characterization of a regulatory RNA of mammalian cells for the first time. However, from the point of view of therapeutic applications, the need for understanding outweighs any preferences with regard to mechanism. For this reason, there is a considerable responsibility on the part of those interested or involved in immune RNA therapy to see that these rigorous RNA characterization methods are applied as soon and as efficiently as possible.

*Editor's Note: Dr. Roeder did not submit a manuscript for publication in this volume. See Roeder, R. G., L. B. Schwartz and V. E. F. Sklar. Structure, function and regulation of eukaryotic RNA polymerases. *Hormones and Molecular Biology*: the 34th Symposium for the Society for Developmental Biology. John Papaconstantinou, Ed. Academic Press, New York, in press.

REFERENCES

1. Dickson, E., W. Prensky and H. D. Robertson (1975). Comparative studies of two viroids: Analysis of potato spindle tuber and citrus exocortis viroids by RNA fingerprinting and polyacrylamide-gel electrophoresis. *Virology 68*:309-316.
2. Diener, T. O. (1971). Potato spindle tuber "virus". IV. A replicating, low molecular weight RNA. *Virology 45*:411-428.
3. Diener, T. O., D. R. Smith and M. J. O'Brien (1972). Potato spindle tuber viroid. VII. Susceptibility of several plant species to infection with low molecular-weight RNA. *Virology 38*:844-846.
4. Prensky, W., D. M. Steffenson and W. L. Hughes (1973). Use of iodinated RNA for gene localization. *Proc. Nat. Acad. Sci. USA 70*:1860-1864.

5. Robertson, H. D., E. Dickson, P. Model and W. Prensky (1973). Application of fingerprinting techniques to iodinated nucleic acids. *Proc. Nat. Acad. Sci. USA* 70:3260-3264.

6. Robertson, H. D. and E. Dickson (1974). RNA processing and the control of gene expression. *Brookhaven Symposia in Biology* 26:240-266.

SESSION II
What Is Immune RNA?

Chairman, Frank L. Adler

CURRENT STATUS OF IMMUNE RNA

M. Fishman and F. L. Adler

Division of Immunology
St. Jude Children's Research Hospital
Memphis, Tennessee 38101

Two principal sources exist for the stream of literature dealing with the role of ribonucleic acids in the immune response in general and in antibody formation in particular. One of these originates in the area of studies concerned with the biosynthesis of proteins and centers on certain classes of RNA such as transfer (t-RNA) and messenger (m-RNA) ribonucleic acids in the formation of immunoglobulins (1,2,3,4). Possible modifications of the general mechanisms to meet the more specific requirements of antibody formation have recently been discussed by Haurowitz (5). The other point of departure for studies on the role of RNA in immunogenesis was the important finding of Garvey and Campbell (6) followed by a series of other reports from this group of workers (7,8,9,10) which implicated metabolically stable complexes of antigen fragments and RNA-like material in immunogenesis. Formation and persistence of such complexes in the liver, retention of immunogenicity, and appearance of such complexes in serum and urine under specified conditions have been studied in great detail.

It seems important to stress that this dichotomy in approach does indeed reflect a diversity in biologically active substances which unfortunately have often been lumped together under the designation "immune RNA." This paper shall attempt to differentiate clearly between two classes of RNA. One has frequently been given the descriptive designation of "informational RNA (i-RNA)" and we hope to show that it has all the characteristics of m-RNA. The other is a complex of RNA and antigen or antigen fragments which resembles in many aspects the material investigated by Garvey and Campbell in their *in vivo* studies. We shall take advantage of the fact that most laboratories active in relevant research are represented at this meeting and thus can and will present their own data and conclusions directly. This allows us to lean heavily on our own data without the risk of slighting the important contributions of others.

In order to explain the rationale of this approach to non-immunologists is to say that our studies are based on an entirely *in vitro* system of organ and tissue culture which supports the initiation of an immune response and the *de novo* synthesis of specific antibody. Using bacteriophages as our antigens, we found that the sequential action of two classes of cells was required – first a reaction between the antigen and peritoneal macrophages and then the response of lymphoid cells to the products of the reaction just described. These reaction products were contained in the RNA fraction obtained from antigenically stimulated macrophages and consisted of two classes of active materials, namely informational RNA (i-RNA) and RNA-antigen complexes. The evidence in support of the conclusion that two physically and functionally distinct classes of RNAase-sensitive material were obtained is described in the following paragraph.

Physical separation of fractions rich in one and deprived of the other RNA fraction could be attained by chromatography on columns of methylated bovine serum albumin-Kieselguhr from which i-RNA was eluted with peak II (estimated 10-12 s) and which yielded the RNA-antigen fragment with fraction III (23 s) (11). Separation could also be attained through the use of potent antisera specific for T2 bacteriophage, our test antigen (11). Such antisera would cause the precipitation of the RNA-antigen complex, leaving i-RNA in the supernatant. Additional criteria for the distinction of i-RNA and RNA-antigen complexes are the following: (a) Actinomycin D added to the peritoneal macrophage suspension prior to the addition of the antigen inhibits the formation of i-RNA but not that of RNA-antigen complexes (12). (b) The formation of effective amounts of RNA-antigen complexes requires relatively large doses of antigen (100 phage particles per macrophage) while the induction of i-RNA is triggered by minimal amounts of antigen (1 phage particle per 1000 macrophages) and appears to be independent of antigen concentration above this threshold level (13). (c) Treatment of the crude RNA preparation with pronase destroys the activity of the RNA-antigen complex but leaves intact the activity of the i-RNA. (d) Although both activities are destroyed by the action of ribonuclease, i-RNA is significantly more labile than the RNA-antigen complex. (e) Under the standard conditions of our experiments the i-RNA evokes an antibody response that is primarily IgM while the RNA-antigen complex causes formation of IgG antibodies (13). In cultures of lymph node fragments these two responses are temporally separated with the IgG response following the IgM response. If one used as the source of RNA not macrophages from normal animals, as we have done, but lymphatic tissue from immunized animals, as others have done, one would expect to obtain i-RNA which reflects the immunoglobulin class of antibody that is being produced at that time, and that, indeed, has been verified experimentally (14). (f) It has been possible to show that the target cells for the RNA-antigen complex are lymphocytes which possess recognition for the antigen component of the complex. Cultures deprived of such cells by appropriate absorption on antigen-coated glass bead columns nevertheless react in an undiminished fashion to i-RNA (15).

The contention that i-RNA initiates antibody formation independently of antigen was not readily accepted. Arguments that there may be undetectable yet immunogenic traces of antigen in the preparation could not be answered experimentally because it is technically impossible to provide rigid proof for the absence of something. Our finding that i-RNA transferred from one cell to another the ability to produce immunoglobulin molecules which possessed both the antibody and the allotypic specifities of the cells from the donor animal appeared to settle this argument in favor of the transfer of information (16). Whether the transfer of allotypic specificities rigidly proves the transfer of structural information or, alternatively, could be explained by a de-repressor role for i-RNA is still debatable; but we believe that the weight of the evidence favors the concept of transfer of structural information.

Further support for the dual nature of RNA classes involved in the immune response comes from a series of recent studies which suggest that different cells are responsible for the formation of i-RNA and that of RNA-antigen complexes. Fractionation of the peritoneal exudate cell preparations (80-90 percent macrophages) on Ficoll (17) or BSA (18) density gradients has led to data which show that the active cells can be concentrated in a sub-population which contains only 15 percent of the original population and that refinement of the gradients can then separate the active cells into two bands of which one contains the cells responsible for i-RNA formation, the other the cells that produce RNA-antigen complexes.

With regard to those cells among the peritoneal macrophages that produce i-RNA, the addition of just 1 phage particle per 1000 macrophages appears sufficient to activate these cells. This implies the presence of an efficient recognition or trapping mechanism which, most likely, would be membrane-bound antibody reactive with T2 phage. Since the presence of such (actively synthesized) antibody would imply possession of relevant m-RNA by the cell, it is possible that the cells which produce i-RNA in our system belong to a sub-class of peritoneal macrophages which are producers but not necessarily secretors of specific antibodies. While macrophages are not generally considered to be producers of antibody, there exists evidence to the contrary, as for example in the recent work by Bussard (19).

In collaboration with colleagues at the Roche Institute, we have recently come closer to identifying i-RNA with m-RNA (20). Briefly summarized, available evidence in support of this notion is the following: (a) The addition of actinomycin D or of rifampicin to cultures of spleen cells fails to inhibit their responsiveness to i-RNA under conditions that prevent completely their response to antigen or antigen-RNA complexes. (b) Polysomal RNA from macrophages that had been incubated with antigen for 30 minutes appears to possess greater specific activity than does total cellular RNA. (c) A cell-free extract from mouse L cells was shown to be capable of supporting the translation of i-RNA from rabbit cells. The formation of rabbit IgM antibody with the allotypic specificity of the (rabbit) cell donor and specificity for the test antigen (T2 phage) could be shown both by binding and by neutralization assays. Besides their relevance to the m-RNA nature of the activity, these findings showing transfer of information pertaining to amino acid sequence across species lines support similar observations of others on trans-species transfer (21,22). Thus questions as to whether the genes whose activities are recognized in allotyping are structural or regulatory appear not to be very relevant to the problems under discussion here.

Turning to a discussion of the RNA-antigen complex, we have had some success in our attempts to sort out the cells in which such complexes may originate and we have evidence that the target cells are lymphocytes that possess antigen-specific recognition. While enzyme-inactivation studies clearly have shown that both the antigen and the RNA moieties are essential for biological activities the function of the nucleic acid component still remains unclear. It has been shown experimentally that the formation of *immunogenic* complexes of RNA and antigen require a class of RNA that is restricted in its organ and cellular distribution. In addition, cellsap factors, presumably enzymes, play an important role in the formation of such complexes (23). The existence of electrostatic binding between nucleic acids and many substances, including proteins, is a real phenomenon but should not becloud the existence of more specific and biologically active linkages between RNA and antigen.

Stressing once again that our experience is confined to *in vitro* systems, the next question is how the transfer of RNA from the donor cell (macrophage) to the recipient cell (lymphocyte) is brought about. Several years ago we observed that when radioactively-labeled RNA was added to cultures containing macrophages and lymphocytes (24), the radioactivity was concentrated in macrophages and in those lymphocytes that immediately surrounded them (peripolesis). One possible explanation was that the macrophage took in the RNA and then "fed" it to the adherent lymphocytes. It was equally possible that lymphocytes involved in the peripolesis phenomenon might be activated in some manner and therefore be more receptive to the RNA. In more recent work, we have observed that these morphological observations can be verified by functional tests (15). Spleen cell cultures deprived of adherent cells will not respond to the addition of RNA as if the RNA extracted from macrophages had to be recycled through such cells for appropriate presentation to lymphocytes. The significance of these findings to the *in vitro* system and their relevance to events *in vivo* remains to be established.

One final series of observations is worthy of discussion. It appears that the entry of exogenous RNA into lymphoid cells causes a "shutdown" of synthesis of function of the homologous endogenous RNA. The evidence is based on the results obtained when normal cells of one allotype ($a_2 a_2 / b_6 b_6$) were exposed to a mixture of RNA preparations from macrophages of a rabbit ($a_1 a_1 / b_4 b_4$) that had been exposed to the coli phage T2 and from a second donor rabbit ($a_3 a_3 / b_5 b_5$) exposed to the *B. subtilis* phage SP 82. Antibodies reactive with T2 as well as those reactive with SP82 contained both the b_4 and the b_5 markers, suggestive of the formation of hybrid molecules in response to both exogenous RNA preparations. Neither the anti T2 nor the anti-SP 82 antibodies appeared to contain the b_6 marker native to the cells in which they were produced. It seems possible, therefore, that our RNA preparations act as some viral RNA preparations do in that they inhibit expression of some endogenous RNA functions.

Still confining ourselves to the *in vitro* system as source of information, we believe that we may extrapolate to the *in vivo* system in several respects. We strongly believe that the demonstration of at least two active moieties – the i-RNA and the RNA-antigen complex, substances that are physically and functionally distinct – arise in different cells under different conditions, and act upon distinct cells. This calls for discontinuation of the terms "immune RNA" or "immunogenic RNA," terms which are not only ambiguous but inaccurate in that they convey the impression of designating one substance instead of at least two. This consideration goes beyond pedantic reasons because our *in vitro* data as well as the *in vivo* data of some but not all other workers suggest that i-RNA conveys and transfers information beyond and in addition to that pertaining to antibody specificity. The transfer of allogeneic or xenogeneic antigenic markers by i-RNA may, of course, lead to complications in therapeutic applications. It would seem of importance to know, in situations where RNA from immunized donors can be shown to be therapeutically effective, whether it is i-RNA or the complex of some tumor-associated antigen and RNA.

That macromolecular RNA can enter cells and biologically express itself has been clearly demonstrated with infectious viral RNA and HeLa cells, with immune RNA and lymphoid cells, and more recently with globin m-RNA in toad oocytes. In our *in vitro* system, the entrance of immune RNA into lymphocytes is probably a very inefficient process; and it may even be less efficient in the *in vivo* studies, where RNA is injected intravenously. We would suggest that if i-RNA or RNA-antigen play significant roles in the course of an immune response to the conventionally administered antigen, they may well have to be presented to responder cells by others (helper) cells such as the macrophages described earlier. It has also been said that RNA could not convey information because extracellular RNA would be promptly destroyed by ubiquitous ribonucleases. This argument ignores two important factors, namely the existence of inhibitors of ribonucleases in biological fluids and the experimental demonstration that RNA does reach target cells in active form.

It has been pointed out elsewhere that the major biological significance of i-RNA may come from its ability to enter into lymphoid cells that lack specific receptors for the antigen toward which a response is to be made. Thus it has the potential for expanding a pool of committed cells or of introducing or reintroducing responsiveness to a given antigen where such is missing. If specific unresponsiveness to a given antigen were due to clonal deletion or elimination, i-RNA would be expected to restore responsiveness.

While we are reasonably confident that the identity of i-RNA and m-RNA is established, many questions remain about the mechanism by which such RNA finds its way to the polysomes and about the manner in which information or product are integrated into cellular RNA or protein, respectively. More puzzling remains the mode of action of the RNA-antigen complex. We believe that the demonstrated metabolic stability of these complexes is an important attribute which in itself may be sufficient to bestow on the complex the qualities of the superantigen. It is possible, of course, that the RNA portion of the complex may contribute adjuvant effects in some unknown manner.

Progress in the study of the nature and the mode of action of RNA from antigenically stimulated cells has been slow but steady. Some of our colleagues felt that it could not work and then, sometime later, that it should not work. They are contending now that while it works *in vitro* it could not possibly play a significant role in the immune response *in vivo* (25). While this remains a matter of opinion which may some day be resolved experimentally, we have moved into the era of the application of the principles to cancer therapy. Reports of success (26,27) in this area provide encouragement and, at the same time, reinforce the need for further basic studies in this field.

REFERENCES

1. Askonas, B. A. and A. R. Williamson (1968). Interchain disulphide bond formation in the assembly of immunoglobulin G. Heavy-chain dimer as an intermediate. *Biochem. J. 109*:637-643.

2. Baumal, R., M. Potter and M. D. Scharff (1971). Synthesis, assembly and secretion of gamma globulin by mouse myeloma cells. III. Assembly of the three subclasses of IgG. *J. Exp. Med. 134*:1316-1334.

3. Stevens, R. H. and A. R. Williamson (1973). Isolation of messenger RNA coding for mouse heavy-chain immunoglobulin. *Proc. Natl. Acad. Sci. 70*:1127-1131.

4. Cowan, N. J., D. S. Secher and C. Milstein (1974). Intracellular immunoglobulin chain synthesis in non-secreting variants of a mouse myeloma: Detection of inactive light-chain messenger RNA. *J. Mol. Biol. 90*:691-701.

5. Haurowitz, F. (1973). The role of RNA in antibody formation. Historical perspectives. *Annals. N. Y. Acad. Sciences. 207*:8-16.

6. Garvey, J. S. and D. H. Campbell (1957). The retention of S^{35}-labeled bovine serum albumin in normal and immunized rabbit liver tissue. *J. Exp. Med. 105*:361-372.

7. Garvey, J. S., D. H. Campbell and M. L. Das (1967). Urinary excretion of foreign antigens and RNA following primary and secondary injections of antigens. *J. Exp. Med. 125*:111-126.

8. Yuan, L., J. S. Garvey and D. H. Campbell (1970). Molecular forms of antigen in the circulation following immunization. *Immunochemistry 7*:601-618.

9. Yuan, L. and D. H. Campbell (1971). A unique low molecular weight antigen fragment associated with oligoribonucleopeptide in lymph nodes and spleen. *Immunochemistry 8*:185-199.

10. Yuan, L. and D. H. Campbell (1972). The *in vivo* and *in vitro* immunogenicity of antigenic specificity of lymphoid antigen fragments-oliogoribonucleopeptide conjugated. *Immunochemistry 9*:1-8.

11. Fishman, M. and F. L. Adler (1967). The role of macrophage-RNA in the immune response. *Cold Spring Harbor Symposia on Quan. Biol. 32*:343-348.

12. Fishman, M., J. J. Van Rood and F. L. Adler (1964). The initiation of antibody formation by ribonucleic acid from specifically stimulated macrophages. *Molecular and Cellular Basis of Antibody Formation, Prague Symp.* J. Sterzl, Ed. Czech. Acad. Sci. Prague.

13. Fishman, M. and F. L. Adler (1967). Macrophage RNA and antibody synthesis. *Immunity, Cancer and Chemotherapy.* E. Mihich, Ed. Academic Press, New York.

14. Bell, C. and S. Dray (1972). Conversion of homozygous lymphoid cells to produce IgM antibodies and IgG immunoglobulins of allelic lightchain allotype by injection of rabbits with RNA extracts. *Cellular Immunol. 5*:52-65.

15. Schaefer, A. E., M. Fishman and F. L. Adler (1974). Studies of antibody induction *in vitro*. III. Cellular requirements for the induction of antibody synthesis by solubilized T_2 phage and immunogenic RNA. *J. Immunol. 112*:1981-1986.

16. Adler, F. L., M. Fishman and S. Dray (1966). Antibody formation initiated *in vitro*. III. Antibody formation and allotypic specificity directed by ribonucleic acid from peritoneal exudate cells. *J. Immunol. 97*:554-558.

17. Walker, W. S. (1971). Macrophage functional heterogeneity in the *in vitro* induced immune response. *Nature, New Biology 229*:211-212.

18. Rice, S. G. and M. Fishman (1974). Functional and morphological heterogeneity among rabbit peritoneal macrophages. *Cellular Immunol. 11*:130-145.

19. Lowy, I., R. L. Teplitz and A. E. Bussard (1975). Secretion of IgM antibodies by glass-adherent fraction of immune mouse spleen cells. II. Characteristics of glass-adherent plaque-forming cells. *Cellular Immunol. 16*:36-47.

20. Bilello, P. A., G. Koch and M. Fishman (1975). RNA activity of immunogenic RNA. *Fed. Proc. 34*:1030.

21. Paque, R. E. and S. Dray (1972). Monkey to human transfer of delayed hypersensitivity *in vitro* with RNA extracts. *Cellular Immunol. 5*:30-41.

22. Wang, B. S., P. A. Stuart and J. A. Mannick (1974). Interspecies transfer by immune RNA of lymphocytes proliferative response to specific antigen. *Cellular Immunol. 12*:114-118.

23. Fishman, M. and F. L. Adler (1973). The formation of immunogenic RNA-antigen complexes in a cell-free system. *Cellular Immunol. 8*:221-234.

24. Fishman, M., R. A. Hammerstrom and V. P. Bond (1963). *In vitro* transfer of macrophage RNA to lymph node cells. *Nature 198*:549-551.

25. Lefkovits, I. (1974). Precommitment in the immune system *Current Topics in Microbiology and Immunology 65*:21-58. Springer Verlag, Berlin, Heidelberg, New York.

26. Pilch, Y. H., L. L. Veltman and D. H. Kern (1974). Immune cytolysis of human tumor cells mediated by xenogeneic immune RNA. Implications of immunotherapy. *Surgery 76*:23-43.

27. Veltman, L. L., D. H. Kern and Y. H. Pilch (1974). Immune cytolysis of human tumor cells mediated by xenogeneic immune RNA. *Cellular Immunol. 13*:367-377.

TRANSFER OF CELLULAR IMMUNITY WITH IMMUNE RNA EXTRACTS AND TUMOR IMMUNOTHERAPY

Sheldon Dray

Department of Microbiology
University of Illinois at the Medical Center
Chicago, Illinois

During the last 9 years, studies in our laboratory on the transfer of cellular immunity with RNA extracts of lymphoid tissues have led to the development of an animal model for specific tumor immunotherapy which may be applicable to man (1-17). In all of our experiments, the RNA was extracted by the phenol method from lymph nodes (LN) and/or spleen (Sp) of a donor sensitive to the selected antigen (1-17); this RNA is referred to as "immune" RNA. When the RNA extracts or 8-12S fractions thereof were incubated with nonsensitive lymph node or peritoneal exudate cells (PEC), the cells were converted to a state of specific immunological reactivity as assessed *in vitro* by the inhibition of macrophage migration (Migration Inhibition Factor, MIF, assay) and *in vivo* by the delayed hypersensitivity skin reaction (DHSR) or tumor regression (TuReg) (Table 1). Transfer of cellular immunity has been observed when the donor RNA and recipient cells were of the same species — allogeneic transfers (1-3,7,11,12); of the same inbred line — syngeneic transfers (5,9,13-16); or of different species — xenogeneic transfers (4,6,8,10,13,14) (Table 1). Transfer of cellular immunity by RNA extracts has been observed for a variety of antigens (Table 1).

ALLOGENEIC TRANSFERS

Man

Our investigations began with allogeneic transfers in man (1,2). The RNA was extracted from lymph nodes of patients with skin sensitive to tuberculin purified protein derivative (PPD) and/or histoplasmin. The RNA extract (500 μg/ml) was incubated at 37° C for 15 min with 72 hr-cultured lymph node cells (10^9 cells/ml) from individuals not sensitive to PPD or histoplasmin. When nonsensitive cells were incubated with an RNA extract from lymph nodes of human donors sensitive to PPD, histoplasmin or both, the migration

TABLE 1.
Transfer of Cellular Immunity with Immune RNA Extracts

Exp	Donor RNA Tissue Species		Recipient Cells Tissue Species		Antigen	Assay	References
A. Allogeneic Transfers							
1	LN	Man	LN	Man	PPD	MIF	Thor(1); Thor & Dray (2)
2	,,	,,	,,	,,	Histoplasmin	,,	,,
3a	,,	,,	,,	,,	PPD	,,	Thor & Dray (7)
4a	,,	,,	,,	,,	Histoplasmin	,,	,,
5	LN+Sp	GP	PEC	GP	PPD	,,	Jureziz et al. (3)
6	,,	,,	,,	,,	Coccidioidin	,,	,,
7	,,	,,	,,	,,	ARS-NAT	,,	Schlager et al. (11)
8	,,	,,	,,	GP-S13	PPD	,,	Paque et al. (12)
9	,,	,,	,,	,,b	α, DNP-OL	,,	,,
10	,,	,,	,,	,,b	ε, DNP-OL	,,	,,

B. Syngeneic Transfers

#							
	LN	GP-S13	PEC	GP-S13	PPD	DHSR	Jureziz et al. (5)
11	LN	GP-S13	PEC	GP-S13	PPD	DHSR	Jureziz et al. (5)
12	"	"	"	"	Brain	"	"
13	"	GP-S2	"	GP-S2	α, DNP-OL	"	"
14	"	"	"	"	ε, DNP-OL	"	"
15	LN–Sp	"	"	"	Line-1 TSA	MIF	Paque et al. (9)
16	"	"	"	"	Line-10 TSA	"	Schlager et al. (13)
17[a]	"	"	"	"	Line-1 TSA	"	Paque (16)
18[a]	"	"	"	"	Line-10 TSA	"	"
19	"	"	"	"	Line-10 TSA	TuReg	Schlager et al. (13); Schlager et al. (14,15)
20	Sp	BALB/c	"	BALB/c	MOPC-300 TSA	MIF	Braun & Dray (17)
21	"	"	"	"	MOPC-315 TSA	"	"
22	"	"	"[c]	"[d]	MOPC-315 TSA	"	"

C. Xenogeneic Transfers

#							
23	LN+Sp	Monkey	PEC	GP	PPD	MIF	Paque & Dray (4,8)
24	"	"	"	"	Coccidioidin	"	"
25	"	"	"	GP-S2	Line-10 TSA	"	Schlager et al. (13)
26	"	"	"	"	Line-10 TSA	TuReg	Schlager et al. (13); Schlager & Dray (14,15)
27	"	"	LN	Man	Coccidioidin	MIF	Paque & Dray (6,8)
28	"	"	"	"	KLH	"	"
29[a]	"	"	"	"	KLH	"	Paque & Dray (10)

[a] 8-12 S fractions of RNA extracts used.
[b] Strain 13 guinea pigs are nonresponders to α, or ε, DNP-oligolysines.
[c] BALB/c bearing 5 day tumor whose PEC respond to tumor specific antigen in MIF test.
[d] BALB/c bearing 14 day tumor whose PEC are unresponsive to challenge with tumor specific antigen in MIF test.

of these cells from capillary tubes was inhibited only by the corresponding antigen(s). Neither the antigen alone nor RNA from nonsensitive donors were able to convert nonsensitive cells to a state of sensitivity. Thus, the RNA conversion of nonsensitive cells to sensitive cells is antigen specific, a characteristic which has been confirmed by all subsequent experiments (Table 1).

By sucrose density gradient centrifugation, the RNA extracts have characteristic 28S, 18S and 4S peaks (Fig. 1). The biuret test indicated less than 3% contamination with protein. When the RNA extracts were treated with DNase or trypsin, the sucrose density gradient centrifugation patterns were essentially unaltered and the RNA extracts retained their cell conversion activity. However, when the RNA extracts were treated with pancreatic RNase, virtually all of the high molecular weight RNA species were degraded as indicated by the sucrose density gradient centrifugation pattern which exhibited essentially only 4S material or smaller and this RNA extract was inactive in cell conversion. When the RNA extracts were fractionated by sucrose density gradient centrifugation, the cell conversion activity was localized to the 8-12S fraction (Fig. 1) (7). Thus, the conversion of nonsensitive cells by RNA to sensitive cells requires intact RNA, a characteristic which has been tested in most other transfers and confirmed (Table 1). That the effective RNA is 8-12S has been confirmed for monkey RNA and guinea pig RNA in two other systems (Exps. 17,18,29, Table 1).

Guinea Pig

An animal model for the transfer of cellular immunity with RNA extracts was developed with guinea pigs. The specificity of the reaction was first shown for the relatively crude PPD and coccidioidin antigens and then for the chemically defined low molecular weight antigens, α, DNP-oligolysines, ϵ, DNP-oligolysines, and mono-(p-azobenzene-arsonate)-N-chloracetyl-L-tyrosine (ARS-NAT) (Table 1).

The use of ARS-NAT made it possible to use an exquisitely sensitive quantitative assay for the presence of antigen in the active RNA extracts. Although we had already shown that the PPD or histoplasmin antigen could not be detected in the RNA extracts by inhibiting the migration of sensitive cells (2), this test is not quantitative and the argument has been put forth than antigen is present in RNA extracts but may not be exposed. The use of radioactive labels on crude antigens such as PPD or histoplasmin would not be informative since the argument would be made that antigen fragments free of label would still be present. Thus, ARS-NAT with a molecular weight of 486 and containing arsenic as a chemical marker provided a basis for sensitive assays. RNA extracts from ARS-NAT sensitive donors known to transfer sensitivity to ARS-NAT was assessed for arsenic content by atomic absorption spectroscopy (AAS). AAS assays the arsenic directly, has a sensitivity greater

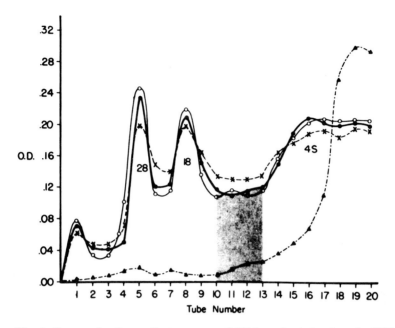

Fig. 1. *Sucrose density gradient patterns of RNA species isolated as the RNA extract from a PPD-sensitive human RNA donor. The RNA species are represented as peaks 4 S (top of tube), 18 S and 28 S (bottom of tube). Intact RNA lymph node extract (solid circles ●); RNA extract after incubation with DNase (open circles ○); RNA extract after incubation with trypsin (X's); RNA extract after incubation with RNase (solid triangles ▲). Shaded area indicates the location of fractions with activity in cell conversion (7).*

than any biological or chemical assay, and since the sample is exhaustively incinerated at 2700°C, "buried antigen" cannot escape detection. AAS, with a sensitivity of 0.1 ng, failed to detect arsenic in 250 μg to 10 mg of ARS-NAT immune RNA (Fig. 2), suggesting that, if arsenic is associated with the RNA-rich extracts, it could be present in an amount of no more than 5 pg in 500 μg of RNA, the amount of RNA usually used for cell conversion. This corresponds to less than 0.0000065% ARS-NAT antigen in the RNA extract. Thus, these results suggest that antigen is not the active component in the RNA extracts. Since we know that the RNA extracts are inactivated by RNase and that the activity resides in an 8 to 12S fraction, it seems reasonable to postulate that the RNA may have a regulatory or an informational role. Regulation might occur by specific derepression of the gene controlling the synthesis of a cell receptor specific for the antigen. Alternatively, the base sequence of the RNA may provide the information for the cell to synthesize the specific cell receptor. The cell bearing the specific cell receptor could then respond to the antigen and release migration inhibition factor. That these hypotheses seem

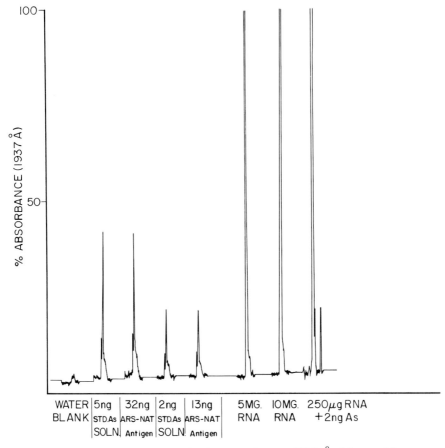

Fig. 2. *Atomic spectroscopic absorbance readings at 1937 Å of 5 ng and 2 ng standard arsenic solutions, 13 ng and 32 ng samples of ARS-NAT antigen, 5 mg and 10 mg samples of RNA extracts prepared from guinea pigs sensitized to ARS-NAT antigen, and a localization control of the background RNA (250 μg RNA) in the presence of arsenic absorbance (2 ng As) (11).*

plausible are also indicated by the observations than RNA extracts convert lymphocytes to secrete allogeneic allotypes or to acquire specific plasmacytoma immunoglobulins on their surface as reviewed by Drs. Fishman and Heller during this workshop.

Delayed hypersensitivity skin reactions to DNP-oligolysines can be induced in inbred strain 2 or individual outbred guinea pigs but not in strain inbred 13 guinea pigs. The ability of guinea pigs to respond immunologically to DNP-oligolysines depends on the inheritance of an autosomal dominant immune response (Ir) gene which is linked to the major histocompatibility locus. Specific sensitivity for α or ϵ, DNP-oligolysines was transferred to nonresponder strain 13 guinea pig peritoneal exudate cells by treatment of the cells with RNA extracts prepared from lymphoid cells of responder strain 2 or Rockefeller guinea pigs as demonstrated by inhibition of macrophage migration (12) confirming an earlier report by Thor (18). Thus, in accordance with the above hypotheses, the RNA may cause the nonresponder strain 13 cell to synthesize a specific gene product capable of recognizing and reacting with DNP-oligolysines. A gene product that might be synthesized could be a membrane receptor which reacts with the DNP-oligolysines and triggers the release of migration inhibition factor. This also fits with the idea that the Ir gene may direct the synthesis of a receptor on T lymphocytes and that a deficiency of these receptors may be the reason for the lack of response in the strain 13 guinea pigs. This could have practical significance for the restoration of genetic defects in the immune system.

SYNGENEIC TRANSFERS

Inbred Guinea Pigs, Strain 2 and Strain 13

We turned to syngeneic systems to determine whether the *in vivo* delayed hypersensitivity skin test could be transferred by adoptive transfer of RNA treated peritoneal exudate cells. Successful transfers were obtained by intraperitoneal injections of approximately 10^9 nonsensitive lymphoid cells previously incubated at $37°$ C for 15 to 30 min with 120 to 300 $\mu g/ml$ of immune RNA (5). The transfer of skin test sensitivity was shown to be specific for PPD, brain antigen, α, DNP-oligolysine and ϵ, DNP-oligolysine (Table 1). Thus, the transfer of cellular immunity with RNA extracts was demonstrated by the classical *in vivo* delayed hypersensitivity skin reaction (5).

The successful transfer of delayed hypersensitivity to various antigens as indicated by the skin test suggested that RNA extracts should transfer delayed hypersensitivity with respect to tumor specific antigens and perhaps have therapeutic effects. For our investigation, we selected the line-1 and line-10 antigenically distinct, transplantable, chemically induced hepatomas in strain 2 guinea pigs. As a first step, we used the *in vitro* MIF test and showed that the transfer with RNA extracts was specific for the line-1 and line-10 antigens (Table 1). Furthermore, as shown for human RNA, the cell conversion activity was localized to the 8-12S fraction (16). These experiments (9,16) will be reviewed by Dr. Paque during this workshop.

When 10^6 line-10 cells are injected intradermally into strain 2 guinea pigs, the tumor is uniformly lethal in 60 to 90 days (19). The tumors are known to metastasize frequently within 6 days after implantation. Guinea pigs injected intradermally with 10^6 line-10 tumor cells admixed with 6×10^6 living BCG organisms never develop an established tumor and are resistant to subsequent challenge with the line-10 tumor (19). When the BCG is injected intratumorally 6 or 12 days after the injection of 10^6 cells, complete tumor regression is observed in approximately 85% or 20% of the animals, respectively (20). This provided us with a syngeneic tumor model to test a variety of immunotherapeutic procedures based on the transfer of cellular immunity with RNA extracts which we have observed *in vitro* by the MIF assay and *in vivo* by the delayed hypersensitivity skin reaction (Table 1). We succeeded in developing an RNA therapeutic regimen which cured all guinea pigs bearing a uniformly lethal 5 to 7 day tumor transplant (13-15). In our procedure, 10^7 syngeneic peritoneal exudate cells were injected first into the area around the tumor, 2.5 mg RNA from the lymphoid cells of tumor-immunized guinea pigs was injected immediately thereafter into the same site, and 1.0 mg of a preparation containing tumor specific antigen (TSA) was injected one hour later into the same site. The rationale for the RNA immunotherapeutic regimen was as follows: to circumvent the possible deficiency of immunocompetent host lymphoid cells in the tumor area, we injected normal peritoneal exudate cells first under the growing tumor; to convert these nonsensitive cells to a specific state of immunological sensitivity to line-10 tumor specific antigen (TSA), the RNA was injected immediately into the same site; to attempt an amplification effect (i.e., blast transformation and cell proliferation) of lymphoid cells that might have been converted by the RNA, line-10 TSA was injected later as a specific immunological stimulant.

The complete and apparent specific regression of the local tumor, survival of all of the tumored animals in which a high frequency of metastases occurs by 5 to 7 days, and the immunity of these animals to further tumor challenge indicated that the RNA immunotherapeutic procedure had systemic effects. To test this further, a lethal dose of tumor cells was injected intradermally into two sites 10 cm apart and only one site was treated. Not only did the treated tumor regress completely as predicted, but so did the untreated tumor and all animals survived. These experiments on the immunotherapy of line-10 tumors with RNA extracts (13-15) will be reviewed by Dr. Schlager during this workshop.

Because strain 2 inbred guinea pigs have been difficult to obtain, we have not had the opportunity to determine the optimal doses and timing for the injection of cells, RNA and TSA in relation to the tumor load, particularly in guinea pigs with increased tumor loads. Since strain 2 guinea pigs have been unavailable to us in sufficient numbers and since it is important to extend our observations to other tumor systems, we chose initially to investigate plasmacytomas in BALB/c mice. Tumor specific antigens were prepared from the MOPC-300 and MOPC-315 plasmacytomas. As in the experiments with the line-1 and line-10 tumors, we found that RNA extracts from tumor-immunized animals transferred cellular immunity to nonsensitive peritoneal exudate cells as assessed by the MIF test (Exp. 20, 21, Table 1).

Cellular immunity by the MIF test was assessed daily in BALB/c mice after receiving a transplant. We found that at 5 to 7 days, the peritoneal exudate cells were specifically inhibited in their migration by MOPC-315 TSA; but that by 12-14 days, the cells were unresponsive to challenge with TSA (Fig. 3). Similar results were obtained with MOPC-300, except that the time frame for the appearance and subsequent suppression of specific cell-mediated antitumor immunity was different. When RNA was extracted from 5-7 day MOPC-315 tumored mice and 200 μg RNA was used to treat 3×10^6 peritoneal exudate cells of 12-14 day tumored mice, the unresponsive 12-14 day cells were converted to be immunologically responsive to the MOPC-315 TSA by the MIF test (Fig. 4). Thus, it appears that lymphoid cells which become unresponsive in tumored animals can be converted by RNA to a state of immunological responsiveness and we will see later that this has important implications for the development of an immunotherapy procedure for man.

XENOGENEIC TRANSFERS

Monkey to Guinea Pig

Since the development of an immunotherapy procedure using lymphoid RNA extracts in man might depend on the availability of a tumor immunized RNA donor of another species, we examined the possibility of transferring cellular immunity with RNA extracts across species barriers (Table 1). We found that RNA extracts from immunized monkeys transferred immunological

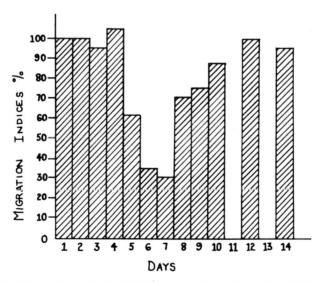

Fig. 3. *Cellular immunity in BALB/c mice after implantation of 10^6 MOPC-315 plasmacytoma tumor cells. Migration inhibition factor (MIF) assay of peritoneal exudate cells challenged with MOPC-315 tumor specific antigen (TSA).*

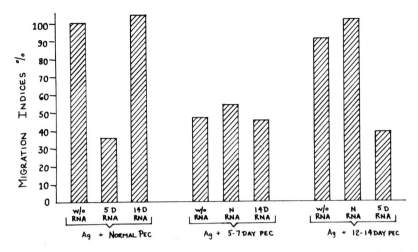

Fig. 4. *Transfer of cellular immunity with immune RNA extracts of 5-7 day (MOPC-315) tumored BALB/c mice to unresponsive peritoneal exudate cells of 12-14 day (MOPC-315) tumored BALB/c mice. Release of MIF was tested with MOPC-315 TSA. Controls without RNA (w/o), with RNA from normal BALB/c mice (N RNA), with normal PEC and 5-7 day PEC are shown.*

reactivity for PPD, coccidioidin and line-10 TSA to guinea pigs as assayed *in vitro* by the MIF test (4,8,13) and indeed was equally effective in the *in vivo* regression of the line-10 tumor as syngeneic RNA extracts (13-15).

Monkey to Man

Finally, we have shown that RNA extracts from immunized monkeys can convert human peripheral lymphocytes so that they release MIF upon challenge by the specific antigen, i.e., coccidioidin or keyhole limpet hemocyanin (KLH) (Table 1). The effective RNA was in the 8 to 12S fraction (10) as had been shown previously for human and guinea pig RNA. Thus, this provides a rational basis for experimenting with the use of xenogeneic RNA for immunotherapy in man. Also, the ability to transfer cellular immunity with xenogeneic RNA favors the hypothesis of an informational role for the RNA since the postulated cell receptors are presumably characteristic of the RNA donor as shown by the experiments with allotypes and plasmacytoma immunoglobulins.

CONSIDERATIONS FOR TUMOR IMMUNOTHERAPY IN MAN WITH IMMUNE RNA

Immune RNA

Based on our experience thus far in the use of immune RNA for the treatment of line-10 lethal hepatomas in guinea pigs, the development of similar procedures for tumor immunotherapy may be considered for man. In principle, the use of syngeneic or allogeneic immune RNA would be preferable for man. Suitable techniques may develop for the deliberate immunization of a normal human donor with the tumor or tumor specific antigen without risk and make this immunization procedure practical in selected circumstances. However, at the present time, experimental trials with xenogeneic immune RNA seem like the most practical approach since donors for tumor immunization are plentiful and adjuvants required for hyperimmunization can be used. Xenogeneic immune RNA had no obvious side effects when given to guinea pigs in a therapeutic procedure which cured the guinea pigs of lethal doses of the line-10 tumor. Moreover, in this workshop, Dr. Pilch will refer to his experience of injecting sheep immune RNA into man without yet observing untoward effects. Finally, with evidence accumulating that the active component of the immune RNA extracts is in the 8 to 12S fraction, only that fraction might be needed for therapeutic purposes. Nevertheless, since xenogeneic RNA could, in principle, have adverse effects on the host's normal tissue antigens as well as the antitumor effect, the use of xenogeneic RNA in man must proceed with considerable caution until more information is available from tests on experimental animals.

Immunocompetent Cells

From our work with the line-10 tumor in guinea pigs, it appears that an important requirement for immunotherapy is the availability of large numbers of immunocompetent cells at the tumor site relative to the tumor load. Tumor bearing individuals may have a deficiency in immunocompetent cells capable of responding to the specific tumor as illustrated by the tumored mice (Fig. 3). In man, external sources of immunocompetent cells are limited to identical twins or to allogeneic donors matched for the major histocompatibility antigens. In view of our experiments which show that immune RNA may convert cells of nonresponder normal guinea pigs to responder cells with respect to DNP-oligolysines (Exps. 9,10, Table 1) and that immune RNA may convert nonresponder cells from tumored mice to responder cells with respect to MOPC-315 TSA (Exp. 22, Table 1), it seems reasonable to attempt the use of autologous lymphocytes obtained from the tumored patient by lymphophoresis.

Tumor Specific Antigen

Our work in the guinea pig also suggests that tumor specific antigen has an important role as part of the immunotherapy regimen with immune RNA (13-15). In many situations for man, sufficient tumor tissue is removed at surgery so that the preparation of tumor specific antigen is feasible.

Tumor Immunotherapy

Even with the availability of immunocompetent cells, immune RNA, and tumor specific antigen, the development of protocols for tumor immunotherapy in man and their evaluation seem formidable. Initially, it might be desirable to select patients with a relatively large localized primary tumor which has a relatively high incidence of metastases. The logistics would require 1 to 3 weeks for immunization of an animal with the tumor or tumor specific antigen. It would not take long to prepare the RNA extracts from lymph nodes and spleen. Immune RNA should be tested before use for their ability to convert lymphoid cells *in vitro* so that they respond to tumor specific antigen and release migration inhibition factor. *In vitro* cytotoxicity tests could also be used for this purpose. Since the tumor would have been removed, the cells, RNA, and TSA could not be injected into a tumor site as was done in the guinea pig model. However, this may not be necessary in view of the observation in the guinea pig that the treatment of one line-10 tumor site was effective in developing systemic tumor immunity in sufficient time to cause complete tumor regression of a small but lethal second tumor 10 cm away (14,15). Hopefully, the *in vivo* interaction of cells, RNA and TSA at nontumor sites might also be sufficient to induce a sufficient degree of systemic tumor immunity. This latter question can be easily tested in the guinea pig when sufficient strain 2 animals become available.

In our experience with the guinea pig, we have only succeeded thus far in an immune RNA immunotherapy regimen which is based on the *in vivo* interaction of immune RNA with immunocompetent cells followed by TSA at the tumor site. We have not been successful in other experiments in which immune RNA was injected alone or in which immunocompetent cells were incubated *in vitro* with immune RNA and then injected. However, by varying the dose and route of injection, conditions may be found in which immune RNA alone or immunocompetent cells incubated with immune RNA *in vitro* might be more effective. For the development of optimal conditions for immune RNA therapeutic procedures which may become applicable to man, the line-10 hepatoma model in the guinea pig seems highly suitable but other tumor systems should also be investigated.

REFERENCES

1. Thor, D. E. (1967). Delayed hypersensitivity in man: A correlate *in vitro* and transfer by an RNA extract. *Science 157*:1567-1569.
2. Thor, D. E. and S. Dray (1968). The cell-migration-inhibition correlate of delayed hypersensitivity. *J. Immunol. 101*:469-480.
3. Jureziz, R. E., D. E. Thor and S. Dray (1968). Transfer with RNA extracts of the cell migration inhibition correlate of delayed hypersensitivity in the guinea pig. *J. Immunol. 101*:823-829.
4. Paque, R. E. and S. Dray (1970). Interspecies "transfer" of delayed hypersensitivity *in vitro* with RNA extracts. *J. Immunol. 105*:1334-1338.
5. Jureziz, R. E., D. E. Thor and S. Dray (1970). Transfer of the delayed hypersensitivity skin reaction in the guinea pig using RNA-treated lymphoid cells. *J. Immunol. 105*:1313-1321.

6. Paque, R. E. and S. Dray (1972). Monkey to human transfer of delayed hypersensitivity *in vitro* with RNA extracts. *Cell. Immunol.* 5:30-41.
7. Thor, D. E. and S. Dray (1973). Transfer of cell-mediated immunity by immune RNA assessed by migration-inhibition. *Ann. N. Y. Acad. Sci.* 207:355-368.
8. Paque, R. E. and S. Dray (1973). Interspecies "transfer" of delayed hypersensitivity *in vitro* with RNA extracts. *Ann. N. Y. Acad. Sci.* 207:369-379.
9. Paque, R. E., M. S. Meltzer, B. Zbar, H. J. Rapp and S. Dray (1973). Transfer of tumor-specific delayed hypersensitivity *in vitro* to normal guinea pig peritoneal exudate cells using RNA extracts from sensitized lymphoid tissues. *Cancer Res.* 33:3165-3171.
10. Paque, R. E. and S. Dray (1974). Transfer of delayed hypersensitivity to nonsensitive human leukocytes with Rhesus-monkey lymphoid RNA extracts. *Transpl. Proc.* 6:203-207.
11. Schlager, S. I., S. Dray and R. E. Paque (1974). Atomic spectroscopic evidence for the absence of a low-molecular-weight (486) antigen in RNA extracts shown to transfer delayed-type hypersensitivity *in vitro*. *Cell Immunol.* 14:104-122.
12. Paque, R. E., M. Ali and S. Dray (1975). RNA extracts of lymphoid cells sensitized to DNP-oligolysines convert nonresponder lymphoid cells to responder cells which release migration inhibition factor. *Cell. Immunol.* 16:261-268.
13. Schlager, S. I., R. E. Paque and S. Dray (1975). Complete and apparently specific local tumor regression using syngeneic or xenogeneic "tumor-immune" RNA extracts. *Cancer Res.* 35:1907-1914.
14. Schlager, S. I. and S. Dray. Tumor regression at an untreated site during immunotherapy of an identical distant tumor. *Proc. Nat. Acad. Sci. USA*, in press.
15. Schlager, S. I. and S. Dray. Complete regression of a guinea pig hepatocarcinoma by immunotherapy with "tumor-immune" RNA or antibody to fibrin fragment E. *Israel J. Med. Sci.,* in press.
16. Paque, R. E. (1976). An *in vitro* model for tumor specific sensitivity with "tumor-immune" RNA extracts and localization of immunologically active RNA fractions. *The Macrophage in Neoplasia.* Mary A. Fink, Ed. Academic Press, New York.
17. Braun, D. P. and S. Dray. "Tumor-immune" RNA conversion of lymphoid cells of tumored mice from being unresponsive to responsive upon challenge with tumor specific antigen. *Proc. Amer. Ass. Cancer Res. Abs.,* in press.
18. Thor, D. E. (1969). Discussion of "Elaboration of effector molecules by activated lymphocytes." *Mediators of Cellular Immunity.* H. S. Lawrence and M. Landy, Eds. Academic Press, New York.
19. Rapp, H. J. (1973). A guinea pig model for tumor immunology. *Israel J. Med. Sci.* 9:366-374.
20. Zbar, B., I. D. Bernstein, G. L. Bartlett, M. G. Hanna and H. J. Rapp (1972). Immunotherapy of cancer: Regression of intradermal tumors and prevention of growth of lymph node metastases after intralesional injection of Mycobacterium bovis. *J. Natl. Cancer Inst.* 49:119-130.

TRANSFER FACTOR: IS IT RELATED TO IMMUNE RNA?

Fred T. Valentine, M.D.

Department of Medicine
New York University Medical Center
New York, New York 10016

Transfer factor has not been found to be a form of immune RNA. Rather, it is being discussed because it is a substance present in extracts of sensitive lymphoid cells that is able to transfer delayed hypersensitivity or cell-mediated immunity. The identification of the exact biochemical nature of transfer factor remains an elusive problem.

Not only will this paper not answer the question "What is immune RNA?," but for the moment it will be left to others to decide whether transfer factor might be an immune RNA. I will attempt to provoke a good discussion rather than to present a review. The subject of transfer factor has been thoroughly reviewed elsewhere (1,2).

Investigations of transfer factor have been in progress for many years, and have proceeded independently from those on immune RNA. Transfer factor (TF) is an operational designation for nonimmunoglobulin substances present in extracts of sensitive blood lymphocytes that can introduce specific cell-mediated immunity into nonsensitive humans. These investigations, performed largely by H. S. Lawrence and his collaborators, have delineated many biological characteristics of an immunological phenomenon mediated by substances whose chemical nature still is not known. The direction that these studies have taken is to some extent opposite from that taken with immune RNA. Transfer factor might be said to be an immunological phenomenon in search of a molecule. There are many similarities between these problems, and between types of experiments encountered during investigations of these two types of immunological reagents. Those seeking the nature and mode of action of transfer factor, however, must also contend with the additional conceptual hazard of the apparently unprecedented ability of a small molecule to convey immunologically specific effects to recipient cells. Also, until very recently, TF has not been demonstrated in any species other than man.

The following discussion includes certain crucial biological and immunological observations about TF; what is known about the biochemical nature of this substance; a consideration of recent investigations whose goal is to describe a reproducable effect of TF *in vitro* or in animals that might serve as a bioassay for the characterization of this material; and the mention of a few examples of clinical applications of TF that have given great impetus to the field.

A most important observation is that the donor whose lymphocytes are used to prepare TF must have exceedingly high reactivity to antigen, otherwise the extract from his cells will not be able to transfer specific immunity. Extracts of lymphocytes from nonsensitive individuals or from those with only slight sensitivity will not suffice. The immunological reactivity that is transferred can immediately be detected in the recipient, a characteristic associated with passive transfers. The recipient acquires the immunological sensitivities of the donor as measured by delayed type hypersensitivity reactions elicited by the injection of the appropriate antigens into the recipient's skin. He becomes positive not only at the site into which the TF was injected but throughout his body. The immunological reactivity that has been transferred can be detected even when the skin test is not applied until several months after the injection of TF. The sensitivity introduced by TF lasts for several years, far longer than can be explained by the persistence of a molecule of conventional biological half-life. To further investigate this question a serial transfer was performed (3). The recipient of TF was used as a donor for a second successful transfer. If the recipient behaved as though sensitive because of the passive introduction of a unique long lasting immunological molecule, then one would suspect that this molecule would have been diluted out during the serial passage. Thus, recipients of TF seemed to have synthesized an increased amount of TF to the particular antigen, so that they in turn can serve as donors. It should be noted that in these experiments recipients were skin-tested with antigen prior to serving as donors of TF. It is possible that this antigenic stimulus induced a proliferation of the newly sensitized cells in the recipient, and that these new cells accounted for the increased amount of TF detected in the serial passage.

The recipient of TF does not have to be skin-tested to determine if he has acquired specific immunity. Several investigators have found that the introduction of immunological reactivity can be detected in recipients of TF by means of tests for cell-mediated immunity *in vitro*. Recipient lymphocytes most frequently have been noted to acquire the ability to form migration inhibitory factor when stimulated by specific antigen *in vitro*, but an immunological transfer also has been detected by measurements of lymphocyte-proliferation and lymphocyte-mediated cytotoxicity *in vitro*.

One of the most important issues in the study of TF, and that point which is most frequently questioned, is whether TF transfers immunologically specific immune reactivity. Several lines of evidence suggest that this occurs. TF has been successfully used to transfer delayed type hypersensitivity to many microbial antigens. With rare exceptions only, those sensitivities possessed by the donor were successfully transferred into the recipient. Delayed cutaneous reactivity has been successfully transferred to the new antigenic determinants on etheline oxide-denatured serum protein by means of TF (4). In these studies the recipients became sensitive to antigenic determinants that they had never met before. A similar experiment has recently been performed in which the sensitivity to keyhole lymphoid hemocyanin (KLH) has been successfully transferred in 10 of 10 consecutive tries (5). TF prepared from donors not sensitive to KLH but sensitive to tuberculin PPD transferred reactivity only to PPD. It should be noted that the recipients were not skin-tested prior to receiving the TF. Only 5 to 10% of normal individuals have been reported to be sensitive to KLH.

Perhaps the most elaborate evidence for the specificity of transfer factor is provided by studies on the transfer of the accelerated homograft rejection (6). In these experiments an individual A was sensitized by the repeated application of skin grafts from individual B. TF was then prepared from individual A and injected into recipient C. Recipient C was then challenged by the application of skin grafts from individual B, and control skin grafts from individuals D and E. Only the skin graft from individual B who had been used to sensitize the donor of TF was given an accelerated rejection. Prior to skin grafting or after he had received only one or two skin grafts, individual A was not able to transfer the specific immune reaction. These experiments demonstrate that TF can transfer specific cell-mediated immunity, and also indicate that the transfer factor for a particular immunological reactivity is not present in a donor until he has been specifically sensitized.

The immunological specificity of transfer factor also has been reported in studies of patients with cancer. When patients afflicted with osteogenic sarcoma were injected with TF from donors sensitive to the tumor, recipient lymphocytes temporarily acquired the ability to destroy this tumor *in vitro* (7). If these individuals were subsequently injected with TF prepared from donors unreactive to the tumor, the recipient lymphocytes did not acquire the ability to destroy this tumor. A repeat injection of TF from sensitive donors again leads to the acquisition by recipient lymphocytes of the ability to destroy the tumor *in vitro*.

In another study, a patient with chronic mucocutaneous candidiasis was injected with TF from a donor sensitive to candida (8). The patient's lymphocytes acquired the ability to produce migration inhibitory factor (MIF) after receiving the TF from the sensitive donor. In addition, a long remission in the disease was induced. When the disease relapsed, the administration of TF from a donor not sensitive to candida was unable to initiate prolonged clinical improvement nor to induce the recipient lymphocytes to respond to the candida antigen *in vitro*. When the patient again was treated with transfer factor from a sensitive donor, the patient experienced a prolonged clinical remission and her lymphocytes again acquired the ability to form MIF when cultured with candida antigen *in vitro*. This cross-over study suggests that TF transfers specific reactivity as judged by an *in vitro* test for cell-mediated immunity, and also that the clinical remission induced depends on specific sensitization of the donor of TF.

Finally, experiments performed with several animal models that have been recently described also provide evidence that TF can transfer immunologically specific reactivity. Another intriguing property of TF is that, when examined, TF has not been documented to transfer the formation of specific antibodies even in recipients who have acquired high levels of delayed type hypersensitivity (9).

The possibility that antigen might be present in some preparations of immune RNA has been discussed in several presentations at this meeting. A similar question has been raised about TF. Several observations however suggest that antigen is not present in TF. First, the production of antibody has not been detected in recipients of TF. It seems unlikely that the putative antigen in TF might be of a configuration that renders it unable to induce the formation of antibody even while inducing cell-mediated immunity. The rapidity with which the recipient of TF develops cell-mediated immunity would seem to provide little time for active sensitization to occur in response to any antigen that might be present in TF. It also should be noted that in the majority of experiments the donors of TF have been normal individuals who were sensitized to microbial antigens at a very remote time. They do not have large amounts of antigen present anywhere at the time that they donate. In studies on the administration of TF to patients afflicted with overwhelming infectious diseases, it has been documented that TF is able to induce specific reactivity to the microbe involved in spite of the fact that the patient has enormous amounts of microbial antigen already in his system. The reason why this occurs is not known, but it indicates that transfer factor possesses immunological properties that cannot be attributed to antigen. Transfer factor from a tuberculin sensitive donor added to cultures of lymphocytes *in vitro* will not stimulate lymphocytes from a highly sensitive individual whose cells normally will respond to as little as 1 nanogram of tuberculin PPD per ml. Taken together these observations suggest, but do not prove, that the presence of antigen in transfer factor does not account for its biological activity.

A brief listing of the biochemical properties of TF is especially relevant to a consideration of the relation between TF and immune RNA. Attempts have been made to degrade the biological activity of TF with several enzymes. In each case appropriate substrate markers were included to show that the enzyme was active in the presence of TF. TF activity proves to be resistant to DNase and also to pancreatic RNase. To the best of my knowledge, no other nucleases have been tested in experiments where a valid bioassay was used to determine the results. TF has also been found to be resistant to the action of trypsin. A limited number of experiments suggest that the activity may be degraded by pronase.

Perhaps the most puzzling and exciting observation about transfer factor was the discovery that the active component in extracts of leucocytes was of a sufficiently small molecular weight to be dialysable (10). The biological activity passed through a cellulose membrane normally considered to retain molecules above 12-14,000 Daltons. While some of the experiments previously discussed were conducted with whole extracts of sensitized leucocytes, the majority of the recent work has been performed with dialyzable TF. In addition, it has been found that biological activity of TF is inactivated by heating at 56 degrees for 30 minutes. To my knowledge, the resistance of TF to pancreatic RNase is quite unlike the characteristics of immune RNA. On the other hand, it is theoretically possible that the active component might consist at least in part of RNA protected by protein or containing unusual bases that would render it resistant to this enzyme.

It is not surprising that leucocyte dialysates have been found to contain polynucleotides, polypeptides and orsinol positive material. Ribonucleoprotein molecules also have been detected (11). Preliminary attempts at fractionation of dialysable TF in several laboratories, however, have not provided conclusive evidence as to the biochemical nature of the active component (10-14). Work on the chemical characterization of TF has been plagued by the lack of easily performed assays of biological activity. Little agreement exists between different investigators as to the position at which the biological activity of dialysable TF elutes from sephadex gel when the ability to transfer delayed type hypersensitivity in humans is used as an end point. This difficulty may result in part from the reported adsorption of the biologically active substances to the sephadex (13). The likelihood that the biologically active component of TF adsorbs to sephadex also makes the estimation of molecular weight by this method unsatisfactory.

Strong evidence previously discussed indicates that TF can mediate specific immunological reactivity. However, other evidence suggests that dialysates from lymphocytes can also have immunologically nonspecific effects. These observations almost exclusively have been made in patients with diseases associated with immunological difficiencies of unknown cause. For example, it has been found that some patients, who could not be sensitized by repeated applications of dintrochlorobenzene, acquired the ability to manifest a positive skin test to this antigen after receiving TF from donors who were considered to be nonsensitive (15). It is apparent that these patients had been exposed to abundant amounts of antigen, and that TF may have enabled them to develop or express a sensitization to the antigen to which they had been exposed rather than nonspecifically introducing reactivity which the donor did not possess. In another study, patients with immunological difficiencies and low numbers of circulating T-lymphocytes were found to have an increase in the number of T cells following the injection of TF (16). In a group of patients whose lymphocytes were unable to manifest mixed leucocyte culture reactivity, the injection of TF restored the ability of their lymphocytes to engage in this reaction (17). Fractions of TF obtained from sephadex columns and from thin layer chromatography have been found to restore the ability of patients with sarcoidosis to manifest delayed type hypersensitivity responses (18).

In these examples TF has an immunologically nonspecific effect. In each case, however, TF could be described as having circumvented a block in immunocyte differentiation or in the expression of sensitivities previously present in the donor but not expressed because of his disease. It is obvious that in addition to containing specific TF, dialysates of human lymphocytes contain many small molecules of the cell, some of which might be expected to be biologically active. For example, dialysates of leucocytes have been found to contain substances that elevate cyclic GMP in monocytes (19). Recent studies on extracts of the thymus have disclosed that a dialysable polypeptide, thymopoeitin, is able to facilitate the differentiation of T-lymphocytes *in vitro* (20,21). A second polypeptide, ubiquitin, with a molecular weight of 8,451 and a known amino acid sequence is present in a wide array of living things and is able to facilitate the differentiation of both T and B lymphocytes (22,23). It is not yet clear whether the nonspecific effects of transfer factor in patients with immunological deficiency diseases are attributable to nonspecific molecules of this sort.

Attempts to detect an effect of TF *in vitro* that might be used as an assay for this material have been especially plagued by problems of nonspecific effects. One example will illustrate this point. Lymphocytes from donors sensitive to an antigen undergo a vigorous proliferative response when cultured *in vitro* with that antigen (24). Lymphocytes from donors not sensitive to an antigen manifest little or no proliferation in response to the antigen. Dialysate of leucocytes obtained from a donor sensitive to a given antigen and added along with that antigen to lymphocytes from a nonsensitive donor will enable those lymphocytes to undergo a significant proliferative response (25). However, in a large series of experiments, it became apparent that dialysates obtained from donors with no sensitivity to an antigen worked just as well in enabling nonsensitive lymphocytes to respond to that antigen as did those obtained from donors who were sensitive to the antigen (26,27). Furthermore, the significant proliferative response of nonsensitive lymphocytes in the presence of dialysates and antigen was dependent upon the small and sometimes barely detectable response of the nonsensitive lymphocytes to antigen alone. Thus, if the small clone of cells responding to a given antigen in a population of lymphocytes from a nonsensitive donor was eliminated by first culturing those lymphocytes with that antigen in the presence of high specific activity thymidine, the hot pulse technique, then further reactivity to that antigen in presence of leucocyte dialysates was not observed (27). Dialysable substances in leucocytes therefore act in this *in vitro* system as nonspecific adjuvants that enhance a minimal proliferative response to antigen. The predominant effect of unfractionated dialysates of leucocytes on the proliferation of lymphocytes *in vitro* depends on the minimal background response of the cells used to detect this effect rather than on the specific sensitivities of the donor of the dialysate.

The effects of leucocyte dialysates in several other assays of cell-mediated immunity *in vitro* are now being investigated for their suitability as an assay for TF. Preliminary studies of the ability of TF to enable nonsensitive lymphocytes to respond to antigen by the production of migration inhibitory factor seem promising.

Recently TF has been found to be effective in the transfer of cell-mediated reactions in several experimental animals. These effects seem sufficiently reproducible so that they might serve as the much needed assay for further studies on TF. Dialysable TF has been found to transfer specific immunological reactions from man to monkey and from monkey to monkey (28-30). In these animals, reactivity was tested either by the biopsy of specific skin tests or by *in vitro* tests for cell-mediated immunity. It is perhaps not surprising that a substance as small as TF is able to transfer immunity between species. In a recent study, marmosets were protected by means of a prior injection of TF against lethal disease caused by herpes viruses (30). Animals receiving TF from humans with high levels of cell-mediated immunity against herpes simplex virus were protected when subsequently challenged with this virus. Animals receiving no TF or TF from baboons sensitized to herpes virus saimiri died from a lethal systemic infection and encephalitis when challenged with herpes simplex virus. Several guinea pig models and a mouse model also are under investigation. Dialysable TF has been reported to transfer immunity in rats to a coccidian parasite (31) and a similar system is under study in cows (32).

For many years few investigators studied the transfer of delayed type hypersensitivity with TF. With the increased awareness that cell-mediated immunity plays an important role in defending the host against a variety of intracellular parasites and against neoplastic disease (33), a number of physicians have employed TF in attempts to introduce cell-mediated immunity into patients afflicted with these diseases (2,34-36). The initial reports of clinical improvements following the administration of TF have stimulated many investigators to look again at this immunological phenomenon. TF is now being studied not only because of the immunological and biochemical puzzles that it presents but also because of its potential usefulness as a therapeutic agent in a variety of diseases. The greatest clinical efficacy of TF has been seen in patients with the Wiscott-Aldrich syndrome and in patients with chronic mucocutaneous candidiasis. In each of these groups approximately 50% of patients respond to injections of TF. TF also has been used in attempts at the therapy of neoplastic diseases (37,38). While clinical remissions have been reported, no controlled studies currently have been completed to prove the efficacy of this form of immunotherapy in neoplastic disease. However, it has been clearly demonstrated that large amounts of TF can be administrated repeatedly over long periods of time without any detrimental effects on the patient (39). A list of other diseases in which TF is being tested for its therapeutic benefit is long and includes chronic infectious diseases such as chronic active hepatitis and diseases of unknown etiology such as Bechets disease. The apparent ability of TF made in one species to be effective in other species suggests the possibility of sensitizing animals against the etiologic agents of human infectious disease or against human tumors and using these sensitized animals as donors of TF for the treatment of patients. Since the biological activity of TF is able to pass a dialysis membrane, there is little chance that viral genome might inadvertantly be transferred into the patient. The fact that until recently TF could only be studied in humans has slowed the progress of investigations, however, it has also facilitated the early testing of this material for its efficacy in human disease.

In summary, TF is a dialysable material present in the lymphocytes of sensitive individuals that has been demonstrated by many investigators to transfer specific cell-mediated immunity to nonsensitive humans. The biochemical nature of this substance and the mechanisms by which it works are unknown. It is hoped that the development of *in vitro* systems for detecting TF and the development of models in animals soon will enable us to elucidate these problems. If the immunological and biochemical enigmas of TF do not provide reasons for its intensive investigation then surely the effectiveness of TF in the treatment of certain human disease will stimulate additional work. Everyone is aware that this phenomenon cannot easily be explained by current biochemical concepts and that makes it even more interesting.

REFERENCES

1. Lawrence, H. S. (1969). Transfer Factor. *Advances in Immunology, Vol. 11*. Academic Press, New York.
2. Lawrence, H. S. (1974). Transfer Factor. *Cellular Immunity, Harvey Lecture Series 68*. Academic Press, New York.
3. Lawrence, H. S. (1955). The transfer in humans of delayed skin sensitivity to streptococcal M substance and to tuberculin with disrupted leucocytes. *J. Clin. Invest.* 34:219-230.

4. Maurer, P. H. (1961). Immunologic studies with ethylene oxide-treated human serum. *J. Exp. Med. 113*:1029-1040.

5. Zuckerman, K. S., J. A. Neidhart, S. P. Balcerzak and A. F. LoBuglio (1974). Immunologic specificity of transfer factor. *J. Clin. Invest. 54*:997-1000.

6. Lawrence, H. S., F. T. Rapaport, J. M. Converse and W. S. Tillet (1960). Transfer of delayed hypersensitivity to skin homografts with leucocyte extracts in man. *J. Clin. Invest. 39*:185-198.

7. Levin, A. S., V. S. Byers, H. H. Fudenberg, J. Wybran, A. J. Hackett, J. O. Johnson and L. E. Spitler (1975). Osteogenic sarcoma. Immunologic parameters before and during immunotherapy with tumor-specific transfer factor. *J. Clin. Invest. 55*:487-499.

8. Littman, B. H., R. E. Rocklin, R. Parkman, J. T. Potts and J. R. David (1976). Combination transfer factor – Amphotericin B therapy in a case of chronic mucocutaneous candidiasis: A controlled study in transfer factor. *Transfer Factor. Basic Properties and Clinical Applications.* Academic Press, New York.

9. Lawrence, H. S. and A. M. Pappenheimer, Jr. (1956). Transfer of delayed hypersensitivity to diphtheria toxin in man. *J. Exp. Med. 104*:321-336.

10. Lawrence, H. S., S. Al-Askari, J. R. David, E. C. Franklin and B. Zweiman (1963). Transfer of immunological information in humans with dialysates of leucocyte extracts. *Trans. Ass. Amer. Physicians 76*:84-89.

11. Gottlieb, A. A., L. G. Foster and S. R. Waldman (1973). What is transfer factor? *Lancet 2*:822-823.

12. Baram, P. and M. M. Mosko (1965). A dialysable fraction from tuberculin-sensitive human white blood cells capable of inducing tuberculin-delayed hypersensitivity in negative recipients. *Immunology 8*:461-474.

13. Neidhart, J. A., R. S. Schwartz, P. E. Hurtubise, S. G. Murphy, E. N. Metz, S. P. Balcerzak and A. F. LoBuglio (1973). Transfer factor: Isolation of a biologically active component. *Cell. Immunol. 9*:319-323.

14. O'Dorisio *et al.*; Tomar; Goust, *et al.*; Grob, *et al.*; Gottlieb, *et al.*; Krohn, *et al.*; Andron, *et al.*; Baram, *et al.*; Klesius, *et al.*; Burger, *et al.*; and Khan, *et al.* (1976). *Transfer Factor: Basic Properties and Clinical Applications.* Academic Press, New York.

15. Griscelli, C., J. P. Revillard, H. Betuel, C. Herzog and J. L. Touraine (1973). Transfer factor therapy in immunodeficiencies. *Biomedicine 18*:220-227.

16. Wybran, J., A. S. Levin, L. E. Spitler and H. H. Fudenberg (1973). Rosette-forming cells, immunologic deficiency diseases and transfer factor. *New Eng. J. Med. 288*:710-713.

17. Dupont, B., M. Ballow, J. A. Hansen, C. Quick, E. J. Yunis and R. A. Good (1974). Effect of transfer factor therapy on mixed lymphocyte culture reactivity. *Proc. Nat. Acad. Sci. 71*:867-871.

18. Krohn, K., A. Uotila, P. Grohn, J. Vaisanen and K-M Hiltunen (1976). A component in transfer factor with immunologically nonspecific activity. *Transfer Factor: Basic Properties and Clinical Applications.* Academic Press, New York.

19. Kirkpatrick, C. H. and T. K. Smith (1976). Effects of dialyzable transfer factor on accumulation of cyclic nucleotides by leucocytes. *Transfer Factor: Basic Properties and Clinical Applications.* Academic Press, New York.

20. Bach, J-F, A. L. Dardenne, A. L. Goldstein, A. Guha and A. White (1971). Appearance of T-cell markers in bone marrow rosette-forming cells after incubation with thymosin, a thymic hormone. *Proc. Nat. Acad. Sci. 68*:2734-2738.

21. Scheid, M. P., M. K. Hoffmann, K. Komuro, U. Hammerling, J. Abbott, E. A. Boyse, G. H. Cohen, J. A. Hooper, R. S. Schuloff and A. L. Goldstein (1973). Differentiation of T-cells induced by preparations from thymus and by non-thymic agents. *J. Exp. Med. 138*:1027-1032.

22. Goldstein, G., M. S. Scheid, U. Hammerling, E. A. Boyse, D. H. Schlessinger, and H. D. Naill (1975). Isolation of a polypeptide that has lymphocyte-differentiating properties and probably represented universally in living cells. *Proc. Nat. Acad. Sci.* 72:11-15.

23. Schlesinger, D. H., G. Goldstein and H. D. Niall (1975). The complete amino acid sequence of ubiquitin, an adenylate cyclase stimulating polypeptide probably universal in living cells. *Biochemistry 14*:2214-2218.

24. Valentine, F. T. (1971). The transformation and proliferation of lymphocytes *in vitro*. *Cell-Mediated Immunity. In Vitro Correlates.* J-P Revillard, Ed. S. Karger, New York.

25. Ascher, M. S., W. J. Schneider, F. T. Valentine and H. S. Lawrence (1974). *In vitro* properties of leucocyte dialysates containing transfer factor. *Proc. Nat. Acad. Sci.* 71:1178-1182.

26. Erickson, A. D., R. S. Holzman, F. T. Valentine and H. S. Lawrence (1976). *In vitro* comparison of TF_{DM} obtained from skin test positive and negative individuals. *Transfer Factor: Basic Properties and Clinical Applications.* Academic Press, New York.

27. Cohen, L., R. S. Holzman, F. T. Valentine and H. S. Lawrence (1976). Leucocyte dialysates require precommitted, antigen-reactive cells to augment lymphocyte proliferation. *Transfer Factor: Basic Properties and Clinical Applications.* Academic Press, New York.

28. Maddison, S. E., M. D. Hicklin, B. P. Conway and I. Kagan (1972). Transfer factor: Delayed hypersensitivity to schistosoma mansoni and tuberculin in Macaca Mulatta. *Science 178*:757-759.

29. Zanelli, J. and W. H. Adler (1975). Transfer factor – transfer of tuberculin cutaneous sensitivity in an allogeneic and xenogeneic monkey model. *Cell. Immunol.* 15:475-478.

30. Kniker, W. T., R. W. Steele, J. W. Eichberg, R. L. Heberling and S. S. Kalter (1976). The efficacy of transfer factor in nonhuman primates including immunotherapy for herpes virus induced infection and malignancy. *Transfer Factor: Basic Properties and Clinical Applications.* Academic Press, New York.

31. Liburd, E. M., H. F. Pabst and W. D. Armstrong (1972). Transfer factor in rat coccidiosis. *Cell. Immunol.* 5:487-489.

32. Klesius, P. H., F. Kristensen and T. T. Kramer (1976). Bovine transfer factor: Isolation and characteristics. *Transfer Factor: Basic Properties and Clinical Applications.* Academic Press, New York.

33. Valentine, F. T. and H. S. Lawrence (1971). Cell-mediated immunity. *Adv. Internal Med.* 17:51-93.

34. Lawrence, H. S. (1974). Selective immunotherapy with transfer factor. *Clin. Immunol.* 2:115-152.

35. Spitler, L. E., A. S. Levin and H. H. Fudenberg (1975). Transfer factor II, results of therapy. *Immunodeficiency in Man and Animals. Birth Defects: Original Article Series 11.* D. Bergsma, Ed. Sinauer Associates, Sunderland, Mass.

36. See also several reports in *Transfer Factor: Basic Properties and Clinical Applications*, Academic Press, New York, 1976.

37. Valentine, F. T. (1974). Transfer factor in the immunotherapy of cancer. *Interaction of Radiation and Host Immune Defense Mechanisms in Malignancy.* Brookhaven National Laboratory, Associated Universities, Inc.

38. Neidhart, J. A. and A. F. LoBuglio (1974). Transfer factor therapy of malignancy. *Seminars in Oncology* 1:379-385.

39. Oettgen, H. F., L. Old, J. Farrow, F. T. Valentine, H. S. Lawrence and L. Thomas (1974). Effects of dialysable transfer factor in patients with breast cancer. *Proc. Nat. Acad. Sci.* 71:2319-2323.

MOLECULAR STUDIES OF IMMUNOPATHOLOGY IN PLASMACYTOMA. POSSIBLE ROLE OF INTRACISTERNAL A PARTICLES*

Dario Giacomoni and Jerry Katzmann

Department of Microbiology
University of Illinois at the Medical Center
Chicago, Illinois

INTRODUCTION

Patients with multiple myeloma have a diminished immune response which is accompanied by a change in the character of the lymphocyte surface immunoglobulin (SIg) (1,2,3). An excellent animal model for this disease and for the study of the immunopathology of myeloma is provided by BALB/c mice bearing plasmacytoma. The growth of plasmacytoma is accompanied by a depression of the primary immune response (4) and by a change of the normal SIg of lymphocytes to new SIg with the idiotypic specificity of the immunoglobulin (Ig) produced by the plasmacytoma (5). This change in the character of the SIg has been called "cell conversion" and was detected by immunocytoadhesion reaction (5). When normal lymphocytes enclosed in a nitrocellulose chamber were implanted in tumor bearing mice, they underwent cell conversion and showed a reduced primary immunological response (5,6). These experiments led to the conclusion that filterable, humoral factors are responsible for the immunodepression and cell conversion observed in plasmacytomatous mice.

Earlier work in our and Heller's laboratory has suggested that these humoral factors owe their activity to an RNA molecule; in fact, it is possible to experimentally induce these two immunopathological events (immunodeficiency and cell conversion) by injecting mice with RNA-rich extracts (200-500 µg/mouse) obtained from the plasma of plasmacytomatous mice or from the tumor cells themselves (7,8). Cell conversion was achieved also *in vitro* when plasmacytoma RNA was incubated with normal mouse spleen cells

---

*Supported by a grant from the Chicago Leukemia Research Foundation

(7,9). The cell converting activity of the RNA extracts was inactivated by RNase under conditions that caused relatively few breaks in marker RNA and was unaffected by DNase or pronase treatment under conditions that resulted in marked degradation of marker DNA or protein (9). Similarly RNase but not DNase or proteolytic enzymes destroyed the immunosuppressive activity of tumor RNA (8). The cell converting activity of tumor RNA is highly specific: lymphocytes converted with RNA from one tumor failed to react when assayed for the presence of SIg with the idiotype of the Ig secreted by another tumor (7,10). RNA from normal mouse liver or spleen had no effect on the primary immune response, nor did it change the character of the SIg (8,9).

The RNA extracts that caused cell conversion and immunodeficiency did not induce tumors, thus allowing one to study the mechanisms of the immunopathology of this tumor system independent of the presence of the tumor itself. In the experiments reported here, we have attempted to isolate and characterize the factors responsible for the two phenomena of immunodeficiency and cell conversion. Since plasmacytoma cells contain virus-like intracisternal A particles of yet unknown activity, we were particularly interested in the isolation of these particles in order to assess their effect on the immune response of the mouse. Other studies on the immunopathology of plasmacytomatous mice will be presented during this workshop by Heller *et al*.

RESULTS AND DISCUSSION

In an effort to isolate the plasma humoral factors that cause cell conversion and immunodeficiency in mice, blood from tumor carrying mice was fractionated by electrophoresis and the cell converting activity was found in the α-fraction (11). RNA with cell converting activity could be extracted from this fraction into which free RNA does not migrate. These data suggested that RNA with cell-converting activity is not present in the plasma as a "naked" molecule but is bound to other components that may protect it from degradation. To gather more information on some physical properties of this RNA-containing fraction, the plasma of plasmacytomatous mice was fractionated in sucrose step gradients. The plasma from 10-15 mice was distributed over sucrose step gradients (65, 40, and 20% w/w), centrifuged at 40,000 rpm for 90 min (Spinco rotor SW41) and the material obtained from each interface was injected intraperitoneally into 3 to 4 mice. In one set of experiments, the plasma from MOPC-300 carrying mice was fractionated and the different fractions tested for cell converting activity (12). The percentage of lymphocytes carrying plasmacytoma specific SIg (converted cells) was tested 3 days after injection of the fractions. The cell converting activity was found in the interface between 20 and 40% sucrose layers (20/40 fraction)(Table 1). No activity

TABLE 1.
*Cell Converting and Immunosuppressive Activity of Fractions from Plasma of Plasmacytomatous Mice**

Fraction	Converted Cells (%)	Immune Response to SRBC (% PFU of Control)
Supernatant	4 ± 2	75 ± 15
20/40	19 ± 4	48 ± 25
40/65	4 ± 2	77 ± 25
Control (Saline)	2 ± 1	100 ± 16

*Each experimental value is the average of 6-8 mice tested in two separate experiments.

was found in the supernatant or 40/65 fraction nor in similar fractions isolated from normal plasma. Control experiments showed that under these conditions of centrifugation no free RNA migrated to the 20/40 fraction, again suggesting that the active RNA is complexed to other materials to form a "particulate" fraction with a density between 1.08 and 1.18 g/ml. Similar results were obtained when the plasma from mice bearing LPC-1 was fractionated.

In a second series of experiments, plasma from mice bearing MOPC-104E was fractionated as described above and the different fractions tested for immunosuppressive activity. Mice were injected intraperitoneally and three days later were challenged with sheep red blood cells (SRBC). Five days later their spleen cells were tested in a hemolytic plaque assay to assess the number of cells producing antibodies against SRBC (PFU/10^6 cells). The immunosuppressive activity of the plasma was found in the 20/40 fraction (Table 1). Thus the cell converting and immunosuppressive activities were found in the same fraction.

We looked for fractions with similar activities in the tumor cells. Tumor cells were fractionated as described by Kuff *et al.* (13) to achieve the separation of polysomes from the virus-like intracisternal A particles present in these tumors. We obtained a nuclei fraction, supernatant A (enriched in nonmembrane bound polysomes), precipitate B (enriched in A particles), and precipitate C (enriched in membrane bound polysomes) (12). In one series of experiments, the subcellular fractions obtained from MOPC-300 were injected into mice and lymphocytes were tested for cell conversion 3 days later. The cell converting activity was consistently found in precipitate B (Table II). Although more UV absorbing material from the nuclei and supernatant A was injected, these fractions caused no cell conversion and precipitate C caused only very little. In a second series of experiments, the same subcellular fractions from MOPC-104E were injected into mice and tested for their effect on the immune response as previously described. The highest immunosuppressive activity was found in the fraction enriched in A particles (precipitate B) (Table II). Similar fractions from normal mouse liver and spleen did not cause

TABLE 2.
Cell Converting and Immunosuppressive Activity of Subcellular Fractions from Plasmacytoma

Fraction	Cell Conversion*		Immune Response to SRBC**	
	mg/mouse	Converted cells (%)	mg/mouse	% PFU of Controls
Nuclei	3.0	4 ± 1	0.5	102 ± 35
Supernatant A	1.0-4.0	3 ± 1	0.5	80 ± 20
Precipitate B	0.2-1.0	21 ± 3	0.5	65 ± 11
Precipitate C	0.2-0.5	7 ± 4	0.5	88 ± 13
Control (Saline)	–	3 ± 1	–	100 ± 15

*Each value is the average of 10-20 mice tested in 4 experiments.
**Each value is the average of 5 mice.

immunodeficiency. The data reported in Tables I and II indicate that both the cell converting and immunosuppressive activities are contained in the same fractions.

When RNA was extracted from the various subcellular fractions, RNA derived from precipitate B was the most active in cell conversion and immunosuppression. The RNA induced immunosuppression is not due to any toxic effect of the RNA. In fact, Heller et al. (14) have shown that mice injected with RNA from MOPC-104E still have a normal immunological response to dextran, a polysaccharide for which the MOPC 104E Ig has antibody activity.

To gather information on the chemical nature of the active RNA, its poly-A content was determined. RNA from precipitate B (MOPC-300) was fractionated on an oligo (dT) cellulose column (15), and poly-A containing RNA was separated from the bulk RNA. In an *in vitro* assay for cell conversion we found that 20-40 μg of poly A containing RNA were sufficient to change SIg of a significant portion of normal mouse spleen cells while 100 μg of RNA not containing poly-A were inactive (Table III). In these experiments RNA was extracted by the method of Perry et al. (16) that insures complete extraction of poly-A containing RNA.

In a separate series of experiments, MOPC-104E was RNA fractionated on an oligo (dT) cellouse colum and tested for immunosuppressive activity, which was also found to reside in the RNA containing poly-A (unpublished observations).

To determine the sedimentation velocity of the RNA active in cell conversion, RNA from the plasma of MOPC-300 bearing mice and from precipitate B of the same tumor was sedimented on a 10% to 40% sucrose gradient (12). Ten fractions were collected from the gradients and each fraction was analyzed for its cell converting activity *in vitro*. The sedimentation profile of the bulk RNA from the plasma and from precipitate B was very different in that

TABLE 3.
Cell Converting Activity of Plasmacytoma RNA Fractionated by Oligo (dT) Cellulose Chromatography

Fraction	Cell Conversion	
	µg	Converted Cells (%)
Poly A containing RNA	40	25
	20	15
Non-poly A containing RNA	200	11
	100	6
	50	2

RNA from precipitate B contained large amounts of 18S and 28S RNA in addition to RNA of lower MW, whereas RNA from plasma consisted mostly of low MW RNA (Figure 1). In both preparations the RNA with cell converting activity was found in two regions (12-18S and 40-50S respectively). Thus, although the bulk of the RNA contained in the plasma and in precipitate B is different, they probably contain small amounts of identical molecules active in cell conversion (Fig. 1). Interestingly, some preparations of RNA

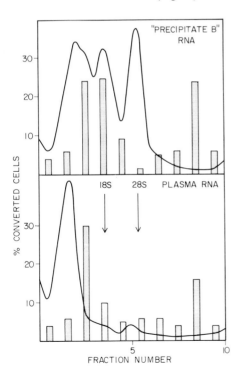

Fig. 1. *Sedimentation velocity of RNA molecules with cell converting activity.*

RNA (200 µg) from precipitate B (MOPC-300) and from plasma of mice bearing MOPC-300 was fractionated on a sucrose gradient (12). RNA from different fractions was precipitated with ethanol and tested for cell covering activity in vitro (bars). The continuous line indicates the sedimentation profile of the bulk of the RNA.

from B fractions with comparable cell converting activity contained very little rRNA, suggesting that polysomal RNA plays no role in this phenomenon.

The size of RNA molecules responsible for immunosuppression was also determined. RNA from precipitate B (MOPC-104E) was fractionated as described above and each fraction was assayed for its immunosuppressive activity. The immunosuppressive activity was found in RNA molecules in similar regions of the sucrose gradients (12-18S and 40-50S) (Figure 2).

To investigate the relationship between the active RNA molecules in the low and high molecular weight regions, we examined the effect of heat on the sedimentation properties of the active RNA molecules. RNA from precipitate B of MOPC-300 (250 μg) in 0.5 ml of 0.01 M Tris-HCl pH 7.4, 0.01 M EDTA buffer were heated in a water bath at 65° C for 2 min and chilled immediately. The RNA was then fractionated on a sucrose gradient and the fractions were analyzed for cell conversion *in vitro*.

Upon heating, the activity in the 40-50S region disappeared and that RNA with cell converting activity was found only in the 12-18S region (Figure 3). In a similar experiment, the high MW RNA (40-50S) with immunosuppressive activity was also found to be thermolabile (unpublished observations). The high MW RNA molecules with cell converting and immunosuppressive activity are probably not single molecules but an aggregate that can be dissociated by moderate heating. Heat sensitive aggregates of RNA molecules have been described (17,18). In particular, Ikawa *et al.* (19) have reported that RNA from Friend leukemia virus grown in globin producing cells contains globin mRNA with sedimentation velocity of 9S and 70S. Upon moderate heating, the globin mRNA was found only in the 9S region (19). These data suggested that cellular mRNA can associate with viral RNA.

Since cell converting and immunosuppressive activities were found in a subcellular fraction enriched in intracisternal A particles, we have purified intracisternal A particles from MOPC-300, MOPC-104E and J606 by sucrose gradient equilibrium sedimentation as described by Kuff *et al.* (13). The prep-

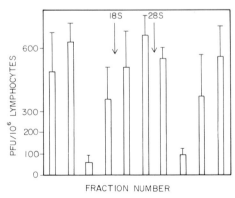

Fig. 2. *Sedimentation velocity of RNA molecules with immunosuppressive activity.*

RNA (200 μg) from precipitate B (MOPC-104E) was fractionated on a sucrose gradient. RNA from each fraction was injected into 4 mice and tested for its effect on the immune response to SRBC (see text).

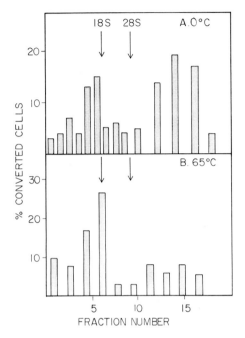

Fig. 3. *Effect of heating on the sedimentation velocity of cell converting RNA.*

RNA from precipitate B (MOPC-300) was heated (see text) and fractionated on a sucrose gradient. The activity of the different fractions was tested as in Figure 1 (3B). The unheated preparation was similarly tested (3A).

arations obtained were analyzed by electron microscopy. In addition to A particles (60-80% of the material), cytoplasmic membranes were present but few polysomes could be detected. All three preparations were immunosuppressive when injected into normal mice (100-125 µg/mouse) (Table IV). Interestingly, smaller amounts of A particles than precipitate B were needed to cause comparable immunodepression to SRBC (see Table II).

The RNA extracted from type A particles was also very effective in causing immunodeficiency (Table IV). We have consistently found that RNA from preparations rich in intracisternal A particles was effective at lower concentrations than RNA from unfractionated plasmacytoma or from precipitate B, indicating a correlation between intracisternal A particle RNA and immunosuppressive activity.

By these experiments we have tried to determine the origin and some physicochemical characteristics of the factors found in plasmacytomatous mice with cell converting and immunosuppressive activities. The results obtained may be summarized as follows:

    a) The plasma of plasmacytomatous mice contains a particulate fraction (specific gravity between 1.08 and 1.18 g/ml) with both cell converting and immunosuppressive activities.

    b) Plasmacytoma subcellular fractions enriched in intracisternal A particles also contain both activities.

TABLE 4.

Immunosuppressive Activity of Intracisternal A Particles and A Particle RNA

Fraction	Amount Injected (μg)	% PFU of Controls
MOPC-300 A particles	125	60 ± 3*
MOPC-104E A particles	125	50 ± 12*
J606 A particles	100	70 ± 18*
Control I**	140	100 ± 15*
MOPC-300 A particle RNA	20	5 ± 2****
MOPC-104E A particle RNA	20	10 ± 3*
Control II***	100	100 ± 20

*Average of 4 mice
**Mice injected with precipitate B from normal mouse spleen and liver
***Mice injected with normal mouse liver RNA
****Average of 10 mice

c) Two RNA species (14-18S and 40-50S) can cause both phenomena.

d) RNA with cell converting and immunosuppressive activity contains only A.

These data strongly suggest that the same factors are responsible for both cell conversion and immunosuppression, but the possibility cannot be ruled out that two different factors with similar physicochemical properties are independently responsible for these two phenomena.

In our cell fractionation studies we have found a positive correlation between the presence of A particles and immunosuppressive and cell converting activities suggesting that A particles play a role in these phenomena. The RNA moiety of the A particles seems to be responsible for these activities. The poly A content of active RNA and the thermolability of 40-50S RNA are consistent with the hypothesis that an active mRNA (12-18S) is partially associated with A particle RNA. The chemical nature of the plasma factor is at present unknown, but conceivably it is related to the subcellular fraction with the same biological activities.

In conclusion, it is conceivable that a humoral factor, containing the mRNA for the plasmacytoma Ig, is released by the tumor cells and is incorporated into normal lymphocytes causing cell conversion. Converted cells may not act as antigen receptors, thus leading to immunodeficiency. Alternatively, converted cells might elicit a suppressor response in the host, leading to a suppression of the immune response. Further studies are in progress to determine the exact nature of the humoral factor and assess its relationship with the intracisternal A particle.

ACKNOWLEDGEMENTS

We are grateful to Dr. S. Chandra for taking electron micrographs of several preparations and to Drs. S. Dray and P. Heller for valuable advice and discussion.

REFERENCES

1. Fahey, J. L., R. Scoggins, J. P. Utz and C. F. Szwed (1963). Infection, antibodies response and gamma globulin components in multiple myeloma and macroglobulinemia. *Am. J. Med.* 35:698-707.
2. Chen, Y., N. Bhoopalam, V. Yakulis and P. Heller (1975). Changes in lymphocyte surface immunoglobulin in myeloma and the effect of an RNA containing plasma factor. *Ann. Int. Med.*, in press.
3. Mellstedt, H., S. Hammarstrom and G. Holm (1974). Monoclonal lymphocyte population in human plasma cell myeloma. *Clin. Exp. Immunol.* 17:371-384.
4. Fahey, J. L. and J. M. Humphrey (1962). Effect of transplantable plasma cell tumors on antibody response in mice. *Immunol.* 5:110-115.
5. Yakulis, V., N. Bhoopalam, S. Shade and P. Heller (1972). Surface immunoglobulins of circulating lymphocytes in mouse plasmacytoma. I. Characteristics of lymphocyte surface immunoglobulins. *Blood* 39:453-464.
6. Tanapatchaiyapong, P. and S. Zolla (1974). Humoral immunosuppressive substance in mice bearing plasmacytomas. *Science* 186:748-750.
7. Bhoopalam, N., V. Yakulis, H. Costea and P. Heller (1972). Surface immunoglobulins of circulating lymphocytes in mouse plasmacytoma. II. The influence of plasmacytoma RNA on surface immunoglobulins of lymphocytes. *Blood* 39:465-472.
8. Yakulis, V., V. Cabana, D. Giacomoni and P. Heller (1973). Surface immunoglobulins of circulating lymphocytes in mouse plasmacytoma. III. The effect of plasmacytoma RNA on the immune response. *Immunol. Commun.* 2:129-139.
9. Giacomoni, D., V. Yakulis, S. R. Wang, A. Cooke, S. Dray and P. Heller (1974). *In vitro* conversion of mouse lymphocytes by plasmacytoma RNA to express idiotypic specificities on their surface characteristic of the plasmacytoma immunoglobulin. *Cell. Immunol.* 11:389-400.
10. Bhoopalam, N., V. Yakulis, D. Giacomoni and P. Heller (1975). Immunoglobulins of lymphocytes in mouse plasmacytoma IV. Evidence for the persistence of the effect of plasmacytoma RNA on the surface immunoglobulins of normal lymphocytes. *Clin. Exp. Immunol.*, in press.
11. Heller, P., N. Bhoopalam, V. Cabana, N. Costea and V. Yakulis (1973). The role of RNA in the immunological deficiency of plasmacytoma. *Ann. N. Y. Acad. Sci.* 207:468-480.
12. Katzmann, J., D. Giacomoni, V. Yakulis and P. Heller (1975). Characterization of two plasmacytoma fractions and their RNA capable of changing lymphocyte surface Ig (cell conversion). *Cell. Immunol.* 18:98-109.
13. Kuff, E. L., N. A. Wivel and K. Leuders (1968). The extraction of intracisternal A particles from a mouse plasma cell tumor. *Cancer Res.* 28:2137-2143.
14. Heller, P., N. Bhoopalam, Y. Chen and V. Yakulis (1976). The relationship of myeloma "RNA" to the immune response. *Immune RNA in Neoplasia.* Mary A. Fink, Ed. Academic Press, New York.

15. Swan, D., H. Aviv and P. Leder (1972). Purification and properties of a biologically active messenger RNA for a myeloma light chain. *Proc. Natl. Acad. Sci. USA* 69:1967-1971.

16. Perry, R. P., J. LaToree, D. E. Kelley and J. R. Greenberg (1972). On the lability of poly (A) sequences during extraction of messenger RNA from polyribosomes. *Biochim. Biophys. Acta* 262:220-226.

17. Canaani, E. and P. Duesberg (1972). Role of subunits of 60 to 70S Avian tumor virus RNA in its template activity for the viral DNA polymerase. *J. Virol.* 10:23-31.

18. Schechter, I. (1973). Biologically and chemically pure mRNA coding for a mouse immunoglobulin L-chain prepared with the aid of immobilized oligo thymidine. *Proc. Natl. Acad. Sci. USA* 70:2256-2260.

19. Ikawa, Y., J. Ross and P. Leder (1974). An association between globin mRNA and 60S RNA derived from Friend Leukemia Virus. *Proc. Natl. Acad. Sci. USA* 71:1154-1158.

I-RNA: SYNTHESIS AND MECHANISMS OF ACTION

D. Jachertz

*Institut für Hygiene und Medizinische Mikrobiologie
der Universität Bern*

In 1959 and 1961, Fishman (1,2) experimentally separated the phase of antigen recognition from the subsequent phase of antibody formation. In 1966, our investigations (3) revealed that in a primary immune response the antigen recognizing cell is polyvalent and does not synthesize antibodies. On the other hand, in a primary response, the antibody synthesizing cells cannot recognize the antigen. Accordingly, information must flow from the antigen recognizing to the antibody synthesizing cells. This information is transmitted by an informational RNA (i-RNA). Using different experimental techniques, our work was recently corroborated by Schaeffer, Fishman and Adler (4).

The synthesis of i-RNA by the antigen recognizing cells which are capable of inducing a specific immunologic reaction — i.e., antibody synthesis and cell-bound immunity — is not only of great scientific interest but also opens up new possibilities in the field of preventive medicine, namely the use of i-RNA as a vaccine.

SYNTHESIS OF I-RNA

A. *Synthesis of i-RNA in a cellular system.* Adherent cells are obtained from induced peritoneal exudates, peritoneal washings, blood leucocytes or spleen cells from various laboratory animals as well as from human blood (3,5,6-18).

The main properties of the cells which are capable of recognizing an antigen are summarized in Table 1.

Antigen is added to an appropriately prepared cell suspension (5-7,13). Optimal amounts of i-RNA synthesis can only be obtained with a previously carefully selected optimal dose of antigen (5,6). In our hands, one antigen (molecule) particle per 100 cells in most instances was the ideal antigen to cell ratio. The use of vastly varying antigen to cell ratios by the different investigators may explain the different yields and quality of RNA (2,20-22).

The use of live cell suspensions for the stimulation of i-RNA synthesis has one great disadvantage in that the i-RNA is contaminated with ribosomal, transfer and different types of messenger RNAs. The physical and chemical

TABLE 1.
Properties of Antigen Recognizing Cells

	References
polyvalent	3, 14
long lasting competence	3
synthesis of m-RNA coding for AAP*	3, 6
synthesis of AAP	5, 6
synthesis of i-RNA	1, 2, 3, 4
amplification of i-RNA by replicase	13, 17, 19

**AAP: *A*ntibody-*A*nalogous *P*roduct

properties of the i-RNA from such a system, however, are identical to the i-RNA obtained from an antigen stimulated cell-free system which is described in the following section.

B. *Synthesis of i-RNA in a cell-free system.* The advantages of a cell-free system for the synthesis of i-RNA are manyfold. The required components can be prepared in a purified way and can be reassembled in optimal amounts. The i-RNA synthesized in such a system is not contaminated with other types of RNAs or with i-RNA of different immunological specificity. The cell-free system used for antigen recognition and i-RNA synthesis also does not synthesize protein.

Cell-free systems are prepared from spleen cells, blood leucocytes or lymphnode cells (3,5,9-18). The optimal concentrations of the components of an i-RNA synthesizing system and its properties are shown in Tables 2 and 3.

The method used for the isolation of i-RNA from such a system after antigenic stimulation is schematically shown in Fig. 1.

The method is as follows: After interaction of the antigen with the cell-free components, i-RNA synthesis is interrupted with sodium-dodecyl-sulfate and the i-RNA is removed by centrifugation through a saccharose solution

TABLE 2.
Components and Optimal Concentrations of the Antigen Recognizing Cell-Free System

DNA	10^{-4} µg/ml
S 100	0.3 µg N_2/ml
rXTP	10^{-6} M
^{3}H UTP	10^{-7} M spec. activity 20 Ci/mmol
antigen	10^5 particles or molecules

TABLE 3.
Properties of the Antigen Recognizing Cell-free System

	References
polyvalent	7, 13, 14
synthesis of i-RNA	1, 2, 4, 12
amplification of i-RNA by replicase	13, 17
synthesis of i-DNA by reversed transcriptase	13, 17

into a solution of CsCl. The amount of i-RNA synthesized is expressed in terms of ^{3}H UTP incorporated during various time intervals. The results of such an experiment are shown in Fig. 2.

After removal of the CsCl by dialysis the i-RNA is isolated by sedimentation through a continuous Cs_2SO_4 gradient as is shown in Fig. 3.

The dialyzed material can be further fractionated by equilibrium centrifugation in Cs_2SO_4 gradient. As in shown in Fig. 4, this procedure yields 3 types of RNAs with densities of 1.72, 1.70 and 1.68 in regard to Cs_2SO_4.

Figures 5, 6 and 7 are electron microscopic pictures of single stranded, replicative intermediate and of double stranded i-RNAs corresponding to the 3 types of i-RNA separated by Cs_2SO_4 gradient centrifugation.

ACTIONS OF I-RNA

A. *Actions of i-RNA in a cell-free system.* For the preparation of the cell-free system the same cells as previously mentioned were used. Since the preparations are freed of DNA (11,13), there will be no DNA dependent transcription of i-RNA. Thus, accidental contact with antigen that might induce i-RNA synthesis is prevented.

However, since the system contains ribosomes which are usually prepared from the same cell source (11,13,18), protein synthesis can occur. And indeed, the addition of i-RNA to such a system will be followed by the synthesis of antibody. The concentration of the various components which we used and have found to be optimal for an i-RNA induced antibody synthesis are shown in Table 4.

The actions of i-RNA in this cell-free system are summarized in Table 5.

The outstanding feature, namely, the synthesis of specific antibody is illustrated with the following two experiments. I-RNA was obtained either by stimulation of a cell-free system with live poliovirus (type I) or with benzoyl-penicilloyl-formyl-L-lysine (BPoFLys). The i-RNAs were extracted and were used in a second cell-free system for the induction of antibody synthesis. The

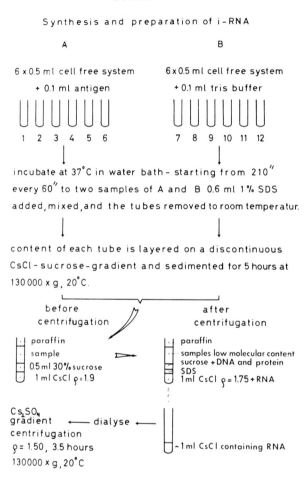

Fig. 1. *Procedures for formation and isolation of i-RNA.*

total amount of protein synthesized after addition of i-RNA in the cell-free system was estimated by trichloroacetic acid (TCA) precipitation and scintillation counting. The antibody was consecutively adsorbed to and eluted from antigen containing sepharose columns. The results are shown in Table 6 and Fig. 8.

B. *Actions of i-RNA in cellular systems.* Cell cultures are prepared from blood leucocytes or from suspensions of spleen cells. After addition of i-RNA in the culture medium, antibody and in certain instances newly synthesized i-RNA can be found (8,10,13). The presently known biological activities of i-RNA in lymphoid cell cultures are summarized in Table 7.

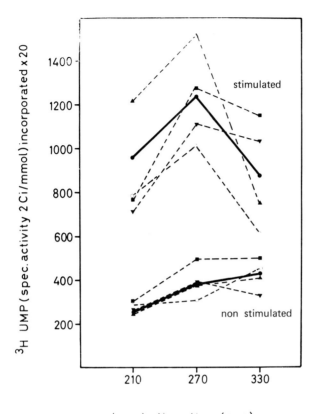

Fig. 2. Incorporation of ^3H UTP in RNA after antigenic stimulation of a cell-free system with BPoFLys.
As stimulating antigen 10^5 molecules of BPoFLys in 0.1 ml Tris buffer were added to 0.5 ml aliquots of 0.1 M Tris-HCl buffer, pH 8.0 containing 0.04 M KCl; 0.005 M Mg-acetate; 10^{-4} µg DNA; 10^{-6} M of each UTP, ATP, GTP; 10^{-7} M ^3H UTP (specific activity 20 Ci/mmol, Radiochemical Center Amersham) and the S-100-system in a concentration of 0.15 µg N_2. After incubation in a waterbath of $37°$ C for 210, 270 and 330 seconds, the RNA synthesis was stopped with 0.6 ml of a 1 % solution of sodiumdodecylsulfate. After mixture the samples were put on a discontinuous CsCl-sucrose gradient and sedimented at 20° C for 5 hours at 130,000 g. The CsCl cushion was withdrawn by puncturing the bottom of the centrifuge tube and 0.1 ml was used for scintillation counting.

As already indicated, i-RNA not only instructs the lymphoid cells to synthesize antibody but also the cell's own i-RNA. This latter effect is achieved by means of a regulator protein. I-RNA exists within the cells in either an active or a latent form. The synthesis of i-DNA by a specific reverse transcriptase and the synthesis of i-RNA by a specific i-RNA dependent i-RNA

TABLE 4.
Components of a Cell-free Antibody Synthesizing System

S 100	0.6	$\mu g\ N_2/ml$
monosomes	10^{-3}	$\mu g\ N_2/ml$
^{3}H-L-Leucine	10^{-6}	M, spec. activity 38 Ci/mmol
^{3}H-L-Valine	10^{-6}	M, spec. activity 19 Ci/mmol
i-RNA	10^6	molecules or $10^{-7}\ \mu g$

polymerase opens two pathways for extrachromosomal genetics. Both the i-RNA and the i-DNA may be objects of selection, and the i-RNA may be incorporated into the cellular genome. Synthesis of i-DNA in an antibody synthesizing cell alone is not necessarily the result of mitosis.

C. *Action of i-RNA in microorganisms.* Experiments were performed using measles virus and i-RNA. We initiated the synthesis of i-RNA against the measles virus in a cell-free system using as antigen an attenuated vaccine strain (Moraten Berna, Swiss Serum and Vaccine Institute Berne). The cell-free system for the production of i-RNA was derived from guinea pig spleen cells.

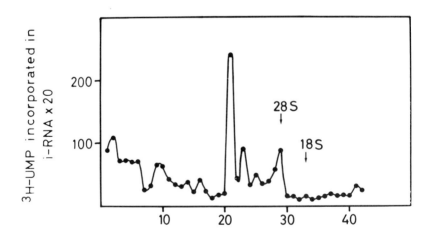

Fig. 3. *Sedimentation of i-RNA in Cs_2SO_4.*
A volume of 0.3 ml of a dialyzed RNA preparation (obtained as described in the legend to Fig. 2) was put on top of 3.0 ml of Cs_2SO_4 solution with a density of 1,540. After centrifugation at 20° for 3.5 hours at 130,000 g, the bottom of the centrifuge tube was punctured and fractions of 0.15 ml were collected. A volume of 0.15 ml aqua dest. was added to each fraction, mixed, and 0.05 ml were used for scintillation counting.

TABLE 5.
Actions of i-RNA in a Cell-free System

		References
translation	— synthesis of antibody	9, 10, 11, 13, 15, 18
	synthesis of regulator protein	10
replication	— synthesis of i-RNA by replicase	13, 17, 19, 21
reversed transcription	— synthesis of i-DNA	13, 17, 21

After isolation, the i-RNA was injected i.p. into normal guinea pigs in amounts of 10^{-2} to 10^2 p.g. The animals were bled at various time intervals after the injection of the i-RNA and the amount of specific antibody was determined. In parallel, a second group of animals was immunized with Moraten. The neutralizing antibody titers obtained with the two immunizing procedures are shown in Figs. 9a and 9b.

The antibody titers obtained with 10^{-1} p.g. i-RNA are comparable to those obtained with an optimal dose of the live attenuated measles virus (Moraten) in a dilution of 10^{-3}, i.e., 10 $TCID_{50}$. This exemplifies that i-RNA can be used instead of the viral vaccine. The advantages of using i-RNAs instead of potentially harmful antigens as immunizing agents are evident.

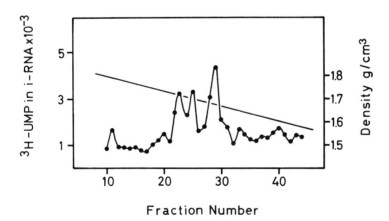

Fig. 4. *Equilibrium centrifugation of i-RNA in a Cs_2SO_4 gradient.*
0.7 ml of the dialyzed RNA preparation (obtained as described in the legend to Fig. 2) was mixed with 2.3 ml of Cs_2SO_4 solution to give a density of 1.7000. After centrifugation for 46 hours at $20°C$ and 100,000 g the bottom of the centrifuge tube was punctured and fractions of 0.07 ml were collected. The radioactivity of each fraction was estimated by scintillation counting.

Fig. 5. *Electron micrograph of single-stranded i-RNA. Magnification ca. 63,000 x. Length of the molecule 2,650 μ.*

As further examples of the induction of cellular immunity, the following data concerns the treatment of animals by i-RNA after transplantation of tumor cells. Different types of tumor cells were used. The mastocytoma 815 (16) was transplanted in DBA/2 mice. Polyomavirus transformed tumor cells were injected into Lewis rats. First we stimulated the correspondent cell-free

TABLE 6.
Antibody Synthesis in a Cell-free System after Addition of i-RNA *

RNA	protein ^{3}H-L-Leucine and ^{3}H-L-Valine incorporated		
	TCA (cpm)	antibody adsorbed on column (cpm)	
		polio	BPoFLys
i-RNA polio	$1.5 \cdot 10^4$	$1.1 \cdot 10^4$	ϕ
control	$3.2 \cdot 10^3$	ϕ	ϕ
i-RNA$_{BPoFLys}$	$9.0 \cdot 10^3$	ϕ	$5.2 \cdot 10^3$
control	$1.2 \cdot 10^2$	ϕ	ϕ

*in collaboration with S. Neuenschwander, Thesis, Berne 1976

To three 0.5 ml aliquots of 0.1 M Tris-HCl buffer, pH 8.0 containing 0.04 M KCl and 0.005 M Mg-acetate, 0.3 μg N_2, S-100-system from guinea pig spleen cells 10^{-6} M ^{3}H-L-Valine (specificity 40 Ci/mmol) (Radio-chemical Center Amersham) and 10^{-3} μg N_2 of ribosomes from guinea pig spleen cells 0.1 ml of a solution containing 10^6 molecules of i-RNA was added. The samples were incubated for 45 minutes in a waterbath at 37°C. Thereafter 0.1 ml of a 0.01 % bovine serum albumine solution and 0.2 ml of a 10 % Trichloroacetic acid solution (TCA) were added to 0.1 ml of each of three incubated mixtures. The precipitate thus obtained was washed with 5 % TCA on a filter disc. The filter discs were dried and radioactivity was estimated by scintillation counting.

systems with live tumor cells. After incubation at 37° C as described in Fig. 1, i-RNA synthesis was stopped by filtration of the sample through a millipore filter with a diameter of 0.45 μ in a 1 % SDS solution in order to separate the tumor cells from the cell-free system. Further steps of isolation of i-RNA were the same as described in Fig. 1-3. The biological activity of a given i-RNA preparation can be tested by a cytotoxicity test using the i-RNA coded protein synthesized in a cell-free system. As target cells we used the correspondent tumor cells grown in culture and labeled with ^{51}Cr. The results of such an experiment are demonstrated in Fig. 10.

Preparations of i-RNA which initiated the synthesis of a cytotoxic protein in cell-free system were used to treat animals after transplantation of correspondent tumor cells. The results of such experiments are demonstrated in Tables 8 and 9.

To groups of DBA/2 mice 1 day after implantation of P815 cells, i-RNA in concentrations as indicated was injected intraperitoneally. The i-RNA was obtained by stimulation of a cell-free system derived from spleen cells of DBA/2 mice with P815 tumor cells. Details of isolation of i-RNA are published (16).

I-RNA in concentrations as indicated was injected intraperitoneally into groups of Lewis rats 1 day after implantation of tumor cells. The i-RNA was

TABLE 7.
Biological Activities of i-RNA in Lymphoid Cell Cultures

		References
translation	— synthesis of antibody	11, 13, 17, 18
	synthesis of regulator protein	10
gene activation	— transcription of i-RNA from the cell genome initiated by action of regulator protein	10, 13
replication	— i-RNA dependent synthesis of i-RNA	13, 19, 21
reversed transcription	— probable	17, 21
integration of i-DNA in the genome in culture	— ?	
repression	— cell own i-RNA is not translated in the first 2-3 days	8
	— after antibody synthesis i-RNA remains in latency	8, 13
derepression	— after nonspecific and specific stimuli i-RNA dependent i-RNA synthesis	8, 13

TABLE 8.
Survival Rates of P815 Implanted DBA/2 Mice After Treatment with i-RNA$_{P815}$

Group No.	Treatment**	Surviving mice/number of mice tested, 48 days after implantation of P815 cells
1	no RNA	1/9
2	day 1 control RNA 10^8 molecules = $5 \cdot 10^2$ pg	0/6
3	day 1 control RNA 10^6 molecules = 5 pg	0/16
4	day 1 i-RNA influenza* 10^6 molecules = 5 pg	0/11
5	day 1 i-RNA$_{P815}$ 10^8 molecules = $5 \cdot 10^2$ pg	0/6
6	day 1 i-RNA$_{P815}$ 10^7 molecules = $5 \cdot 10^1$ pg	0/6
7	day 1 i-RNA$_{P815}$ 10^6 molecules = 5 pg	14/16
8	day 1 i-RNA$_{P815}$ 10^5 molecules = $5 \cdot 10^{-1}$ pg	0/6

*Influenza = Influenza A/Turkey/England/63/L30 virus used as antigen for stimulation of the cell-free i-RNA-synthesizing system.

**All animals received $2 \cdot 10^3$ P815 cells i.c. on day 0.

Fig. 6. *Electron micrograph of the replicative intermediate of i-RNA. Magnification ca. 61,000 x. Length of the double-stranded molecule 2,870 µ; length of the single-stranded molecules 2,300 µ.*

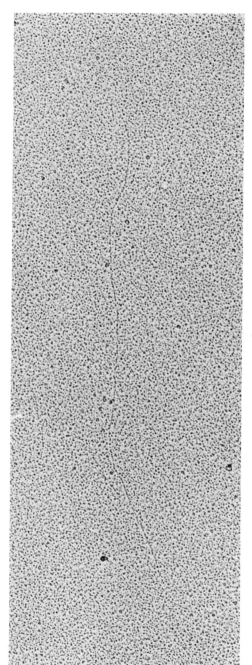

Fig. 7. *Electron micrograph of double-stranded i-RNA. Magnification ca. 73,400 x. Length of the molecule 2,45 μ.*

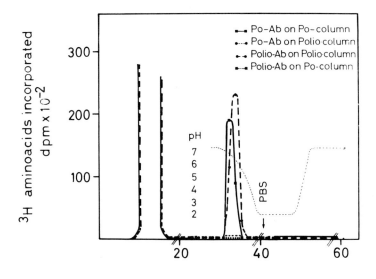

Fig. 8. *Immunadsorption of antibodies to poliovirus or to BPoFLys.*

Poliovirus or bovine serum albumin (BSA) was bound to activated Sepharose 4 B (23). The BSA coated Sepharose was labeled with the penicilloyldeterminant (24).

With both Sepharose-preparations, columns of 10.0 cm length and a diameter of 1 cm were prepared. After incubation with either i-RNA$_{Polio}$ or i-RNA$_{BPoFlys}$, 1.0 ml of a cell-free system was passed through the corresponding column. After elution of all radioactivity at pH 7.2 with PBS, the column was washed with a solution of 0.05 n HCl. After reaching a pH of 3.0 in the effluent, the column was equilibrated with PBS pH 7.2. Starting from the application of the cell-free system, fractions of 2.0 ml were collected of which 0.1 ml were used for scintillation counting.

obtained by stimulation of a cell-free system derived from spleen cells of a Lewis rat with polyomavirus induced tumor cells. Details of isolation of i-RNA were the same as used for the experiments described in Table 8.

It can be seen that in all experiments treatment of animals with i-RNA inhibits the tumor growths. But it can also be demonstrated that certain variations in the amount of transplanted cells, the concentration of i-RNA injected, and the time of treatment lead to enhancement rather than diminution of tumor growth. So it is necessary to find the optimal therapeutic dose of a given i-RNA preparation and in this regard our experimental results are preliminary.

SUMMARY

We have presented evidence in support of Fishman's view of a cellular compartmentalization of the immunologic process culminating in antibody synthesis. We attempted to isolate and characterize the subcellular materials

TABLE 9.
Incidence and Growth of Polyomavirus Induced Transplanted Tumors in Lewis Rats After Treatment with i-RNA

Group	Treatment	Number of cells transplanted	Incidence of tumors	Weight of tumors in the animal group
1	i-RNA 5 • 10^{-6} µg	1 • 10^3	8/10	136.7 g
2	no RNA	1 • 10^3	6/8	27.0 g
3	i-RNA 5 • 10^{-6} µg	5 • 10^3	7/10	4.7 g
4	i-RNA 5 • 10^{-7} µg	5 • 10^3	6/9	23.4 g
5	no RNA	5 • 10^3	3/6	16.8 g
6	tumor cell RNA 4 • 10^{-1} µg	5 • 10^3	7/8	28.7 g
7	i-RNA 5 • 10^{-6} µg	1 • 10^4	1/15	1.78 g
8	no RNA	1 • 10^4	5/9	68.6 g

needed to trigger the events leading to antibody synthesis. We have shown that an i-RNA is synthesized when subcellular materials contact antigen and that this i-RNA provokes the synthesis of antibodies when added to another subcellular material. Of particular interest is the fact that i-RNA can be used in man and animals for the stimulation of specific antibody production as well as cell bound immunity.

REFERENCES

1. Fishman, M. (1959). Antibody formation in tissue culture. *Nature 183:*1200-1201.
2. Fishman, M. (1961). Antibody formation *in vitro. J. Exp. Med. 114*:837-856.
3. Jachertz, D. und H. Noltenius (1966). Antikörper-Synthese *in vitro*. V. Die Informationsübertragung von der Antigenverarbeitenden Zelle auf die Antikörperproduzierende Zelle. *Z. med. Mikrobiol. u. Immunol. 152*:112-133.
4. Schaefer, Annette E., M. Fishman and F. L. Adler (1974). Studies on antibody induction *in vitro*. III. Cellular requirements for the induction of antibody synthesis by solubilized T_2 phage and immunogenic RNA. *J. Immunol. 112*:1981-1986.
5. Drescher, J. und D. Jachertz (1966). Antikörper-Synthese *in vitro*. III. Bildung eines Antikörper-analogen Produktes (AAP) nach Stimulierung mit Influenzavirus-Antigen. *Zbl. Bakt. I. Abt. Orig. 199*:315-328.
6. Jachertz, D. (1966). Antikörper-Synthese *in vitro*. I. Spezifische Stimulierung einer Zellkultur durch Phagen-Receptor-Partikel. *Z. med. Mikrobiol. u. Immunol. 152*:1-19.
7. Jachertz, D. (1966). Antikörper-Synthese *in vitro*. II. Stimulierung eines zellfreien Systems. *Z. med. Mikrobiol. u. Immunol. 152*:20-32.
8. Jachertz, D. (1966). Antikörper-Synthese *in vitro*. VI. Die manifeste und die latente Funktion der informatorischen RNS in der Milzzelle. *Z. med. Mikrobiol. u. Immunol. 152*:262-272.

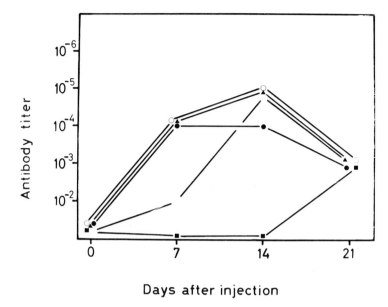

Fig. 9a. *Immunization of guinea pigs against measles by i-RNA. Animals treated with i-RNA in following dilutions:*
○——○ 10^2 pg ▲——▲ 10^1 pg ●——●1 pg □——□ 10^{-1} pg ■——■ 10^{-2} pg.
To groups of 5 guinea pigs i-RNA $_{Moraten}$ in concentrations of 10^2 pg to 10^{-2} pg/ml was injected intraperitoneally. The i-RNA was obtained by stimulation of a cell-free system derived from guinea pig spleen cells according to the procedure outlined in Fig. 1. Control groups of guinea pigs were injected with dilutions of Moraten Berna (live measles vaccine). At various time intervals blood was drawn. Antibody concentrations were estimated by neutralization tests.

9. Jachertz, D. (1967). Antikörpersynthese *in vitro*. VII. Ueber die Wirkung informatorischer RNS-Moleküle in zellfreien Systemen aus Milzzellen. *Z. med. Mikrobiol. u. Immunol.* 153:250-268.

10. Jachertz, D. (1968). Antikörpersynthese *in vitro*. VIII. Ueber die Wirkung der informatorischen RNS-Moleküle auf die Milzzellkultur. *Z. med. Mikrobiol. u. Immunol.* 154:1-22.

11. Jachertz, D. (1968). Antikörpersynthese *in vitro*. IX. Synthese von informatorischer RNS und Antikörpern in einem zellfreien System aus menschlichen peripheren Leukocyten. *Z. med. Mikrobiol. u. Immunol.* 154:245-266.

12. Jachertz, D. (1969). Antikörpersynthese *in vitro*. X. Information zur Antikörpersynthese in der DNS von nicht-kompetenten Zellen. *Z. med. Mikrobiol. u. Immunol.* 154:300-311.

13. Jachertz, D. (1973). Flow of information and gene activation during antibody synthesis. *Ann. N. Y. Acad. Sci.* 207:122-144.

14. Jachertz, D. und J. Drescher (1966). Antikörper-Synthese *in vitro*. IV. Multiple Zustandigkeiten von Genom und Zelle fur die immunologische Spezifität des AAP. *Z. med. Mikrobiol. u. Immunol.* 152:33-44.

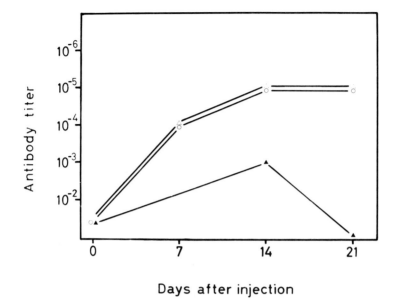

Fig. 9b. *Antibody responses to injection of live, attenuated measles vaccine, Moraten Berna, in guinea pigs (0.5 ml/animal i.p.).*
Animals treated with Moraten Berna in following dilutions:
△——△ 10^{-2} ○——○ 10^{-3} ▲——▲ 10^{-4}

15. Jachertz, D. and J. Drescher (1970). Antibody response in Rhesus monkeys and guinea pigs to inoculation with RNA derived from antigenically stimulated cell-free systems. *J. Immunol.* 104:746-752.
16. Jachertz, D. and Marliese Egger (1974). Treatment of P815 mastocytoma in DBA/2 mice with RNA. *J. Immunogenetics* 1:355-362.
17. Jachertz, D., Uta Opitz and H.-G. Opitz (1972). Gene amplification in cell-free systems. *Z. Immun.-Forsch.* 144:260-272.
18. Jachertz, D., H. Trachsel and Ada Hoida-Werner (1975). Antibody synthesis *in vitro*. XII. Activity of reconstituted and hybrid recombined ribosomes in cell-free synthesis of antibody. *J. Immunogenetics* 2:239-251.
19. Neuhoff, V., W.-B. Schill und D. Jachertz (1970). Nachweis einer RNA-abhängigen RNA-Replicase aus immunologisch kompetenten Zellen durch Mikro-Disk-Elektrophorese. *Hoppe-Seyler's Z. physiol. Chem.* 351:157-162.
20. Gottlieb, A. A. and R. H. Schwartz (1972). Antigen-RNA interactions. *Cell Immunol.* 5:341-362.
21. Mitsuhashi, S., K. Saito and S. Kurashige (1973). The role of RNA as amplifier in the immune response. *Ann. N. Y. Acad. Sci.* 207:160-171.
22. Roelants, G. E. and J. W. Goodman (1969). The chemical nature of macrophage RNA-antigen complexes and their relevance to immune induction. *J. exp. Med.* 130:557-574.
23. Borris, Elsa and G. Koch (1975). Isolation of poliovirus neutralizing antibody by affinity chromatography. *Virology*, in press.
24. Schneider, C. H. and A. L. de Weck (1965). A new chemical aspect of penicillin allergy: The direct reaction of penicillin with ε-amino-groups. *Nature (Lond.)* 208:57-59.

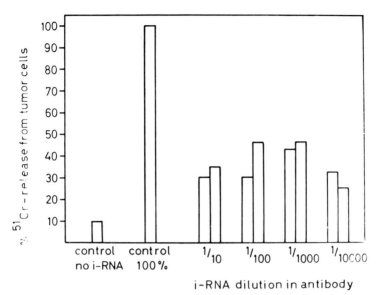

Fig. 10. ^{51}Cr release from Lewis rat tumor cells.
Tumor cells were grown in Medium 199 containing 0.04 µg Gentamycin/ml, fortified with 5 % fetal calf serum. To the medium 0.25 µCi of ^{51}Cr, specificity 652 mCi/mg Cr (Radioisotopenproduktion Würenlingen, Switzerland) was added. After incubation for 1 hour at 37° C, the cells were washed – usually 5 times – with PBS until the supernatant was free of radioactivity. Then 10^5 cells in a volume of 0.1 ml were incubated with 0.1 ml of i-RNA coded and cell-free synthesized protein. After this, 0.5 ml PBS was added, mixed, and the cells were sedimented by centrifugation at 500 g for 2 min. The clear supernatant was estimated in a gamma counter for the content of free ^{51}Cr. As 100 % was calculated the total amount of ^{51}Cr on the cells and the supernatant after incubation for 1 hour at 37° C; as background we used the radioactivity in the supernatant of cells incubated with a protein coded by a heterologous i-RNA.

INFORMATIONAL RNA DIRECTED SYNTHESIS OF VIRUS NEUTRALIZING PROTEINS IN CELL-FREE EXTRACTS PREPARED FROM RAT SPLEEN AND MOUSE L CELLS

G. Koch*, P. Bilello*, M. Fishman[†],
R. Mittelstaedt*[††] and E. Borriss[+]

*Roche Institute of Molecular Biology
Nutley, New Jersey 07110

[†]Public Health Research Institute
City of New York, Inc.
New York, New York

[+]Abteilung fuer Molekularbiologie
Medizinische Hochschule, Hannover
Hannover, W. Germany

INTRODUCTION

Two different RNA species isolated from antigen-exposed, peritoneal exudate cells can induce a specific immune response in spleen cell cultures. One type is an RNA-antigen complex (1-3) and the second is an informational RNA (iRNA) which is free of detectable antigen (4-6). This presentation will deal only with the latter kind of RNA. The iRNA has been reported to induce the formation of allotype and antigen specific 19S immunoglobulins in spleen cultures (4,5). It has been suggested that iRNA itself acts as mRNA (4-6). However, it has not yet been clearly established whether iRNA contains all the coding capability to direct the synthesis of light, heavy and J chains of 19S immunoglobulins.

We have applied two systems to delineate the role and functions of iRNA: (a) iRNA isolated from phage- and poliovirus-exposed mouse peritoneal exudate (PE) cells was used to direct the synthesis of proteins in rat spleen

---

[††]Present Address: Department fuer Innere Medizin
Medizinische Hochschule, Hannover
Hannover, W. Germany

cell-free extracts and (b) iRNA isolated from phage-exposed, rabbit PE cells was used to direct the synthesis of proteins in extracts prepared from mouse L cells. The properties of the *in vitro* synthesizing proteins were subsequently analyzed by the use of affinity chromatography on virus-agarose columns, virus-neutralizing assays, allotype characterization and determination of monomeric and polymeric size.

RESULTS

Isolation of virus neutralizing antibodies of affinity chromatography.

Poliovirus can be covalently bound to cyanogen bromide activated agarose with high efficiency and with at least partial retention of its biological activity, i.e., its infectivity and its capacity to bind neutralizing antibodies (7). Poliovirus agarose columns specifically retain poliovirus neutralizing antibodies. The elution of these antibodies from the columns requires a two step procedure: (a) exposure to pH 2.0 or 3.0 and (b) adjustment of the column eluting fluid to neutrality. Table I summarizes the relevant data. Comparable observations have been made with MS2 phage coupled to cyanogen bromide activated agarose (Mittelstaedt and Koch, unpublished data). The elution of the MS2-phage antibodies retained by this column also occurs only after exposure to dilute acid with subsequent return of the eluting fluid to neutrality. The retention of proteins at low pH on virus-agarose columns is a special characteristic of agarose columns. If poliovirus is coupled to cyanogen-activated Sephadex G-200, the virus-neutralizing antibodies retained on the column are eluted by dilute acid alone (Borriss, unpublished results).

Stimulation of synthesis of poly(A)-containing RNA in murine PE cells by exposure to poliovirus or MS2 phage.

One characteristic property of mammalian mRNA is its poly(A)-containing 3' end which provides a unique tool for the rapid separation of

TABLE 1.
Purification of Poliovirus Neutralizing Antibodies on Poliovirus Agarose

Protein fraction derived from:	protein µg/ml	PFU titer (% of control)	% inhibition of PFU formation
Serum IgG preparation	200	10-15	85-90
Affinity chromatography			
a) Flow through	∼200	85-90	10-15
b) Wash fraction	1	95-100	<5
c) pH 2.0 eluate	<1	95-100	<5
d) pH 2.0 concentr. eluate	1-10	95-100	<5
e) pH 7.4 eluate	<1	5-10	90-95

mRNA from other cellular RNA species by chromatography on poly(dT)-cellulose (8). Exposure of murine PE cells to poliovirus or MS2 phage induces a 2 to 3 fold increase in the incorporation of ^{3}H uridine into total, but especially into poly(A)-containing RNA in PE cells within 30 min at 37°C. We assumed that part of the poly(A)-containing RNA is iRNA. Therefore, the coding activity of iRNA and poly(A)-containing iRNA was determined by following the incorporation of radioactive amino acids into proteins in cell-free extracts prepared from rat spleen cells and by characterization of the newly synthesized proteins.

Characterization of the proteins synthesized in rat spleen cell-free extracts in response to murine iRNA.

We analyzed the *in vitro* synthesized proteins in response to iRNA in a virus neutralizing assay. We found that the plaque forming titer of both poliovirus and MS2 phage increased specifically after a 60 min incubation at 37°C of virus with the *in vitro* reaction product (Table 2). Poliovirus infectivity was enhanced only by incubation with the *in vitro* product synthesized in response to iRNA isolated from poliovirus exposed PE cells but not from MS2 phage exposed cells. Likewise, the same specificity was found in the enhancement of MS2 phage infectivity. Therefore, the proteins in the extract were fractionated and concentrated by precipitation with 40% ammonium sulfate, gel filtration on Sephadex G-200 and affinity chromatography on virus-agarose columns. The data shown in Table 2 show that both fractionation by ammonium sulfate and sephadex as well as isolation by affinity chromatography yielded proteins which specifically neutralized either MS2 phage or poliovirus. When the ammonium sulfate precipitated proteins were subjected to sephadex gel filtration, only the void volume fractions contained virus neutralizing activity. Like-

TABLE 2.
Effect of iRNA Directed In Vitro Synthesized Proteins on the Infectivity of Poliovirus and MS2 Phage

Protein fraction derived from:	PFU titer % of control*	
	MS2	Poliovirus
crude cell-free extract	120-140*	130-150*
sephadex 19S fraction of cell-free extract	–	30-40*
eluate from affinity chromatography	5-40*	15-50*

– not done
* variation seen in different experiments
Incubation of virus and protein fraction as described in Table 1

wise, isolation of these proteins by affinity chromatography yielded virus neutralizing activity only in fractions in which specific antibodies elute (Table 2). The protein fractions which neutralized one virus were always found to be specific. That is, iRNA isolated from PE cells exposed to MS2 phage directed the synthesis of proteins which specifically neutralized MS2 phage but not poliovirus and vice versa. As a further control, anti-phage MS2 protein was passed through poliovirus-agarose columns. None of the fractions contained proteins with neutralizing activity against poliovirus. All these data indicate that iRNA induces in spleen cell-free extracts the synthesis of virus-specific neutralizing proteins which resemble antibodies. However, they do not prove that iRNA contains genetic information for the synthesis of complete antibody molecules since the cell-free system itself contains mRNA which might contribute to the synthesis of complete antibody molecules.

Characterization of proteins synthesized in L cell-free extracts in response to rabbit iRNA.

On the assumption that tissue culture cells are normally not involved in antibody synthesis, mouse L cells were selected as a source for a cell-free system. This system has been used by other investigators previously for the translation of exogeneous mRNA (9). The iRNA synthesized in rabbit PE cells in response to exposure to T_2 phage was used since this RNA has been shown to initiate a donor, allotype specific immune response in spleen cell cultures (4,5). Therefore, the use of this iRNA provided a possible method for the characterization of its cell-free translation product not only for antigen specificity but also for allotype. PE cells from either 11/44 or 11/55 rabbits were incubated with either low (1 phage per 100 cells) or high (100 to 1) input of T_2 phage for 30 min at 37°C and the RNA isolated. Either total RNA (20 to 60 µg/ml), sucrose fractionated RNA or mRNA eluted from dT-cellulose columns (20-20 µg/ml) were used to stimulate protein synthesis in cell-free extracts from L cells. The specificity of the T_2 iRNA directed protein synthesis was analyzed by an allotype amplified T_2 phage neutralization assay. The data are summarized in Table III.

Using 11/44 iRNA, the neutralizing activity was amplified by anti-allotype-4 but not with anti-allotype-5 serum. Titers of up to 40% neutralization at a 1:8 dilution were obtained. The translation products of 11/55 T_2 iRNA gave similar neutralization titers. They also amplified the neutralizing activity with anti-allotype-5-serum, but not with anti-allotype-4.

Binding of phage to cell-free system products was carried out in an effort to further characterize the L cell translation product. The data are shown in Table IV. [^{35}S] methionine-labeled aliquots of the *in vitro* reaction product were dialyzed, then precipitated with 40% ammonium sulfate. The precipitates were incubated with phage one hour at 37°, then overnight at 4°. Separation

TABLE 3.
Amplification of Neutralization of T_2 Phage with Anti-Allotype-4-Serum

L-Cell-Free System Translation Product	% Neutralization anti-4	anti-5
11/44 T_2 iRNA (Hi)	37*	0
11/44 T_2 iRNA (Lo)	32	9
dT-cellulose poly (A) T_2 iRNA (Hi)	35	–
sucrose 16S T_2 iRNA (Hi)	25	–
sucrose 16S T_2 iRNA (Lo)	33	–
sucrose 18S T_2 iRNA (Hi)	28	–
sucrose 18S T_2 iRNA (Lo)	40	–
11/44 Normal RNA	13	0
Endogenous	16	0
23/55 T_2 iRNA (Lo)	2	25

A 1:8 dilution of the translation product was incubated with T_2 phage for 30' at 37°, followed by incubation with anti-4 for 30' at 37° and 1 hr at 4°. Aliquots were plated on *E. coli* B.

T_2 iRNA (Hi): PE cells exposed to T_2 at a ratio of 100 T_2/cell.

T_2 iRNA (Lo): PE cells exposed to T_2 at a ratio of 1 T_2/100 cells

– not done

* Neutralization of more than 20% is significant

of the antigen-antibody complexes formed during this incubation from free translation products was carried out by Sepharose 2B column chromatography. The column was equilibrated with a Tris-Mg^{++} buffer containing either 0.5 or 1.0 M NaCl at pH 7.5. Phenol red was used as a marker. Phage T_2 and SP82 elute in the void volumn of sepharose 2B whereas 19S immunoglobulins appear in later fractions clearly separated from both phage and 7S immunoglobulins. One to three percent of the labeled proteins recovered from sepharose were found in the 19S region of the eluate, but only when the ammonium sulfate precipitate of iRNA stimulated *in vitro* synthesized proteins was fractionated. The 19S proteins were converted to 7S size proteins by exposure to 100 mM mercaptoethanol for 60 min at room temperature. The endogeneous cell-free extracts contained no proteins of the 19S size but almost exclusively proteins of small molecular weight eluting with the phenol red marker. After incubation with phage, proteins from both unstimulated and iRNA stimulated *in vitro* reactions eluted with the void volume together with

TABLE 4.
Fractionation of In Vitro Synthesized Proteins After Incubation with Phage T_2 or SP82 by Sepharose 2B Gel Filtration

	T$_2$ iRNA Stimulated				Unstimulated	
	incubated with T$_2$		incubated with SP82		incubated with T$_2$	
	% Labeled Protein Recovered from Sepharose-in					
	void vol.	19S fract	void vol.	19S fract	void vol.	19S fract
Ammonium sulfate precipitation	9-11*	1-2*	3-6*	1-3*	1-3*	0
Pooled phage from void volume incubated with mercaptoethanol	6	0	0	0	0	0
Pooled phage from void volume incubated at pH 2.4 †	0	0	—	—	—	—

* variation seen in different experiments
† all label present in 7S size proteins after exposure to pH 2.4.

the phage. Two observations, however, may indicate that phage specific labeled antibodies can be identified also with this method: (a) more labeled proteins are bound to T_2 phage than to SP82 with the T_2 iRNA stimulated reaction products and (b) after dissociation of polymeric proteins with mercaptoethanol only T_2 phage retained 50% of the labeled proteins. The remaining proteins can be dissociated from T_2 by exposure to pH 2.4. The mercaptoethanol as well as the acid released proteins are of 7S size.

DISCUSSION

Exposure of PE cells to antigen results in increased incorporation of radioactive uridine into poly(A)-containing iRNA. Addition of this RNA to cell-free extracts prepared from rat spleen cells or mouse L cells stimulated the incorporation of [^{35}S]-methionine into proteins 2 to 10 times above the incorporation in the endogenous reaction. Affinity chromatography on columns of virus coupled to cyanogen-bromide activated agarose was used to characterize the proteins synthesized *in vitro* in response to iRNA. A fraction of these proteins binds specifically to virus-agarose but also to virus in suspension and specifically neutralizes that virus which was used to stimulate the synthesis of poly(A)-containing iRNA in PE cells. The virus neutralizing activity is found in proteins of 19S size only and is amplified specifically by anti-allotype serum. The 19S proteins dissociate into 7S proteins upon exposure to 100 mM mercaptoethanol at room temperature for 60 min. Upon incubation of the T_2 phage-protein complex under this condition, 50% of the labeled proteins remained T_2 bound. The remaining T_2 phage bound proteins are released from T_2 phage as 7S proteins by exposure to pH 2.4.

The size of the iRNA was determined by glycerol gradient centrifugation with ribosomal RNA as marker. iRNA recovered in fractions corresponding in size to 16S and 18S is able to induce the synthesis of virus neutralizing activity in L cell-free extracts. Our results support the conclusion of others (5,6,12,13) that iRNA functions as mRNA. They further indicate that iRNA may contain the coding capability to direct the synthesis of complete 19S antibodies.

Ten mg of a human IgG preparation in 50 ml of phosphate buffered saline pH 7.4 were loaded on a poliovirus-agarose column (0.9 cm X 10 cm) as described (7). The column was washed first with buffered saline, pH 7.4, then with 0.15 M NaCl in 0.01 N HCl and again with buffered saline pH 7.4. The plaque test was performed as described (10). The activity of poliovirus neutralizing proteins was assayed after incubation of poliovirus (500 PFU/ml) with protein fractions for 3 hr at 37°C followed by at least 1 hr at 4°C. Protein concentrations were determined by a modified Lowry method (11). (For further details see reference 7.)

SUMMARY

Virus neutralizing antibodies can be highly purified by affinity chromatography on virus coupled to cyanogen-bromide activated agarose. This method was applied to characterize proteins synthesized *in vitro* in response to iRNA obtained from antigen (MS2 phage or poliovirus) stimulated murine peritoneal exudate (PE) cells. Exposure of PE cells to these antigens enhances the synthesis of poly(A)-containing RNA which directs the synthesis of proteins in cell-free extracts from rat spleen. These proteins specifically neutralize the virus used for stimulation of the PE cells. Characterization of these proteins by sizing and allotype specificity suggests that they are similar to IgM. Likewise, poly(A)-containing RNA isolated from T_2 phage-exposed rabbit PE cells directs the synthesis of antigen and allotype specific antibodies in cell-free extracts prepared from mouse cells. These results indicate that iRNA functions as mRNA and contains the information required to code for the synthesis of IgM antibodies.

REFERENCES

1. Garvey, J. S. and D. H. Campbell (1957). The retention of S^{35}-labeled bovine serum albumin in normal and immunized rabbit liver tissue. *J. Exptl. Med. 105*:361-372.
2. Fishman, M. and F. L. Adler (1967). The role of macrophage-RNA in the immune response. *Cold Spring Harbor Symposium Quant. Biol. 32*:343-348.

3. Gottlieb, A. A. (1969). Macrophage ribonucleoprotein: Nature of the antigenic fragment. *Science 165*:592.

4. Adler, F. L., M. Fishman and S. Dray (1966). Antibody formation initiated *in vitro*. III. Antibody formation and allotypic specificity directed by ribonucleic acid from peritoneal exudate cells. *J. Immunol. 97*:554-558.

5. Fishman, M., F. L. Adler and S. G. Rice (1973). RNA in the *in vitro* immune response: Macrophage RNA in the *in vitro* immune response to phage. *RNA in the Immune Response*. H. Friedman, Ed. New York Academy of Sciences.

6. Jachertz, D. (1973). Flow of information and gene activation during antibody synthesis. *RNA in the Immune Response*. H. Friedman, Ed. New York Academy of Sciences.

7. Borris, E. and G. Koch (1975). Isolation of poliovirus-neutralizing antibodies by affinity chromatography. *Virology 67*:356-364.

8. Gilham, P. (1964). The synthesis of polynucleotide-cellulose and their use in the fractionation of polynucleotides. *J. Amer. Chem. Soc. 86*:4982.

9. McDowell, M. J., W. K. Joklik, L. Villa-Komaroff and H. F. Lodish (1972). Translation of reovirus messenger RNAs synthesized *in vitro* into reovirus polypeptides by several mammalian cell-free extracts. *Proc. Nat. Acad. Sci.* USA *69*:2649-2653.

10. Bishop, J. M. and G. Koch (1969). Plaque assay for poliovirus and poliovirus specific RNAs. *Fundamental Techniques in Virology.* K. Habel and N. P. Salzman, Eds. Academic Press, New York.

11. Shatkin, A. J. (1969). Colorimetric reactions for DNA, RNA, and protein determinations. *Fundamental Techniques in Virology.* K. Habel and N. P. Salzman, Eds. Academic Press, New York.

12. Mitsuhashi, S., K. Saito and S. Kurashige (1973). The role of RNA as amplifier in the immune response. *RNA in the Immune Response*. H. Friedman, Ed. New York Academy of Sciences.

13. Giacomoni, D., S. R. Wang and S. Dray (1973). Discussion paper: Persistence and size of homologous RNA incorporated by rabbit spleen cells. *RNA in the Immune Response*. H. Friedman, Ed. New York Academy of Sciences

ISOLATION, PURIFICATION AND CELL-FREE TRANSLATION OF IMMUNOGLOBULIN MESSENGER RNAS FROM IMMUNOGLOBULIN PRODUCING CELLS AND VARIANTS

Sidney Pestka, Laura Bailey, Roderich Brandsch, Peter Graves,
Michael Green, Robert Jilka, James McInnes, Akira Okuyama,
Philip Tucker, David Weiss, Lyn Yaffe and Tova Zehavi-Willner

Roche Institute of Molecular Biology
Nutley, New Jersey 07110

INTRODUCTION

Plasmacytoma cell lines provide an experimental system in which to investigate the origin of antibody diversity, the cellular commitment and specialization of lymphocyte populations, and the intracellular control and organization of immunoglobulin biosynthesis. Several immunoglobulin messenger RNAs from a variety of plasmacytoma lines have been isolated and characterized to develop an insight into some of these questions (1-11). Standard methods such as density gradient fractionation and oligo(dT)-cellulose chromatography have been employed to purify these mRNA molecules. In addition, specific techniques such as the immunoprecipitation of immunoglobulin synthesizing polysomes (9,12) or the immunoprecipitation of a complex formed between the immunoglobulin heavy chain mRNA and the complete immunoglobulin molecule have been used (13). These mRNA preparations have been used in initial attempts to distinguish between the major theories pertaining to the generation of antibody diversity, the germ-line theory, the somatic mutation theory, and the somatic recombination theory by direct measurement of the number of immunoglobulin genes present in the cellular DNA by hybridization studies (14-19). The theoretical foundations, predictions, and implications of these theories have been reviewed (20). Additionally, the nucleotide sequence of part of an immunoglobulin light chain mRNA has been reported (21). Such studies should help delineate possible processing or translational control regions in mRNA sequences.

The isolation of immunoglobulin mRNAs from additional plasmacytoma cell lines is useful to develop and expand our understanding of immunoglobulin biosynthesis and gene expression. We thus have concentrated on the isolation, purification, and cell-free translation of mRNAs for both the heavy chain and the light chain of various myeloma proteins in order to examine many of the questions enveloping and relevant to immunoglobulin biosynthesis and gene expression. The methodology employed is described in detail elsewhere (10,11,22).

RNA EXTRACTION AND FRACTIONATION

RNA from MOPC-315 plasmacytoma was obtained by two different procedures as illustrated in Fig. 1. In one procedure (Fig. 1A), the total cellular RNA was obtained by extracting the entire tumor. This procedure had the advantage of not artifically biasing — qualitatively or quantitatively — the composition of the mRNA fraction that was subsequently obtained. In the second procedure (Fig. 1B), polyribosomal RNA was extracted from the membrane-

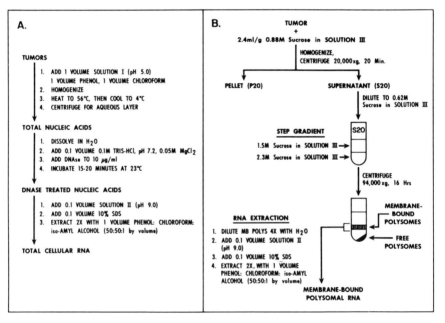

Fig. 1. *Flow diagrams depicting preparation of (A) total cellular RNA and (B) membrane-bound polysomal RNA. Solution I consists of 10 mM sodium acetate (pH 5.0), 1.2 g/l polyvinyl sulfate, 0.3 g/l bentonite, 1.0 g/l 8-hydroxyquinoline, and 5 g/l sodium dodecylsulfate; Solution II, 1.0 M Tris-HCl (pH 9.0), 1.0 M NaCl, and 0.01 M EDTA; Solution III, 50 mM Tris-HCl (pH 7.4), 25 mM KCl, 5 mM $MgCl_2$, and 6 mM 2-mercaptoethanol.*

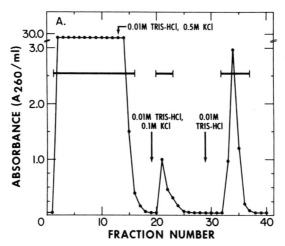

Fig. 2. *Oligo(dT)-cellulose chromatography of membrane-bound polysomal RNA from MOPC-315 tumors. Column profile obtained with 1750 A_{260} units of RNA. Elution with 0.5 M KCl in 0.01 M Tris-HCl (pH 7.4), 0.1 M KCl in 0.01 M Tris-HCl (pH 7.4), or 0.01 M Tris-HCl (pH 7.4) is indicated by the arrows. Column fractions were pooled for further analysis as indicated by the horizontal bars.*

bound polysomes recovered as a distinct band in the sucrose step-gradient employed for polysome fractionation. This method attempted to take advantage of the observation that immunoglobulin biosynthesis takes place predominantly on ribosomes bound to the endoplasmic reticulum (23-25).

The poly A-containing mRNA fractions were obtained by chromatography of the membrane-bound polysomal RNA or the total cellular RNA on oligo(dT)-cellulose (Fig. 2). Immunoglobulin mRNAs, like most eucaryotic mRNAs, contain a poly A segment at their 3′-hydroxyl end and thus bind to oligo(dT)-cellulose at the appropriate ionic strength (3,5). The poly A-containing fraction that eluted with the lowest ionic strength buffer usually comprised 1 to 2 percent of the total material applied to the column whether membrane-bound polysomal RNA or total cellular RNA was used. Column fractions were pooled and concentrated by ethanol precipitation. Analysis of the RNA fractions by centrifugation in sucrose gradients (Fig. 3) demonstrated that the fraction composed of the sample front and the 0.5 M KCl eluate contained tRNA and the 5S, 18S and 28S RNA, and is called the flow-through fraction. The 0.1 M KCl wash fraction contained primarily 18S and 28S rRNA. In initial experiments with oligo(dT)-cellulose chromatography, significant contamination of the poly A-containing fractions by rRNA was observed despite lengthy washing of the column with 0.5 M KCl. It was discovered, however, that introduction of the 0.1 M KCl wash step prior to final elution removed most of the residual rRNA bound to the column. The major

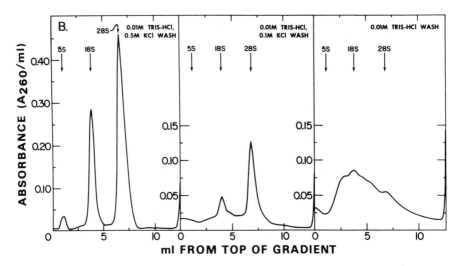

Fig. 3. *Analytical sucrose gradient sedimentation of RNA in pooled column fractions. RNA was analyzed on linear 5% to 20% sucrose (w/v) gradients in 0.02 M sodium acetate (pH 5.0). Centrifugation was performed at 40,000 rpm for 6 hours at 4° in the SW40 rotor of the L2-65B Spinco ultracentrifuge.*

component of this residual RNA is 28S rRNA (Fig. 3). The poly A-containing mRNA fraction obtained from the oligo(dT)-cellulose column in the final elution step is a mixture of several RNA species (Fig. 3).

Poly A-containing mRNA derived from the total cellular RNA and from membrane-bound polysomal RNA was further fractionated on preparative sucrose gradients in 0.02 M sodium acetate (pH 5.0). In order to minimize aggregation of RNA, the poly A-containing mRNA was heated briefly (80° for 30 secs) in this low ionic strength buffer before application to the gradient (22). This procedure results in an RNA profile simpler than that reported previously in sucrose gradients containing SDS (10,11). The fractions from the gradient were pooled as shown in Fig. 4 and the RNA was recovered by ethanol precipitation. As can be seen by comparison of sucrose gradient profiles of heated and unheated mRNA (Fig. 4), heating of the mRNA fraction significantly reduced the material sedimenting greater than 18S. It is very likely that prior heating of the mRNA dispersed aggregates present in the unheated sample. A similar gradient profile for membrane bound poly A-containing mRNA also shows that the short period of heating at 80° reduced the number of sedimenting aggregates greater than 18S (Fig. 5). Furthermore, by a comparison of the heated poly A-containing mRNA from total and membrane bound preparations (Fig. 4B and Fig. 5B, respectively), it can be seen that the membrane-bound preparation was deficient in large mRNA of 18S and greater.

The 12-14S RNA species, fraction B′ (Fig. 4B) and the 16-19S RNA species, fraction C′ (Fig. 4B) were further fractionated on separate sucrose gradients (Fig. 6A,B). The RNA species (fraction L and fraction H, Fig. 6) were recovered by ethanol precipitation. These RNA species were subsequently fractionated by preparative polyacrylamide gel electrophoresis in formamide. Both L and H fractions showed a diffuse gel absorbance profile at 260 nm (Fig. 7). However, at least one major band was evident in these fractions on the polyacrylamide gels. These bands were even more clearly seen on analogous gels which were stained with Stainall rather than scanned for an absorbance profile (Fig. 8). Gel slices were removed as shown in Fig. 7 and the various RNA species were isolated (22).

The molecular weights of the two major bands, L_C and H_C (Fig. 7) were

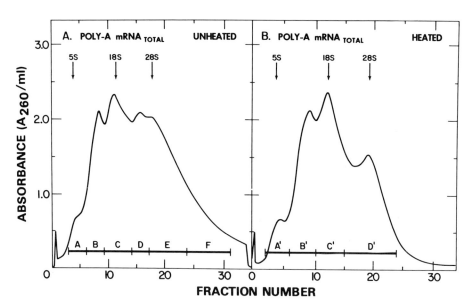

Fig. 4. *Preparative sucrose gradient fractionation of total poly A-containing mRNA extracted from MOPC-315 tumors. Total poly A-containing mRNA (15 to 25 A_{260} units) in 0.01 M sodium acetate buffer (pH 5.0) was heated (80° for 30 sec) and rapidly cooled on ice before loading onto the gradient. Centrifugation was performed in linear 5 to 20% (w/v) sucrose gradients in 0.02 M sodium acetate (pH 5.0) at 4° for 16 hrs at 26,000 rpm in the Spinco SW 27.1 rotor. Fractions were pooled as indicated. The positions of 5S, 18S, and 28S RNA obtained on a parallel gradient are indicated.*

A. Unheated Poly A-containing mRNA from total cellular RNA;
B. Heated (80°, 30 sec) poly A-containing mRNA from total cellular RNA.

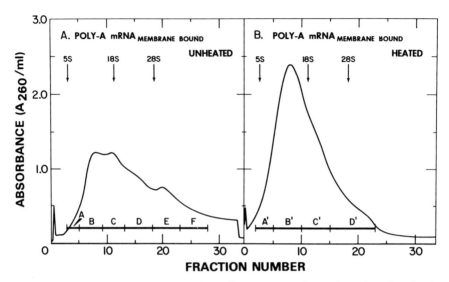

Fig. 5. *Preparative sucrose gradient fractionation of membrane-bound poly A-containing mRNA extracted from MOPC-315 tumors. Centrifugation was performed as indicated in legend to Fig. 4.*

A. Unheated poly A-containing mRNA from membrane-bound polysomes;

B. Heated (80°, 30 sec) poly A-containing mRNA from membrane bound polysomal RNA.

determined by using ribosomal RNA from *E. coli* and from the tumor as well as rabbit globin mRNA as migration standards. The values obtained are 4.0×10^5 daltons (1180 nucleotides; 13.0 S) and 6.8×10^5 daltons (2000 nucleotides; 17.5 S) for the light (L_C) and heavy chain (H_C) mRNA species, respectively. Both these values compare favorably with those found by other workers for immunoglobulin light and heavy chain mRNAs purified from various plasmacytomas (14,15,17,18,21,26,27).

We have prepared mRNA fractions from a number of myeloma tumor lines. These include the following murine plasmacytomas: MOPC-315, MOPC-315 NR, MOPC-315 NP-1, RPC-20, MOPC-104 E, J558, HOPC-1, Y5444, H-2020, MPC-11, MOPC-321, MOPC-41, MOPC-70E, and MOPC-46B. In addition, mRNA has been prepared from the human myeloma line, RPMI 8226.

Immunoglobulin Messenger RNA Fractionation Followed by Cell-Free Translation. Throughout the purification procedure, the different mRNA species were tested for their ability to serve as templates for the synthesis of polypeptide chains in a cell-free system derived from Ehrlich ascites cells. The products synthesized in response to the mRNA fractions released from the formamide gels were compared on SDS-polyacrylamide slab gels (Fig. 9). The primary translation product of the mRNA fractions L_A, L_B, L_C, and L_D (Fig.

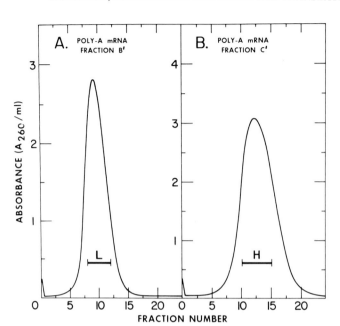

Fig. 6. *Preparative sucrose gradient fractionation of poly A-containing mRNA species. Poly A-containing mRNA (6 to 12 A_{260} units) in 0.01 M sodium acetate buffer (pH 5.0) was heated (80° for 30 sec) and rapidly cooled on ice before loading onto the gradient. Centrifugation and sucrose gradients are described in the legend to Fig. 4.*
  A. mRNA fraction B' (Fig. 4B);
  B. mRNA fraction C' (Fig. 4B).

7) migrated slightly slower than the authentic light chain of MOPC-315 protein (L^{315}). This translation product (pL^{315}) is a precursor of L^{315} (10,11). The majority of the light chain mRNA activity was associated with species L_B and L_C. In the case of mRNA fraction L_E (Fig. 7A), the primary translation product migrated approximately halfway between L^{315} and the dye front. This protein product has not yet been identified, but represents the product coded for by an mRNA or class of mRNAs smaller than L^{315} mRNA, as measured by polyacrylamide gel electrophoresis in formamide (22). With respect to the mRNA fractions H_A, H_B, H_C, and H_D (Fig. 7B), it can be seen that fractions H_B and H_C gave rise to a primary translation product, barely detectable, which migrated slightly faster than the authentic heavy chain of the MOPC-315 protein (H^{315}). This translation product is a protein species antigenically and structurally related to H^{315} (10,11). Interestingly, all mRNA fractions, H_A, H_B, H_C, and H_D directed the synthesis of detectable amounts of H^{315} when globin mRNA was included in the cell-free incubation mixtures (Fig. 10). The reason for this stimulation is unknown at the present time.

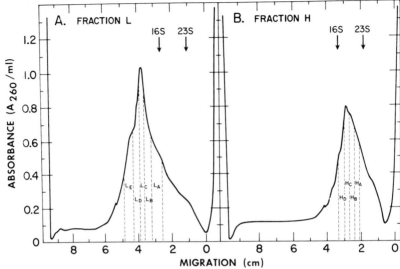

Fig. 7. *Polyacrylamide gel electrophoresis in formamide of poly A-containing mRNA fractions from the second sucrose gradient centrifugations. The RNA samples (10 to 30 µg) were loaded onto cylindrical 4% polyacrylamide gels in 97 to 98% formamide and electrophoresis was performed at 80 volts for 20 hrs at 4°. The gel was scanned at 260 nm in a Gilford spectrophotometer. Gel slices were removed as shown and RNA recovered as described (22). E. coli 16S and 23S RNA was electrophoresed on a parallel gel and the positions obtained with these RNA species are indicated.*
A. mRNA fraction L (Fig. 6A);
B. mRNA fraction H (Fig. 6B).

It has been reported in previous studies (10,11) that L^{315} mRNA was present in the H^{315} mRNA fraction (fraction C', Fig. 4) after the first sucrose gradient centrifugation. Similarly, L^{315} mRNA was present in the H^{315} mRNA fraction (fraction H, Fig. 6B) after the second sucrose gradient centrifugation as shown by the analyses of the translation products on SDS-polyacrylamide slab gels (data not shown). This forward trailing of light chain mRNA activity in sucrose gradients has been reported previously (8). The L^{315} mRNA activity was separated from the H^{315} mRNA fractions, H_A, H_B, H_C, and H_D, only after preparative polyacrylamide gel electrophoresis in formamide (Figs. 9,10).

CELL-FREE TRANSLATION OF IMMUNOGLOBULIN MESSENGER RNA FROM MOPC-315 PLASMACYTOMA AND MOPC-315 VARIANTS

A. *MOPC-315 NR, a Variant Synthesizing Only Light Chain*

Variants of plasmacytoma cell lines with respect to immunoglobulin production have been identified and characterized in several laboratories (28-39). In the course of studies with MOPC-315, a plasmacytoma producing an immunoglobulin of the IgA class with high affinity for dinitrophenyl and trinitrophenyl groups, Hannestad et al. (39) reported the isolation of a variant tumor line lacking cell-bound myeloma protein capable of binding 2,4,6-trinitrophenylated sheep erythrocytes. This variant cell line, MOPC-315 NR (39), secreted only the light chain of the MOPC-315 IgA molecule.

MOPC-315 cells synthesize both heavy (H^{315}) and light (L^{315}) chains (40). The inability of the variant MOPC-315 NR to secrete H^{315} may reside in an altered DNA, a defective, deficient, or absent mRNA, or possibly a specific translational or post-translational lesion. Our studies were designed to localize the level and nature of the defect. We thus examined the cell-free

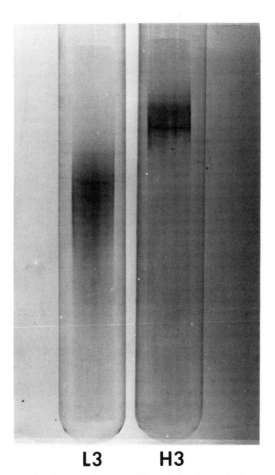

L3   H3

Fig. 8. *Photograph of the polyacrylamide gel fractions similar to those of Fig. 7. The polyacrylamide gels were stained with Stainall.*

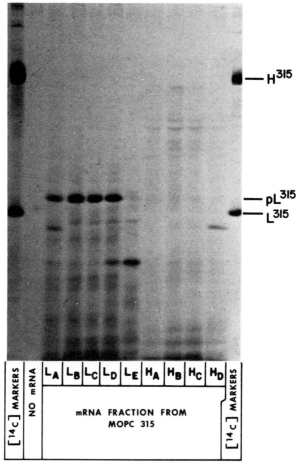

Fig. 9. *Autoradiogram of [^{14}C] labeled cell-free products synthesized in the Ascites cell-free system and analyzed on an SDS-polyacrylamide gel slab. Products were synthesized in the presence of the RNA fractions indicated in the figure. The positions of authentic H^{315} and L^{315} as well as the position of the L^{315} precursor (pL^{315}) are indicated.*

translation of the mRNAs obtained from both MOPC-315 and MOPC-315 NR to localize and define the defect.

Proteins Synthesized by Intact Cells. A comparison of the intracellular and secreted proteins synthesized by cell suspensions prepared from MOPC-315 and MOPC-315 NR tumors in shown in Fig. 11. It is clear that both cell cultures synthesize and secrete substantial amounts of L^{315}, whereas the MOPC-315 NR cells synthesize little, if any, H^{315}.

Cell-Free Protein Synthesis and Analyses of Products. The total poly A-containing mRNA preparations from both MOPC-315 and MOPC-315 NR tumors and the fractions obtained from sucrose gradients were tested for their ability to serve as templates for the synthesis of polypeptide chains in a cell-free system derived from Ehrlich ascites cells. The products synthesized in response to either MOPC-315 or MOPC-315 NR mRNA were compared on SDS-polyacrylamide slab gels (Fig. 12). Substantial synthesis of both H^{315} and L^{315} precursors occurred in the presence of RNA fractions from the

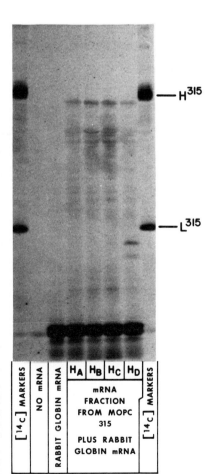

Fig. 10. *Autoradiogram of [^{14}C] labeled cell-free products synthesized in the Ascites cell-free system analyzed on an SDS-polyacrylamide gel slab. Products were synthesized in the presence of the RNA fractions indicated in the figure. The positions of authentic H^{315} and L^{315} are indicated.*

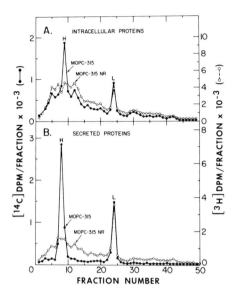

Fig. 11. SDS-polyacrylamide gel electrophoretic analyses of proteins synthesized by MOPC-315 and MOPC-315 NR tumor cell suspensions. MOPC-315 and MOPC-315 NR cells were labeled with $[^{14}C]$leucine and $[^{3}H]$leucine, respectively. The cell cytoplasm and the culture media from the two lines were mixed before electrophoresis.
  A. Intracellular proteins after a 30 min period of labeling;
  B. Secreted proteins after a 3-hr labeling period.

MOPC-315 plasmacytoma. In addition, the L^{315} precursor was synthesized in the presence of mRNA from MOPC-315 NR. It is clear, however, that no detectable H^{315} was synthesized in the presence of any of the mRNA fractions obtained from the variant MOPC-315 NR plasmacytoma. Mixing experiments (data not shown) designed to test for the presence of an inhibitor of H^{315} mRNA translation in the mRNA obtained from MOPC-315 NR were also performed. It was observed that H^{315} mRNA from MOPC-315 could still be successfully translated in the presence of equal amounts of either the total poly A-containing mRNA or the 16-19S mRNA from MOPC-315 NR.

Immunochemical Analysis of the Cell-Free Products. Studies with the mRNAs obtained from the MOPC-315 plasmacytoma had demonstrated that the cell-free products programmed by partially purified H^{315} mRNA fractions (16-19S) resulted in the synthesis of fragments antigenically related to H^{315} in addition to complete H^{315} and L^{315}. To determine whether the 16-19S class of MOPC-315 NR mRNA produced any portion of H^{315}, immunochemical analysis of the cell-free products was performed. The cell-free products of the 16-19S mRNA fractions from both MOPC-315 and MOPC-315 NR were compared with regard to their ability to be precipitated by normal rabbit serum or by rabbit antiserum to L^{315} or H^{315}. The cell-free products directed by the MOPC-315 mRNA fraction contained antigenic determinants and polypeptides related to L^{315} and H^{315} (Fig. 13). The precursors to L^{315} (pL^{315}) and to H^{315} were major products. The cell-free products stimulated by the mRNA fraction from MOPC-315 NR contained the pL 315 protein

(Fig. 14). However, no significant immunoprecipitable heavy chain or H^{315}-related peptides were detected (Fig. 14).

B. *MOPC-315 NP-1, a Variant Synthesizing Only Heavy Chain*

The MOPC-315 NP-1 variant plasmacytoma arose spontaneously during the serial transfer of the MOPC-315 tumor line in the BALB/c strain of mice and appears to be analogous to several of the variants isolated from other tumor cell lines. Specifically, it has lost the ability to synthesize L^{315} and

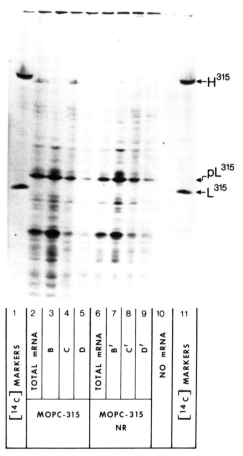

Fig. 12. *Autoradiograms of $[^{35}S]$ methionine-labeled cell-free products analyzed on SDS-polyacrylamide gel slabs. Products were synthesized in the presence of the RNA fractions designated on the figure. The positions of authentic H^{315} and L^{315}, as well as the position of the L^{315} precursor (pL^{315}), are indicated.*

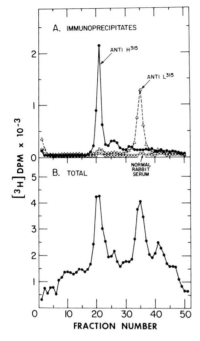

Fig. 13. *Immunoprecipitates of cell-free products of MOPC-315 16-19S mRNA analyzed on continuous SDS-polyacrylamide gels.*

A. *Immunoprecipitation of cell-free products by normal serum (○), anti-H^{315} serum (●), or anti-L^{315} serum (△);*

B. *Total radioactivity in cell-free products.*

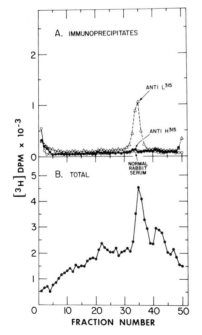

Fig. 14. *Immunoprecipitates of cell-free products of MOPC-315 NR 16-19S mRNA analyzed on continuous SDS-polyacrylamide gels.*

A. *Immunoprecipitation of cell-free products by normal serum (○), anti-H^{315} serum (●), or anti-L^{315} serum (△);*

B. *Total radioactivity in cell-free products.*

synthesizes, but does not secrete, 5-10% of the level of H^{315} synthesized by MOPC-315 cells (L. Bailey and H. Eisen, unpublished results). The inability of the MOPC-315 NP-1 variant tumor line to synthesize L^{315} may reside in an altered DNA, a defective, deficient, or absent mRNA, or possibly a specific translational or post-translational defect. The fact that the MOPC-315 NP-1 cell-line synthesizes, but does not secrete, decreased amounts of H^{315} may be explained by defects at one or several levels of cellular organization. The experiments described here examined some of these possibilities.

Protein Synthesis in Intact Cells. A comparison of the proteins synthesized by cell suspensions prepared from MOPC-315 and MOPC-315 NP-1 tumors is shown in Fig. 15. It is apparent that MOPC-315 NP-1 synthesizes or secretes little, if any, H^{315} or L^{315}. Immunochemical analysis of the intracellular proteins present in the tumor cells is shown in Fig. 16. Both MOPC-315 and MOPC-315 NP-1 cytosols contain H^{315} peptides (Fig. 16A). However, while there was good synthesis of L^{315} in MOPC-315 cells, no detectable L^{315} was synthesized by MOPC-315 NP-1 cells (Fig. 16B).

Cell-free Protein Synthesis and Analyses of Cell-Free Products. Total poly A-containing RNA preparations from both the MOPC-315 and MOPC-315 NP-1 tumor lines and the RNA fractions obtained from the preparative sucrose gradients were tested for their messenger RNA activity in a cell-free

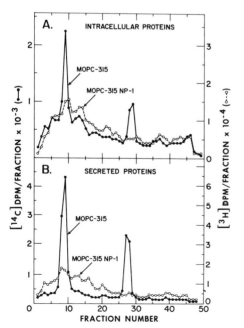

Fig. 15. *SDS-polyacrylamide gel electrophoretic analysis of proteins synthesized by MOPC-315 and MOPC-315 NP-1 tumor cell suspensions. MOPC-315 and MOPC-315 NP-1 cells were labeled with [^{14}C]leucine and [^{3}H]leucine, respectively. The cell lysates and the culture media from the two lines were mixed prior to electrophoresis.*

A. Intracellular proteins after a 30 min period of labeling;

B. Secreted proteins after a 3 hr labeling period.

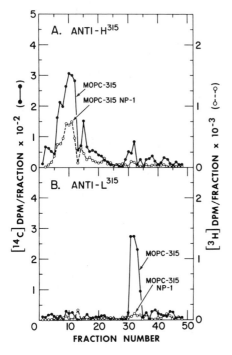

Fig. 16. SDS-polyacrylamide electrophoresis of intracellular proteins synthesized by MOPC-315 and MOPC-315 NP-1 tumor cell suspensions and precipitated with anti-L^{315} and anti-H^{315} sera. Mixtures of the cell lysates shown in Fig. 15A were incubated with anti-L^{315}, anti-H^{315}, or normal rabbit serum, followed by sheep anti-rabbit serum, and the three immunoprecipitates analyzed on separate gels. Background values from the non-immune gel were subtracted from each of the specific antiserum gels. Since the input of $[^3H]/[^{14}C]$ was about 5/1 in the mixtures, the DPM scales have been adjusted accordingly.
  A. Anti-H^{315} immunoprecipitate;
  B. Anti-L^{315} immunoprecipitate.

extract derived from Ehrlich ascites cells (41). The products synthesized in response to either the MOPC-315 or MOPC-315 NP-1 RNA fractions were compared on SDS-polyacrylamide slab gels (10,11) as shown in Fig. 17. It was shown previously that the primary translation product of L^{315} mRNA is a precursor (pL^{315}) longer than L^{315} and that the primary product of H^{315} mRNA translation is a protein species antigenically and structurally related to H^{315} but migrating slightly faster than H^{315} on SDS-polyacrylamide gels (11). Substantial synthesis of both the H^{315} and pL^{315} cell-free products occurred in the presence of the relevant MOPC-315 RNA fractions as reported previously (10,11). However, no detectable pL^{315} and little, if any H^{315} was synthesized in response to any of the RNA fractions obtained from the variant MOPC-315 NP-1 plasmacytoma. Furthermore, it can also be seen that the prominent band migrating approximately halfway between L^{315} and the dye front of SDS-polyacrylamide gels was present in the MOPC-315 cell-free products, but reduced or absent in the MOPC-315 NP-1 cell-free products. Although this product has not yet been identified, it has been shown that it is coded for by an mRNA or class of mRNAs smaller than L^{315} as measured by sucrose gradient sedimentation and electrophoresis in polyacrylamide gels in the presence of 98% formamide (22). Mixing experiments (data not shown) were performed to test for the presence of inhibitors of MOPC-315 mRNA translation in the mRNA obtained from MOPC-315 NP-1. It was observed that H^{315} and L^{315} mRNA from MOPC-315 could still be translated in the presence of equal amounts of the total poly A-containing mRNA from MOPC-315 NP-1.

Immunochemical Analyses of the Cell-Free Products. Previous results (Fig. 13) obtained with the MOPC-315 mRNAs demonstrated the cell-free synthesis of fragments antigenically related to H^{315} in addition to complete

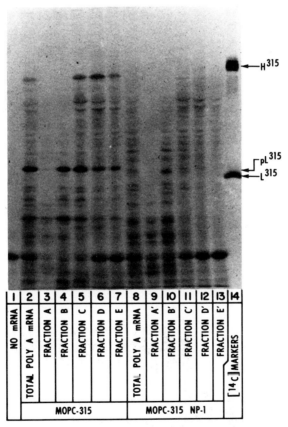

Fig. 17. *Autoradiogram of radioactive cell-free products analyzed on an SDS-polyacrylamide gel slab. Products were synthesized in the presence of $[^{14}C]$alanine, $[^{14}C]$leucine and $[^{14}C]$valine with the RNA fractions indicated in the figure as templates. The positions of authentic H^{315} and L^{315} as well as the position of the L^{315} precursor (pL^{315}) are indicated.*

H^{315} and pL^{315} (10,11). To determine whether the cell-free translation of MOPC-315 NP-1 mRNA yielded any fragments of L^{315} or H^{315}, immunochemical analyses of the cell-free products of the total mRNA (Fig. 18) and of the 16-19S mRNA fraction (Fig. 19) from the MOPC-315 NP-1 tumor line were performed. No pL^{315} or L^{315} related peptides were detected in the products synthesized in the presence of MOPC-315 NP-1 mRNA (Figs. 18,19). Furthermore, H^{315} was detectable only in the cell-free products synthesized in response to the 16-19S mRNA fraction from MOPC-315 NP-1 plasmacytoma. Similarly, L^{315} related peptides were present in the cell-free product synthesized in response to the 16-19S fraction of the MOPC-315 mRNA (Figs. 12,13,17), but absent from that of the analogous fraction of MOPC-315 NP-1 mRNA (Figs. 17-19).

In summary, no L^{315} or immunoprecipitable L^{315} fragments were observed in the cell-free products directed by any of the MOPC-315 NP-1 mRNA fractions. This observation excludes the existence of a specific defect in the translational or post-translational apparatus of the variant as well as the existence of a specific protein inhibitor of L^{315} mRNA translation. An L^{315} mRNA of comparable activity to that of the MOPC-315 tumor line should have been translated in the heterologous Ehrlich ascites system if it existed. Any inhibitor protein, if it existed, would have to survive the rigorous extraction procedures used to prepare the mRNA. It must be noted that possibly fragments of L^{315} that are not recognized and, therefore, not efficiently precipitated by anti-L^{315} serum may be synthesized. The addition of MOPC-315 NP-1 mRNA fractions failed to block the cell-free translation of L^{315} directed by MOPC-315 mRNA fractions. Thus, the existence of an inhibitor RNA species is also unlikely. We conclude, therefore, that no translatable poly A-containing mRNA capable of directing the synthesis of intact pL^{315} can be extracted from the MOPC-315 NP-1 plasmacytoma.

MOPC-315 NP-1 cells synthesize decreased amounts of H^{315} compared to MOPC-315 cells. Analyses of the cell-free products synthesized in response to the mRNAs derived from these tumor lines reflect a similar pattern. Thus, MOPC-315 NP-1 cells contain a decreased amount of and/or an intrinsically less active poly A-containing H^{315} mRNA.

C. Conclusions from Translational Study of Variants

The inability to produce heavy and light chains are stable characteristics of the MOPC-315 NR and MOPC-315 NP-1 variant tumor lines, respectively. Several mechanisms could be postulated to explain these phenomena. It is possible, for example, that the DNA segment coding for the specific chain has been lost, in whole or part. Alternatively, it is possible that the genes have been altered such that they can no longer be properly transcribed or yield upon transcription mRNAs that cannot be properly processed in the nucleus or transported into the cytoplasm. Possibly, modification of the genes has occurred which leads to the transcription of an mRNA sequence that cannot

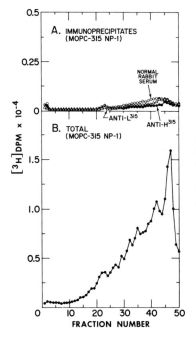

Fig. 18. *Immunoprecipitates of cell-free products of MOPC-315 NP-1 total poly A-containing mRNA analyzed on continuous SDS-polyacrylamide gels.*

A. Immunoprecipitation of cell-free products by normal serum (○); anti-H^{315} serum (●); or anti-L^{315} serum (△);

B. Total radioactivity in cell-free products.

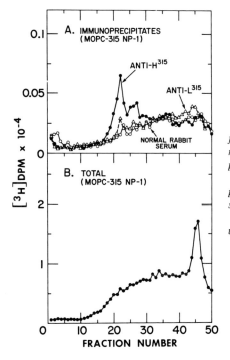

Fig. 19. *Immunoprecipitation of cell-free products of MOPC-315 NP-1 16-19S mRNA analyzed on continuous SDS-polyacrylamide gels.*

A. Immunoprecipitation of cell-free products by normal serum (○); anti-H^{315} serum (●), or anti-L^{313} serum (△);

B. Total radioactivity in cell-free products.

be properly translated. This would be the case if, for example, translation of mRNA was prematurely terminated by a nonsense codon arising either by direct base substitution or as the result of a change in the reading frame of the message. Also, it could be argued *a priori* that the loss of the ability to synthesize the polypeptide chains may be the result of a translational defect which would specifically block H^{315} or L^{315} biosynthesis even if active mRNA were available. Furthermore, conceivably, a specific inhibitor of mRNA transcription or translation might be produced in the variants. Additionally, the presence of one of many post-translational defects could account for the absence of the polypeptide chains. These include increased degradation rate, defective carbohydrate substitution, deficient precursor cleavage enzyme, as well as other alternatives. The above experiments examined some of these possibilities.

No H^{315} or immunoprecipitable H^{315} fragments larger than light chain and no L^{315} or immunoprecipitable L^{315} fragments were observed in the cell-free products programmed by any of the MOPC-315 NR and MOPC-315 NP-1 fractions, respectively. These observations rule out the existence of a specific defect in the translational or post-translational apparatus of these variant plasmacytomas. A specific protein inhibitor of translation is also excluded, for mRNA of comparable activity to that of the MOPC-315 parent line should have been translated in the heterologous Ehrlich ascites system if it existed. To account for the observed results, any inhibitor protein would have to bind to the poly A-containing mRNA tightly enough to survive the rigorous extraction procedures used to prepare the mRNA. It must be noted that small fragments of H^{315} and L^{315} not recognized and, therefore, not efficiently precipitated by the antisera – may be synthesized.

The addition of MOPC-315 NR and MOPC-315 NP-1 mRNA fractions failed to block the cell-free translation of H^{315} and L^{315}, respectively, directed by MOPC-315 mRNA fractions. Thus, the existence of an inhibitor RNA species is unlikely. We conclude, therefore, that no translatable poly A-containing mRNA capable of directing the synthesis of intact H^{315} can be extracted from the MOPC-315 NR variant plasmacytoma; and that no translatable poly A-containing mRNA capable of directing the synthesis of intact L^{315} can be extracted from the MOPC-315 NP-1 variant.

The results described are limited to translational analysis of the mRNAs. Whether the defect is the result of an alteration of DNA or mRNA requires further study. It should be possible, however, to resolve these questions by direct measurement of H^{315} and L^{315} DNA and mRNA by hybridization studies. Highly labeled H^{315} and L^{315} mRNA or a DNA complementary to H^{315} and L^{315} mRNA can be used to probe the DNA of the variants for the presence of H^{315} and L^{315} nucleotide sequences. Additionally, complementary DNA could be used to estimate the amount of H^{315} and L^{315} mRNA sequences in the total cellular RNA.

The data obtained in studies as described here should help elucidate the coordination between the genetic information and the cellular machinery for macromolecular biosynthesis and processing that is necessary for the production of functional immunoglobulin molecules. With this information, it should be possible to develop a cell-free system that will produce an active antibody molecule when programmed by suitable templates. Furthermore, many variant plasmacytoma cells of different types have been found and studied (28-39). The methods outlined here should permit the elucidation of their molecular basis.

D. *Cell-free synthesis of Interferon: Relevance to Immune RNA Phenomena*

One final note is worthy of mention relevant to transfer of biological activity by RNA to intact cells. DeMaeyer-Guignard, DeMaeyer, and Montagnier (42) reported the synthesis of mouse and human interferons by avian cells treated with RNA from mouse and human cells, respectively. This observation was analagous to many other reports of the cellular transfer of information for protein sequence by RNA. Recently, we have synthesized biologically active human interferon *de novo* in a cell-free mouse extract dependent on poly A-containing mRNA from induced human FS-4 fibroblasts (43). The human fibroblasts were induced to produce interferon and interferon mRNA by treatment with poly I·poly C. The synthesis of biologically active human interferon, probably a single polypeptide chain, in a cell-free extract may provide insight into the primary synthesis and post-translational events leading to the production of a biologically active molecule. Immunoglobulins, multichain structures, require a greater number of modifications for the synthesis of biologically active molecules. Cell-free synthesis of all these molecules should permit insight into their total biosynthesis.

REFERENCES

1. Stavnezer, J. and R. C. C. Huang (1971). Synthesis of a mouse immunoglobulin light chain in a rabbit reticulocyte cell-free system. *Nature New Biol.* 230:172-176.

2. Schmeckpeper, B. J., S. Cory and J. M. Adams (1974). Translation of immunoglobulin messenger mRNAs in a wheat-germ cell-free system. *Mol. Biol. Reports* 1:355-363.

3. Swan, D., H. Aviv and P. Leder (1972). Purification and properties of biologically active messenger RNA for a myeloma light chain. *Proc. Nat. Acad. Sci. U.S.A.* 69:1967-1971.

4. Milstein, C., G. G. Brownlee, T. M. Harrison and M. B. Mathews (1972). A possible precursor of immunoglobulin light chains. *Nature New Biol.* 239:117-120.

5. Mach, B., C. Faust and P. Vassalli (1973). Purification of 14S messenger RNA of immunoglobulin light chain that codes for a possible light-chain precursor. *Proc. Nat. Acad. Sci. U.S.A.* 70:451-455.

6. Tonegawa, S. and I. Baldi (1973). Electrophoretically homogenous myeloma light chain mRNA and its translation *in vitro*. *Biochem. Biophys. Res. Commun.* *51*:81-87.

7. Cowan, N. J. and C. Milstein (1973). The translation *in vitro* of mRNA for immunoglobulin heavy chains. *Eur. J. Biochem. 36*:1-7.

8. Harrison, T. M., G. G. Brownlee and C. Milstein (1974). Preparation of immunoglobulin light-chain mRNA from microsomes without the use of detergent. *Eur. J. Biochem. 47*:621-627.

9. Schechter, I. (1973). Biologically and chemically pure mRNA coding for a mouse immunoglobulin L-chain prepared with the aid of antibodies and immobilized oligothymidine. *Proc. Nat. Acad. Sci. U.S.A. 70*:2256-2260.

10. Green, M., P. N. Graves, T. Zehavi-Willner, J. McInnes and S. Pestka (1975). Cell-free translation of immunoglobulin messenger RNA from MOPC-315 plasmacytoma and MOPC-315 NR, a variant synthesizing only light chain. *Proc. Nat. Acad. Sci. U.S.A. 72*:224-228.

11. Green, M., T. Zehavi-Willner, P. N. Graves, J. McInnes and S. Pestka. Isolation and cell-free translation of immunoglobulin messenger RNA. *Arch. Biochem. Biophys.,* in press.

12. Delovitch, T. L., S. L. Boyd, H. M. Tsay, G. Holme and A. H. Sehon (1973). The specific immunoprecipitation of polyribosomes synthesizing an immunoglobulin light chain. *Biochem. Biophys. Acta. 299*:621-633.

13. Stevens, R. H. and A. R. Williamson (1973). Isolation of messenger RNA coding for mouse heavy-chain immunoglobulin. *Proc. Nat. Acad. Sci. U.S.A. 70*:1127-1131.

14. Tonegawa, S., C. Steinberg, S. Dube and A. Bernardini (1974). Evidence for somatic generation of antibody diversity. *Proc. Nat. Acad. Sci. U.S.A. 71*:4027-4031.

15. Premkumar, E., M. Shoyab and A. R. Williamson (1974). Germ line basis for antibody diversity: Immunoglobulins V_H- and C_H-gene frequencies measured by DNA·RNA hybridization. *Proc. Nat. Acad. Sci. U.S.A. 71*:99-103.

16. Delovitch, T. L. and C. Baglioni (1973). Estimation of light-chain gene reiteration of mouse immunoglobulin by DNA-RNA hybridization. *Proc. Nat. Acad. Sci. U.S.A. 70*:173-178.

17. Honjo, T., S. Packman, D. Swan, M. Nau and P. Leder (1974). Organization of immunoglobulin genes: Reiteration frequency of the mouse K chain constant region gene. *Proc. Nat. Acad. Sci. U.S.A. 71*:3659-3663.

18. Stavnezer, J., R. C. C. Huang, E. Stavnezer and J. M. Bishop (1974). Isolation of messenger RNA for an immunoglobulin kappa chain and enumeration of the genes for the constant region of kappa chain in the mouse. *J. Mol. Biol. 88*:43-63.

19. Rabbitts, T. H., J. O. Bishop, C. Milstein and G. G. Brownlee (1974). Comparative hybridization studies with an immunoglobulin light chain mRNA fraction and non-immunoglobulin mRNA of mouse. *FEBS Lett. 40*:157-160.

20. Smith, G. P., L. Hood and W. M. Fitch (1971). Antibody diversity. *Ann. Rev. Biochem. 40*:969-1012.

21. Brownlee, G. G., E. M. Cartwright, N. J. Cowan, J. M. Jarvis and C. Milstein (1973). Purification and sequence of messenger RNA for immunoglobulin light chains. *Nature New Biol. 244*:236-240.

22. McInnes, J., M. Green and S. Pestka, manuscript in preparation.

23. Cioli, D. and E. S. Lennox (1973). Immunoglobulin nascent chains on membrane-bound ribosomes of myeloma cells. *Biochemistry 12*:3211-3217.

24. Lisowska-Bernstein, B., M. E. Lamm and P. Vassalli (1970). Synthesis of immunoglobulin heavy and light chains by the free ribosomes of a mouse plasma cell tumor. *Proc. Nat. Acad. Sci. U.S.A. 66*:425-432.

25. Mach, B., H. Koblet and D. Gros (1968). Chemical identification of specific immunoglobulins as the product of a cell-free system from plasmacytoma tumors. *Proc. Nat. Acad. Sci. U.S.A. 59*:445-452.

26. Bernardini, A. and S. Tonegawa (1974). Hybridization studies with an antibody heavy chain mRNA. *FEBS Lett. 41*:73-77.

27. Faust, C. H., H. Diggelmann and B. Mach (1974). Estimation of the number of genes coding for the constant part of the mouse immunoglobulin kappa light chain. *Proc. Nat. Acad. Sci. U.S.A. 71*:2491-2495.

28. Coffino, P., R. Laskov and M. D. Scharff (1970). Immunoglobulin producton: Method of quantitatively detecting variant myeloma cells. *Science 167*:186-188.

29. Coffino, P. and M. D. Scharff (1971). Rate of somatic mutation in immunoglobulin production by mouse myeloma cells. *Proc. Nat. Acad. Sci. U.S.A. 68*:219-223.

30. Coffino, P., R. Baumal, R. Laskov and M. D. Scharff (1972). Cloning of mouse myeloma cells and detection of rare variants. *J. Cell Physiol. 79*:429-440.

31. Baumal, R., B. K. Birshtein, P. Coffino and M. D. Scharff (1973). Mutations in immunoglobulin-producing mouse myeloma cells. *Science 182*:164-166.

32. Preud-homme, J.-L., J. Buxbaum and M. D. Scharff (1973). Mutagenesis of mouse myeloma cells with "Melphalan." *Nature 245*:320-322.

33. Cotton, R. G. H., D. S. Secher and C. Milstein (1973). Somatic mutation and the origin of antibody diversity. Clonial variability of the immunoglobulin produced by MOPC-21 (mouse myeloma) cells in culture. *Eur. J. Immun. 3*:135-140.

34. Secher, D. S., R. G. H. Cotton and C. Milstein (1973). Spontaneous mutation in tissue culture-chemical nature of variant immunoglobulin from mutant clones of MOPC-21. *FEBS Lett. 37*:311-316.

35. Schubert, D., A. Munro and S. Ohno (1968). Immunoglobulin biosynthesis. I. A myeloma variant secreting light chain only. *J. Mol. Biol. 38*:253-262.

36. Schubert, D. and K. Horibata (1968). Immunoglobulin biosynthesis. II. Four independently isolated myeloma variants. *J. Mol. Biol. 38*:263-271.

37. Schubert, D. and M. Cohn (1968). Immunoglobulin biosynthesis. III. Blocks in defective synthesis. *J. Mol. Biol. 38*:273-288.

38. Lynch, R. G., R. J. Graff, S. Sirisinha, E. S. Simms and H. N. Eisen (1972). Myeloma proteins as tumor-specific transplantation antigens. *Proc. Nat. Acad. Sci. U.S.A. 69*:1540-1544.

39. Hannestad, K., M.-S. Kao and H. N. Eisen (1972). Cell-bound myeloma proteins on the surface of myeloma cell: Potential targets for the immune system. *Proc. Nat. Acad. Sci. U.S.A. 69*:2295-2299.

40. Eisen, H. N., E. S. Simms and M. Potter (1968). Mouse myeloma proteins with anti-hapten antibody activity. The protein produced by plasma cell tumor MOPC-315. *Biochemistry 7*:4126-4134.

41. Green, M., J. McInnes, P. N. Graves, L. K. Bailey and S. Pestka. Cell-free translation of immunoglobulin messenger RNA from MOPC-315 plasmacytoma and MOPC-315 NP-1, a variant synthesizing only heavy chain In preparation.

42. DeMaeyer-Guignard, J., E. DeMaeyer and L. Montagnier (1972). Interferon messenger RNA: Translation in heterologous cells. *Proc. Nat. Acad. Sci. U.S.A. 69*:1203-1207.

43. Pestka, S., J. McInnes, E. Havell and J. Vilček. The cell-free synthesis of human interferon. *Proc. Nat. Acad. Sci. U.S.A.,* in press.

DISCUSSION

F. L. Adler, Chairman

The session served the purpose of bringing before the conferees an up-to-date summary of the more basic immunological studies on "immune RNA." Fishman and Dray presented summaries of studies concerned with the initiation of immune responses both *in vitro* and *in vivo*; the former concentrating on humoral responses, the latter on expressions of cellular immunity. Questions on the possible misinterpretation of data due to contamination of the RNA with either antibody or antigen were raised. With regard to antibody it was pointed out that RNA preparations used in many of these studies have been shown to be free of detectable antibodies by the same techniques that were used to detect antibody formation in response to such RNA; by the resistance of RNA preparations to protease treatment and by the kinetics of the antibody response. Antigen is admittedly present as an essential component of RNA-antigen complexes. Where a response is attributed to informational RNA, however, it can be shown experimentally that the presence or absence of antigen is irrelevant to the production of antibody molecules in response to i-RNA (informational RNA) that conveys to the recipient both the antibody and the allotypic specificities of the donor.

Valentine presented a brief report on transfer factor studies with emphasis on the evidence for specificity of the induced responses. He was followed by Giacomoni who discussed immunosuppression and receptor conversion in plasmacytomas. Evidence was presented to relate these activities to RNA which, in the mouse, appears to be associated with "A" particles in the plasma. The transient nature of the conversion was documented by appropriate studies.

Jachertz presented data based on studies with measles virus as antigen which confirmed and extended his earlier studies in which informational RNA is generated in a cell-free system composed of DNA, antigen, cell sap and RNA precursors. Koch then spoke of work in which i-RNA prepared by Fishman was shown to possess the biochemical characteristics of m-RNA. Also, his data confirmed and extended evidence that cells from one species, in this instance mouse L cells, respond to i-RNA with the synthesis of antibody molecules that possess antigenic markers characteristic of the species of the RNA donor, in this case rabbits.

Finally, Pestka spoke of his studies with m-RNA from myeloma cells which induced synthesis of L and H chains but not of complete molecules with the antibody activity of the myeloma protein in question. He also cited data in support of the trans-species transfer of synthesis of interferon through RNA with messenger activity. In discussion it was brought out that reports of successful transfer of synthesis across species lines reinforced strongly the data based on the transfer of allotypic markers and would override objections based on some recent doubts in the true allelic nature of the genes that determine allotypic markers.

SESSION III
Mediation of Antitumor Immune Responses By I-RNA

Chairman, A. Arthur Gottlieb

THE MEDIATION OF IMMUNE RESPONSES BY I-RNA TO ANIMAL AND HUMAN TUMOR ANTIGENS[1]

Yosef H. Pilch, M.D., Dieter Fritze, M.D.*,
Kenneth P. Ramming, M.D., Jean B. deKernion, M.D.,
and David H. Kern, Ph.D.

*Department of Surgery
Harbor General Hospital
Torrance, California 90509*

*Divisions of Urology and Oncology,
Department of Surgery
University of California
Los Angeles, California 90024
and the Veterans Administration Hospital
Sepulveda, California 91343*

INTRODUCTION

In 1970-71, we reported that immunity to tumor isografts in inbred mice could be mediated by syngeneic spleen cells preincubated with xenogeneic immune RNA extracted from the lymphoid organs of guinea pigs immunized with the specific tumor that was to be treated (1-3). This immunity was manifested when such RNA-incubated spleen cells were injected into syngeneic mice. This response was specific for the particular tumor used to immunize the guinea pig from whose lymphoid organs the I-RNA was prepared. Thus, spleen cells incubated with I-RNA against one tumor inhibited the growth of isografts of that tumor, but not isografts of a syngeneic but antigenically different tumor. Administration of spleen cells incubated with I-RNA against the second tumor inhibited the growth of isografts of that tumor but not isografts of the first tumor. Nonspecific factors associated with guinea pig

---

[1] Supported in part by Grant CA 14846 from the National Institutes of Health, Grant ET-28 from the American Cancer Society, and by grants from the California Institute for Cancer Research and the California Research Coordinating Committee of the University of California.

*Present Address: Dieter Fritze, M.D., D-69 Heidelberg, Medizinische Universitatas-Klinik, Bergheimerstr. 58 (West Germany)

RNA probably did not account for these results, since RNA from guinea pigs immunized with Freund's adjuvant only was without effect. Likewise, injections of untreated syngeneic spleen cells (not incubated with RNA) did not inhibit the growth of tumor isografts. Most important was the finding that I-RNA from the lymphoid organs of guinea pigs immunized with syngeneic normal mouse tissues failed to mediate an antitumor response. This, together with the findings of the specificity of the immune RNA for the immunizing tumor, strongly suggested that the immunity expressed by the spleen cells preincubated with antitumor I-RNA was directed against tumor-associated antigens (TAA).

We then sought a more direct and simpler method of utilizing I-RNA in mediating antitumor immune responses. Such a method would be the direct systemic administration of the I-RNA itself rather than lymphoid cells preincubated with I-RNA. In considering how best to achieve a transfer of immunity *in vivo* by the direct systemic administration of I-RNA, our attention was drawn to the problem of the possible degradation of I-RNA by plasma and interstitial ribonculeases. Ribonuclease is known to inactivate preparations of I-RNA, and, indeed, when the antitumor I-RNA preparations previously studied in our laboratory had been treated with RNase all antitumor immuno-reactivity was abolished. These facts led to the initial assumption that the direct systemic administration of xenogeneic I-RNA without lymphoid cells would be ineffective due to enzymatic degradation and inactivation of I-RNA by interstitial and plasma ribonucleases. We attempted to eliminate this problem by administering our I-RNA in a solution containing a potent RNase inhibitor.

Many polyanions are excellent inhibitors of RNase (4-6) and the RNase inhibitor we selected was the polyanion, sodium dextran sulfate. Previous studies in our laboratory had shown that sodium dextran sulfate of molecular weights 77,500 to 2.0×10^6 were excellent inhibitors of ribonuclease (Pilch and Ramming, unpublished data). We, therefore, elected to administer I-RNA systemically in a vehicle containing sodium dextran sulfate of molecular weights, 77,500 to 500,000 at concentrations known to effectively inhibit RNase at levels approximating those in plasma. When such mixtures of xenogeneic I-RNA and dextran sulfates were administered to mice subcutaneously or to rats by footpad injection every other day for 10 days, such animals were rendered specifically resistant to isografts of that particular tumor used to immunize the I-RNA donor (7,8).

These experiments with xenogeneic immune RNA provided convincing but *indirect* evidence of the specificity of antitumor immune RNA for the tumor associated antigens of the murine tumor. Because of certain obvious difficulties in the interpretation of results obtained in xenogeneic models and so that *direct* evidence could be obtained that Immune RNA could mediate immune responses to TAA, it was desirable to employ a system in which the tumor cells, the lymphoid tissues from which I-RNA was extracted, the spleen cells incubated with the I-RNA, and the recipient host were all from members of a single, inbred strain. This would eliminate the possibility that the inhibition of tumor isograft development observed might be due, in part, to immune responses directed against transplantation antigens. In this model, a syngeneic tumor-host system in female Fischer 344/N rats was used. *Syngeneic* immune RNA was prepared from the spleens of Fischer rats immunized to this tumor by the excision of growing tumor transplants. Control RNA preparations were extracted from the spleens of unimmunized Fischer rats.

We then demonstrated that rats challenged with tumor cells synchronous with the first of three daily intravenous injections of syngeneic spleen cells preincubated with syngeneic antitumor I-RNA extracted from the spleens of immunized rats evidenced a much lower incidence of tumor isograft development than did control rats which received injections of syngeneic spleen cells alone (9). Furthermore, rats receiving spleen cells incubated with the same active I-RNA, but treated with RNase prior to incubation with the spleen cells, manifested a tumor incidence similar to that of control animals. Moreover, RNA from the spleens of unimmunized rats was ineffective in mediating an antitumor immune response.

IMMUNOTHERAPY OF A SPONTANEOUSLY METASTASIZING RAT MAMMARY CARCINOMA WITH IMMUNE RNA

Recently, the successful immunotherapy of the transplantable spontaneously metastasizing mammary adenocarcinoma 13762 in female Fischer 344/N rats with xenogeneic or syngeneic immune RNA has been demonstrated. When growing primary tumor transplants are surgically excised 18 days after transplantation, local recurrence does not occur. However, all animals go on to die of metastases to lungs and other organs if no additional treatment is given. An experiment was performed in which immunotherapy with immune RNA was utilized as an adjuvant to surgical excision of the primary tumor transplant. The data from this experiment is presented in Table 1.
With no treatment at all, all animals died within 70 days. With excision of the primary tumor transplant alone all animals died of metastases within 96 days. Xenogeneic immune RNA was extracted from the lymph nodes and spleens of Hartley guinea pigs two weeks following immunization with mammary adenocarcinoma tissue. This I-RNA was administered at a dose of 1.0 mg in 1 ml of buffer containing 10 mg per ml of sodium dextran sulfate (molecular weight 500,000). 0.25 ml of this mixture was administered into each of the four foot pads of a group of rats every other day for 10 doses (i.e., 20 days) beginning 10 days prior to surgical excision of the primary tumor transplant. Of the animals so treated, 80% survived and remained free of disease for 180 days.

TABLE 1.

Immunotherapy of a Spontaneously Metastasizing Rat Mammary Carcinoma with Immune RNA

Experimental Group	No. of Rats Surviving Free of Disease at 180 Days	p value[1]
No Treatment	0/20[2,3] (0%)	–
Surgery Only at Day 18	0/10[4] (0%)	–
Surgery + Xenogeneic I-RNA Pre & Post-op	4/5 (80%)	<0.005
Surgery + Xenogeneic I-RNA Post-op Only	6/9 (67%)	<0.005
Surgery + Syngeneic I-RNA	3/6 (50%)	<0.05

[1] p value by Fischer's exact test for chi-square with Yates' correction compared to no treatment group.

[2] No. of rats surviving free of disease ÷ total number of rats in group.

[3] All rats were dead by 70 days.

[4] All rats were dead by 96 days.

When xenogeneic immune RNA was administered at a dose of 1 mg every other day for 10 doses beginning at the day of surgical excision of the primary tumor, 67% of animals so treated survived and remained free of disease for 180 days. Syngeneic immune RNA was extracted from the spleens of Fischer rats bearing growing transplants of the mammary adenocarcinoma. When this syngeneic I-RNA was administered at the same dosage schedule pre-and post-operatively to another group of rats, 50% of the animals so treated survived and remained free of disease for 180 days.

MEDIATION OF CYTOTOXIC IMMUNE REACTIONS AGAINST HUMAN TUMOR CELLS *IN VITRO* BY XENOGENEIC IMMUNE RNA

We had previously demonstrated the mediation of immune cytolysis of rodent tumor cells *in vitro* by syngeneic spleen cells preincubated with syngeneic (10,11) or xenogeneic (12) immune RNA. Recently we have reported the mediation of immune cytolysis of human tumor cells *in vitro* by xenogeneic immune RNA (13-15). In these studies, normal, nonimmune human peripheral blood lymphocytes were converted into "killer cells" which were cytotoxic for human tumor cells *in vitro* by incubation with I-RNA extracted from the lymph nodes and spleens of sheep or guinea pigs which had previously been immunized with cells from the particular human tumor being studied. The human tumors employed were (a) a human gastric adenocarcinoma, RI-H, established in tissue culture in our laboratory by primary explant from a surgical specimen, and studied in its 12th to 14th passage; (b) a human adenocarcinoma of the breast, BT-20, obtained from Dr. E. Lasfargues, and studied in its 300th-310th passage; (c) a human malignant melanoma, ED-H, explanted from a surgical specimen in the laboratory of Dr. Donald L. Morton, UCLA, and studied in its 18-25 passage; and (d) a human melanoma, RO-H established by primary explant from a surgical specimen and studied in its 10th-15th passage. Sheep or guinea pigs were immunized separately with viable tumor cells from each of these tissue cultures lines in complete Freund's adjuvant. In some experiments I-RNA was treated with ribonuclease, deoxyribonuclease, or pronase. In early experiments, peripheral blood leukocytes from normal human volunteers were isolated from heparinized specimens of peripheral blood by sedimentation in 10% Plasmagel (Roger Bellon Laboratories, Neuilly, France (13). In subsequent experiments, purified mononuclear cell populations were prepared from heparinized blood specimens on Ficoll-Isopaque gradients suspended in RPMI 1640 with 40% agamma human serum and 10% dimethysulfoxide (DMSO), frozen in liquid nitrogen at 1°C per minute and thawed at 37°C prior to use (13). These lymphocytes were incubated with immune RNA at a concentration of 2 mg/ml for 20 minutes at 37°C. The cells were washed, diluted to the appropriate cell concentration and added to target cells in a standard microcytotoxicity assay in which the tumor target cells were prelabeled with ^{125}I-Iododeoxyuridine (13,15).

In preliminary experiments, it was demonstrated that normal human peripheral blood leukocytes preincubated *in vitro* with xenogeneic I-RNA extracted from the lymphoid organs of either immunized sheep or guinea pigs were markedly more cytotoxic to human tumor cells *in vitro* than were aliquots of the same leukocyte preparation which had been incubated without RNA or incubated with RNA from the lymphoid organs of animals immunized with Freund's adjuvant only. When purified mononuclear cell populations prepared on Ficoll-Isopaque gradients instead of buffy coat leukocytes were incubated with the I-RNA, the magnitude of the cytotoxicity observed was increased and the lymphocyte to target cell ratio could be significantly reduced (from 500:1 to 100:1 or 200:1). Data from two representative experiments are presented in Figures 1 and 2. From the results depicted in Figure 1, it may be seen that at a lymphocyte to target cell ratio of 100:1 a cytotoxicity index of 0.66 ± 0.08 was obtained.

Since lymphocytes normally account for 20-40% of the total peripheral white blood cells in man, this result is quite consistent with what would be expected assuming that one or more population of lymphocytes are the active effector cell population. By effecting approximately a five fold increase in lymphocyte concentration, the effector cell to target cell ratio could be reduced by a factor of 5 with no loss (or even an increase) in cytotoxicity.

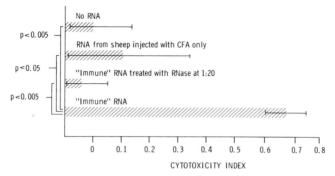

Fig. 1. *Immune cytolysis of RI-H tumor cells mediated by xenogeneic immune RNA extracted from the lymphoid organs of sheep immunized with RI-H tumor cells. The mononuclear effector cells were purified from whole blood on Ficoll-Isopaque gradients. Controls include RNA from sheep injected with complete Freund's adjuvant (CFA) only and active immune RNA after treatment with RNase at an enzyme to substrate ratio of 1 to 20.*

When the active I-RNA preparations were pretreated with deoxyribonuclease (Figure 2) or with pronase (13) (not shown) and then incubated with lymphocytes, the resultant cytotoxicity index was not significantly altered. The fact that deoxyribonuclease and pronase treatment of the I-RNA preparations did not affect the induction of an immune response and that

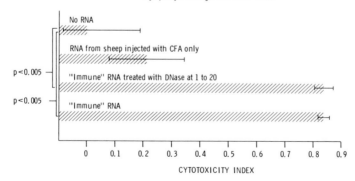

Fig. 2. *Effect of DNase treatment on the activity of immune RNA. Active xenogeneic immune RNA was treated with DNase at an enzyme to substrate ratio of 1 to 20 based on the actual amount of DNA present in the RNA preparations.*

ribonuclease treatment of the I-RNA abrogated the response (Figure 1), again indicated that one or more moieties of RNA were responsible for the immunological activity of these preparations.

Next, the effects of varying the RNA concentration upon the magnitude of the immune response were determined. The concentration of RNA with which the lymphocytes were incubated was progressively reduced so that a dose-response curve might be established and a threshold RNA concentration determined. The results of one experiment are depicted in Figure 3. From these data it can be seen that reducing the RNA concentration from 2.0 mg/ml to 1.0 mg/ml or even to 500 µg/ml had little effect on the response, but reducing the concentration still further to 250 µg/ml resulted in approximately a 50% decrease in cytotoxicity. In other experiments (not shown) it was found that reducing an RNA concentration of 100 µg/ml reduced cytotoxicity still further, but this result remained significantly greater than that obtained with no RNA.

Experiments similar to those described above for the RI-H cell line were performed utilizing the BT-20 breast carcinoma cell line. In these experiments, xenogeneic I-RNA from immunized guinea pigs was used instead of sheep I-RNA. The results are depicted in Figure 4.

Xenogeneic I-RNA extracted from the lymphoid organs of guinea pigs immunized with BT-20 tumor cells induced normal, nonimmune human lymphocytes to become cytotoxic to BT-20 tumor target cells. RNA from guinea pigs immunized with Freund's adjuvant only was inactive. Again, the

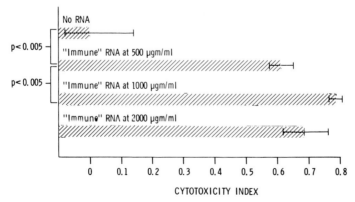

Fig. 3. *Effect of varying the concentration of immune RNA on the cytolysis of RI-H cells.*

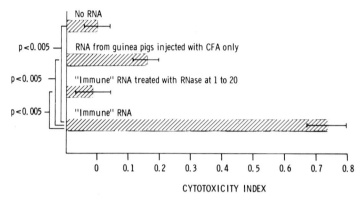

Fig. 4. *Immune cytolysis of BT-20 breast cancer cells mediated by xenogeneic immune RNA extracted from the lymphoid organs of guinea pigs immunized with BT-20 tumor cells.*

activity of the I-RNA was abolished by treatment of the I-RNA preparations with RNase, but was not affected by treatment with DNase or pronase (Figure 5).

Experiments were then designed to extend our studies to a third histologic type of human cancer, malignant melanoma. Melanoma cell lines ED-H and RO-H were previously established in tissue culture from primary explants of surgical specimens from melanoma patients ED and RO, respectively. In the initial experiments, I-RNA was extracted from the lymphoid organs of a sheep immunized with melanoma cells from a fresh surgical specimen obtained from a third melanoma patient, HO. The results of these experiments have been previously published (16). Again, this I-RNA preparation converted normal, nonimmune human peripheral blood lymphocytes to effector cells which were cytotoxic for both RO-H and ED-H melanoma cells *in vitro*. Of particular interest is the fact that RNA from the lymphoid organs of a sheep immunized with melanoma cells from one patient, HO, was effective in mediating cytotoxic immune responses against melanoma cells from two other patients, RO and ED.

In all these experiments, it was obvious, since the effector lymphocytes were allogeneic with respect to the tumor target cells, that the cytotoxic immune responses mediated by the I-RNA preparations were probably directed primarily against the normal transplantation antigens (HL-A antigens) on the surface of the tumor cells although immune responses against TAA might also occur. A critical experiment was then performed in an autologous

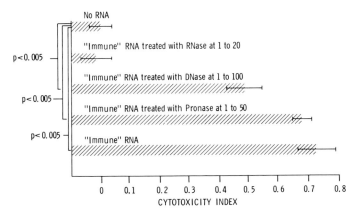

Fig. 5. *Effects of enzyme treatments on the activity of anti-BT-20 immune RNA. RNase was used at an enzyme to substrate ratio of 1 to 20. DNase and pronase were used at enzyme to substrate ratios of 1 to 100 and 1 to 50, respectively, based on the actual amount of DNA and protein in the immune RNA preparation.*

system using autologous lymphocytes obtained from the same patient, ED, from whom the ED-H cell line originated. These lymphocytes were incubated *in vitro* with an anti-melanoma I-RNA extracted from the lymphoid organs of a sheep immunized with fresh melanoma cells excised at surgery from another melanoma patient and reacted with cells from this patient's own tumor *in vitro*. The patient's lymphocytes incubated without RNA served as a control. When incubated with no RNA, the patient's lymphocytes were slightly cytotoxic to his tumor cells with a cytotoxicity index (CI) of 0.22 ± 0.05. However, when this patient's lymphocytes were incubated with anti-melanoma I-RNA, the cytotoxicity index (CI) of his lymphocytes for autologous melanoma cells doubled (CI=0.45 ± 0.02). This increase was significant at $p <$ .005. At the same time, lymphocytes from a normal volunteer were incubated with this I-RNA. The resulting CI increased from 0.18 ± 0.03 to 0.49 ± 0.05. It was apparent that the effect of I-RNA upon *autologous* lymphocytes (autologous with respect to the tumor target cells) was equal in magnitude to its effect on normal, nonimmune allogeneic lymphocytes from a healthy volunteer. When autologous lymphocytes were used as effector cells, cytotoxic immune responses directed specifically against melanoma associated antigens mediated by the I-RNA were almost certainly involved. It is important to note that the anti-melanoma I-RNA was extracted from the lymphoid organs of a sheep immunized with melanoma cells from a *different* melanoma patient, suggesting the mediation of immune responses to common melanoma specific antigens.

To derive additional evidence that immune responses directed specifically against tumor associated antigens could be mediated by I-RNA, a second experiment was performed which included additional controls. I-RNA was extracted separately from the lymphoid organs of two groups of guinea pigs. One group had been immunized with ED's melanoma cells (anti-ED-mel RNA) while the second had been immunized with ED's autologous skin fibroblasts (anti-ED-Fib RNA). Each I-RNA preparation was incubated separately with normal, allogeneic lymphocytes (from a normal healthy donor) and with ED's own autologous lymphocytes. Following incubation, each lymphocyte sample was tested for cytotoxic activity upon ED's own melanoma target cells (ED-H cells). The results of this experiment are depicted in Figure 6. Note that when effector cells which were allogeneic with respect to the target cells were used, both anti-ED-Fib RNA and anti-Mel RNA were equally effective in mediating cytotoxic immune responses to ED-H melanoma target cells. However, when autologous lymphocytes were used as effector cells, only anti-ED-Mel RNA was active and anti-ED-Fib RNA did not cause any increase in the cytotoxic activity of ED's lymphocytes for autologous melanoma cells.

We interpret these results as follows. Both anti-ED-Mel RNA and anti-ED-

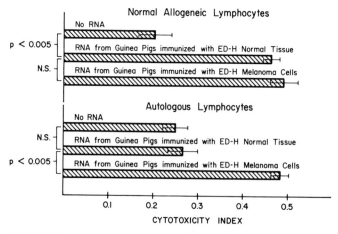

Fig. 6. *Immune cytolysis of ED-H melanoma cells by autologous lymphocytes and by allogeneic lymphocytes (from a normal donor) following incubation with xenogeneic immune RNA extracted from the lymphoid organs of guinea pigs immunized with ED-H melanoma cells or with ED's autologous skin fibroblasts.*

Fib RNA induced allogeneic lymphocytes to effect immune responses directed against normal transplantation antigens (primarily HL-A antigens) as well as tumor associated antigens (in the case of anti-Ed-Mel RNA). When these I-RNA preparations were incubated with autologous lymphocytes, I-RNAs directed against normal transplantation antigens were recognized as "self" by the lymphocytes and no immune responses against these antigens were elicited. Therefore, anti-ED-Fib RNA was inactive when incubated with autologous lymphocytes. Only I-RNAs directed against foreign antigens (i.e., tumor-associated antigens) were capable of eliciting antitumor immune responses from patient ED's own autologous lymphocytes. I-RNA extracted from lymphoid organs of animals immunized with ED's own normal fibroblasts when incubated with ED's autologous lymphocytes failed to mediate cytotoxic immune responses against ED melanoma target cells, indicating a failure to elicit immune responses against "self" cell surface antigens.

In another experiment, aliquots of lymphocytes from a healthy volunteer were incubated separately with either xenogeneic I-RNA extracted from the lymphoid organs of guinea pigs immunized with normal skin tissue from melanoma patient CR or guinea pigs immunized with CR's autologous melanoma cells (Figure 7).

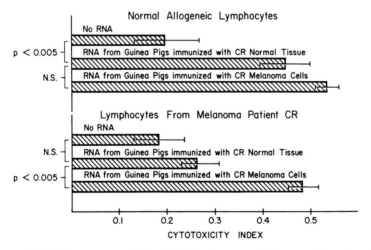

Fig. 7. *Immune cytolysis of ED-H melanoma cells by normal lymphocytes (from an allogeneic donor) and by lymphocytes from melanoma patient CR following incubation with xenogeneic immune RNA extracted from the lymphoid organs of guinea pigs immunized either with melanoma patient CR's melanoma cells or with CR's normal skin fibroblasts.*

Both I-RNA against patient CR's normal tissue and I-RNA against CR's melanoma cells when incubated with normal, allogeneic lymphocytes mediated cytotoxic immune reactions against melanoma target cells. However, when melanoma patient CR's own lymphocytes were incubated separately with these same two I-RNA preparations, only I-RNA extracted from the lymphoid organs of guinea pigs immunized with CR's melanoma cells mediated cytotoxic immune reactions against melanoma target cells. I-RNA extracted from the lymphoid organs of guinea pigs immunized with CR's normal skin tissue when incubated with CR's autologous lymphocytes failed to mediate immune reactions against melanoma target cells.

In another tumor system (Figure 8), it was shown that only I-RNA extracted from the lymphoid organs of guinea pigs specifically immunized with breast cancer cells from patient ME, when incubated with ME's own lymphocytes, mediated cytotoxic immune reactions against breast cancer target cells. RNA extracted from the lymphoid organs of guinea pigs immunized with breast cancer patient ME's normal muscle tissue was not active on ME's own lymphocytes. However, when either of these same two RNA's were incubated with lymphocytes from a normal donor, I-RNA directed against MR's normal tissue as well as I-RNA directed against ME's

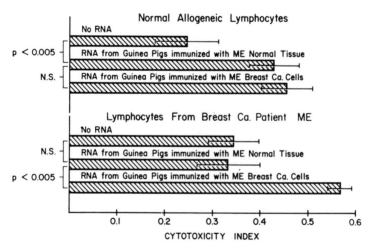

Fig. 8. *Immune cytolysis of BT-20 breast cancer cells by normal lymphocytes (from an allogeneic donor) and by lymphocytes from breast cancer patient ME following incubation with xenogeneic immune RNA extracted from the lymphoid organs of guinea pigs immunized either with patient ME's breast cancer cells or with ME's normal skin fibroblasts.*

breast cancer cells mediated cytotoxic immune responses against breast cancer target cells. It is clear from these several experiments that human peripheral blood lymphocytes have the ability to recognize as "self" I-RNAs against normal, self transplantation antigens. Only I-RNAs against foreign (e.g., tumor-associated) antigens will evoke lymphocytes to effect an active immune response.

MEDIATION OF CYTOTOXIC IMMUNE REACTIONS AGAINST HUMAN CANCER CELLS BY ALLOGENEIC IMMUNE RNA

Additional evidence to confirm that I-RNA could mediate immune responses directed specifically against tumor-associated antigens was derived from another series of experiments involving a different approach to the problem of specificity. Allogeneic I-RNA was extracted separately from the peripheral blood lymphocytes of five "cured" melanoma patients (collected on the continuous flow blood cell separator), one patient with colon carcinoma, one patient with hypernephroma and five normal volunteers. Each I-RNA preparation was incubated separately with aliquots of a single sample of normal lymphocytes collected on the continuous flow blood cell separator from another healthy volunteer, and each aliquot was then tested for cytotoxic activity againt ED-H melanoma target cells. The results are depicted in Figure 9. From this data it is apparent that only I-RNA from the lymphocytes of melanoma patients was effective in mediating a cytotoxic immune response against melanoma target cells. RNA from the lymphocytes of normal

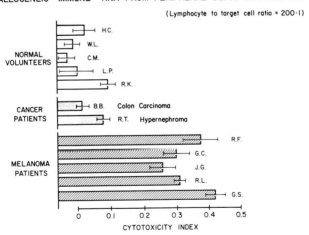

Fig. 9. *Immune cytolysis of ED-H melanoma cells mediated by allogeneic immune RNA extracted from the peripheral blood lymphocytes of "cured" melanoma patients.*

volunteers was invariably inactive, as was RNA from the lymphocytes of the colon cancer patient and the hypernephroma patient. Presumably, the melanoma patients had been sensitized to melanoma associated antigens *in vivo* but not to normal histocompatibility antigens (all were male and none had been multiply-transfused), and I-RNA was synthesized against melanoma-associated antigens.

A similar experiment was carried out using HT-29M colon cancer cells as target cells (Figure 10). When incubated with normal, allogeneic human lymphocytes, I-RNAs extracted from the peripheral blood lymphocytes of three out of four colon cancer patients effected cytotoxic immune reactions against colon cancer target cells. RNAs extracted from the peripheral blood lymphocytes of normal, healthy volunteers were inactive. More importantly, I-RNAs extracted from lymphocytes of cancer patients with cancer other than colon cancer failed to elicit immune reactions directed against colon cancer target cells. These I-RNAs included two specimens from melanoma patients which had been shown to be active in mediating immune responses against melanoma target cells. One of the I-RNAs from a colon cancer patient (B.B.) which was active against colon cancer target cells had previously been shown to be inactive against melanoma cells.

From the results of another experiment (Figure 11), it can be seen that I-RNAs extracted from the peripheral blood lymphocytes of two breast

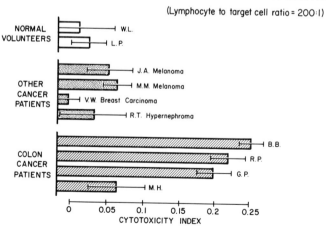

Fig. 10. *Immune cytolysis of colon cancer cells mediated by allogeneic immune RNA extracted from the peripheral blood lymphocytes of "cured" colon cancer patients.*

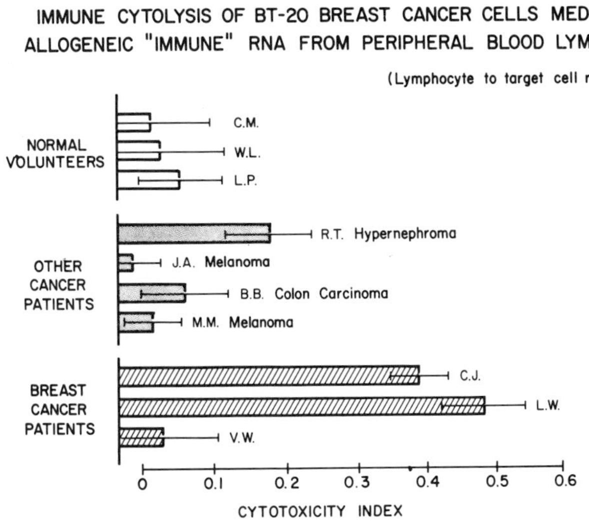

IMMUNE CYTOLYSIS OF BT-20 BREAST CANCER CELLS MEDIATED BY ALLOGENEIC "IMMUNE" RNA FROM PERIPHERAL BLOOD LYMPHOCYTES

(Lymphocyte to target cell ratio·= 200:1)

NORMAL VOLUNTEERS
- C.M.
- W.L.
- L.P.

OTHER CANCER PATIENTS
- R.T. Hypernephroma
- J.A. Melanoma
- B.B. Colon Carcinoma
- M.M. Melanoma

BREAST CANCER PATIENTS
- C.J.
- L.W.
- V.W.

CYTOTOXICITY INDEX

Fig. 11. *Immune cytolysis of BT-20 breast cancer cells mediated by allogeneic immune RNA extracted from the peripheral blood lymphocytes of "cured" breast cancer patients.*

carcinoma patients when incubated with normal lymphocytes from a healthy volunteer mediated immune reactions against BT-20 breast cancer target cells. RNAs extracted from the peripheral blood lymphocytes of normal donors or from patients with cancers other than breast carcinoma (including a colon cancer I-RNA active against colon cancer and 2 melanoma I-RNAs active against melanoma) failed to mediate significant cytotoxic immune responses against breast cancer target cells. RNA extracted from the lymphocytes of a third breast cancer patient was inactive in this assay. A sucrose density gradient profile indicated that this particular RNA was almost completely degraded and was probably inactive as a result of this.

All the cancer patients from whose peripheral blood lymphocytes these I-RNAs were extracted were male except for the breast cancer patients and only one of the breast cancer patients was multiparous (having had four children). None of the patients had been multiply transfused. Therefore, it seems unlikely that they could have been sensitized to HL-A antigens or other normal human transplantation antigens *in vivo*. The patients, therefore, were presumedly sensitized only to tumor-associated antigens of their own tumors. We observed that only I-RNAs extracted from peripheral blood lymphocytes of melanoma patients mediated cytotoxic immune reactions against melanoma cells. Similarly, only I-RNAs extracted from the lymphocytes of colon cancer patients mediated cytotoxic immune responses against colon cancer target cells, and only I-RNAs from the lymphocytes of breast cancer patients mediated immune reactions against breast cancer target cells. In other words, cytotoxic immune reactions mediated by allogeneic I-RNA extracted from the peripheral blood lymphocytes of cancer patients were shown to be directed only against tumor target cells of the same histologic tumor type as the I-RNA donor. It appears that allogeneic I-RNA extracted from the peripheral blood lymphocytes of cancer patients, under the conditions of our experiments, mediates cytotoxic immune reactions which are directed against common tumor-associated antigens specific for the tumor type of the RNA donor.

INITIAL CLINICAL TRIALS OF IMMUNOTHERAPY WITH IMMUNE RNA

Immune RNA offers several theoretical advantages over other methods of adoptive immunotherapy. (a) There is no need to give serum or plasma which has been shown in some instances to contain blocking factors *in vitro* but may facilitate tumor growth *in vivo*. (b) Since it is not necessary to transfuse foreign leukocytes, there is no sensitization of the recipient to foreign histocompatibility antigens (HL-A antigens). Repeated transfusion of allogeneic leukocytes can result in significant toxicity. Immune RNA itself does not contain transplantation antigens. (c) The danger of inducing a graft-versus-host reaction in immunodeficient or immunosuppressed individuals is eliminated. (d) Since immune RNA itself is not a strong antigen, repeated injections of human or even animal RNA would not be expected to sensitize the recipient, cause secondary syndromes, or result in immune elimination of subsequent doses of RNA. (e) Immune RNA does not require the simultaneous administration of adjuvants. (f) Immune RNA would probably be more immunogenic than immunization with tumor antigens alone, assuming that good preparations of human tumor antigens eventually become available. Immune RNA could conceivably make purification and isolation of human tumor specific transplantation antigens unnecessary for effective immunotherapy. (g) Immune RNA might be effective despite certain types of host anergy (defects in the afferent arc of the immune response, i.e., antigen recognition or processing).

The successful interspecies transfer of cytotoxic immune responses to human tumors *in vitro* (see above) provided a rational basis for the immunotherapy of human cancer with I-RNA extracted from the lymphoid tissues of animals specifically immunized with the type of human tumor to be treated. There are several obvious advantages to this form of immunotherapy. First, large quantities of I-RNA could be readily produced without dependence upon human donors. Secondly, since histologically similar human tumors appear to contain common tumor-associated antigens, many patients with the same tumor type could be treated with I-RNA from a sheep sensitized with a single patient's tumor.

We have recently undertaken a preliminary trial of immunotherapy of human cancer with xenogeneic I-RNA. As of September 30, 1975, 51 patients ranging in age from 10 to 69 have been treated. The main intent of this study was to determine the safety and feasibility of immunotherapy with xenogeneic immune RNA and to determine the effects, if any, of this therapy on various tumor specific and nonspecific immunologic functions. However, some interesting observations have been made regarding the effects of immune RNA therapy on certain antitumor immune responses assessed *in vitro*, and some indications of possible therapeutic benefit have been observed.

Two types of patients have been studied: patients with grossly detectable and measurable metastatic disease and patients with "minimum residual disease". Patients in the latter category had no clinically detectable disease following surgical resection of all gross tumor but had a greater than 50% likelihood of developing recurrent and/or metastatic disease within 24 months. The histologic types of tumors treated included malignant melanoma, hypernephroma, sarcoma, breast carcinoma, cholangiocarcinoma, colon carcinoma, carcinoma of the lung, transitional cell carcinoma of the prostate and carcinoma of the stomach. Of the three sarcoma patients, one had an alveolar soft part sarcoma, one a liposarcoma and one an osteogenic sarcoma. Patients with gross disease were accepted for treatment only if (a) they had failed all standard therapy (including standard chemotherapy), (b) no standard therapy of proven efficacy existed, or (c) they had relatively stable disease and standard therapy could be interrupted or withheld for 8 weeks which was the minimum duration of the study. Our 51 patients included 34 patients with gross disease and 17 patients with minimum residual disease. The numbers of patients of each tumor type treated in each category are presented in Table 2.

TABLE 2.
Numbers of Patients Treated with Immune RNA Grouped According to Tumor Type and Extent of Disease

Tumor Types Treated	Gross Disease	Minimum Residual Disease	Total
Malignant Melanoma	11	6	17
Hypernephroma	18	5	23
Sarcoma	3	0	3
Breast Carcinoma	0	1	1
Stomach Carcinoma	0	1	1
Colon Carcinoma	0	3	3
Cholangiocarcinoma	1	0	1
Transitional Cell Carcinoma of Prostate	0	1	1
Carcinoma of the Lung	1	0	1
TOTAL	34	17	51

The patients ranged in age from 10 to 69. All were sensitized and skin tested with DNCB and a battery of 4 common skin test antigens (PPD, monilia, mumps and streptokinase/streptodornase) prior to the initiation of immunotherapy and all were immunocompetent.

Tumor tissue obtained from surgical specimens (either fresh or frozen) was used. A thick suspension of viable tumor cells emulsified in an equal volume of complete Freund's adjuvant was injected intradermally into all 4 extremities of a sheep at weekly intervals for 3 consecutive weeks. Ten days after the last injection the animals were sacrificed and immune RNA was extracted from the spleens and lymph nodes. The I-RNA was reprecipitated twice from solutions made 2 M with respect to potassium acetate, treated with pronase (to remove contaminating protein), and again reprecipitated from 2 M potassium acetate. The RNA was then dialyzed against sterile distilled water, sterilized by passage through a 0.22 micron millipore filter and lyophylized. The RNA was resuspended in normal saline prior to injection. The I-RNA was administered intradermally in 0.1 ml aliquots in the skin of the upper anterior chest near the axillae and lower abdomen near the groins at weekly intervals rotating injection sites. Initially the weekly dose of I-RNA was 1 mg; later 2 mg/week was given and, most recently 8 mg/week has been the usual dose. A few patients have received up to 40 mg/week. Whenever possible, each patient received I-RNA against autologous tumor tissue. Alternatively, I-RNA against allogeneic tumor tissue of the same histologic type was used. Treatment was given for 8 weeks, following which initial responses were evaluated. If progression of disease was noted, immune RNA therapy was discontinued and alternative therapy instituted. The duration of I-RNA therapy to date ranges from 21 months to less than two months.

Responses to therapy in patients with gross disease were evaluated by the following criteria: (a) Patients were considered "improved" if grossly detectable (and measurable) lesions objectively regressed (i.e., decreased in area) and, following regression did not increase in size for at least two months. (b) Patients were considered treatment "failures" if gross lesions significantly increased in area or new lesions appeared. (c) Patients were considered as having achieved "stability" if lesions which had been growing progressively prior to the initiation of therapy ceased all growth and remained stable for at least two months. (d) A few patients were classified as having achieved "possible benefit" when transitory regressions occurred or when prolonged stability was later followed by regrowth while on therapy. (e) Patients with minimum residual disease were considered "free of disease" if no recurrence or metastases developed. If recurrence and/or metastases occurred, this was considered a treatment failure. When progression occurred, immune RNA therapy was discontinued.

No significant local or systemic toxicity has been noted with immune RNA therapy administered weekly for up to 21 months. "Burning" on

TABLE 3.
Summary of Clinical Results – Gross Disease

Tumor Types	Improvement	Stability or Possible Improvement	Failure	Indeterminate	Total
Malignant Melanoma	0	3	7	1	11
Hypernephroma	3	5	7	3	18
Sarcoma	1	0	2	0	3
Carcinoma of the lung	0	0	0	1	1
Cholangiocarcinoma	0	0	0	1	1
TOTAL	4	8	16	6	34

injection was minimal. Local irritation as evidenced by erythema, induration or swelling was absent or very minimal. No nausea, vomiting or febrile reactions of any kind have occurred, and no allergic or anaphylactoid reactions have resulted.

The clinical results in patients with gross disease are summarized in Table 3. The first 26 patients are being reported in detail elsewhere (17,18), and the results obtained in patients with hypernephroma are described in detail in another paper in this volume (19). The clinical status of patients with minimum residual disease is presented in Table 4. The duration of follow up for each group of patients is given in the table. Only one patient has relapsed to date.

In reviewing the results of this study to date, it must be remembered that clinical remissions were not expected at this stage of our work and any clinical responses reported must be considered anecdotal. In this initial Phase I trial, the objectives were: (a) to establish the safety (or toxicity) of sheep I-RNA, (b) to evaluate dosage schedules and routes of administration and (c)

TABLE 4.
Summary of Clinical Results – Minimal Residual Disease

Tumor Types	Number of Patients	Relapsed	Free of Disease	Duration of Follow-up
Malignant Melanoma	6	1 (5 mo.)	5	3-12 mo.
Hypernephroma	5	0	5	4-21 mo.
Transitional Cell Carcinoma of Prostate	1	0	1	4 mo.
Colon Carcinoma	3	0	3	1-6 mo.
Breast Carcinoma	1	1 (9 mo.)	0	–
Stomach Carcinoma	1	1 (5 mo.)	0	–
TOTAL	17	3	14	–

to monitor any possible effects of I-RNA treatment on immunologic parameters, both tumor-specific and nonspecific. The optimum dosage, route and frequency of administration of I-RNA is not known. Certainly, we have not as yet approached a toxic dose of I-RNA.

In our previous animal experiments, it appeared desirable to administer immune RNA in a medium containing a strong ribonuclease inhibitor, the polyanion sodium dextran sulfate. However, since we have not as yet received approval from the Food and Drug Administration to administer dextran sulfate experimentally in man, we have not incorporated a ribonuclease inhibitor into our RNA preparations. Perhaps by so doing we might significantly increase the efficacy of I-RNA therapy.

Although treatment of lympocytes *in vitro* with immune RNA has been shown to induce such lymphocytes to effect antitumor responses *in vivo* and *in vitro*, we have not as yet treated any patients by the intravenous infusion of autologous lymphocytes preincubated *in vitro* with I-RNA. This reluctance has been due to the fear of inducing untoward allergic reactions related to small amounts of sheep protein which contaminate the RNA and might be expected to remain with the lymphocytes even after several washes. However, this method of utilizing immune RNA therapeutically may offer promise of greater efficacy.

Many of our patients had large tumor burdens and may not have been good candidates for immunotherapy of any kind. Perhaps — and this may be true for all types of immunotherapy — the primary usefulness of immune RNA will prove to be in the therapy of minimum residual disease. Our number of patients with minimum residual disease is small. The follow-up period for most of these patients is very short, and we have no concomitantly generated control population. Therefore, few if any conclusions can be drawn at the present time from the results observed in these patients.

In a number of instances, lymphocytes obtained from patients prior to the onset of treatment were incubated *in vitro* with the immune RNA which they were to receive. Following incubation of the lymphocytes with I-RNA for 20 minutes at 37°C, the lymphocytes were tested for cytotoxic effect *in vitro* on allogeneic tumor target cells of the same histologic type. One example is presented in the experiment illustrated in Figure 12. In this experiment, pretreatment lymphocytes from melanoma patient CA were tested against allogeneic melanoma cells (ED-H) before and after incubation with I-RNA extracted from the lymphoid organs of a sheep immunized with CA's own autologous tumor cells and after incubation with I-RNA extracted from the lymphoid organs of an unimmunized sheep. Following incubation with I-RNA from a sheep immunized with CA's tumor cells, the cytotoxicity of CA's lymphocytes for melanoma cells was greatly increased, while incubation with I-RNA from an unimmunized sheep was ineffective.

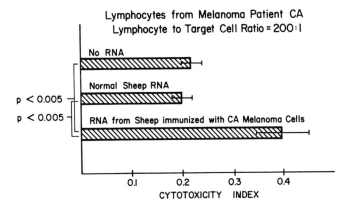

Fig. 12. *Cytotoxicity of pretreatment lymphocytes from melanoma patient CA for ED-H melanoma cells prior to and following incubation of immune RNA extracted from the lymphoid tissues of a sheep immunized with CA's own autologous tumor tissue and (as a control) with immune RNA extracted from the lymphoid organs of a normal sheep.*

Another example is presented in the experiment illustrated in Figure 13. In this experiment, pretreatment lymphocytes from gastric carcinoma patient KI were tested against allogeneic gastric carcinoma cells (RI-H) following (a) incubation without RNA, (b) incubation with I-RNA extracted from the lymphoid organs of a sheep immunized with KI's tumor cells and (c) incubation with I-RNA extracted from the lymphoid organs of a sheep immunized with RI-H tumor cells. These are compared with the cytotoxicity of allogeneic lymphocytes from a normal donor incubated without RNA and with each of the same two I-RNA preparations. Lymphocytes from gastric carcinoma patient KI obtained prior to treatment were significantly cytotoxic for gastric carcinoma target cells. Following incubation with I-RNA from a sheep immunized with KI's tumor cells, the cytotoxicity of KI's lymphocytes for gastric carcinoma cells was greatly increased. The magnitude of this increase was the same as that observed when KI's pretreatment lymphocytes were incubated with I-RNA from a sheep immunized with the gastric carcinoma cells (RI-H) used as target cells in the assay.

In 21 of the 51 patients, peripheral blood lymphocytes from serial bleedings obtained prior to and during I-RNA therapy were tested for cytotoxicity against allogeneic tumor target cells of the same histologic type. The results of these serial lymphocyte cytotoxicity determinations are summarized in Table 5. Of 12 patients in whom significant increases in cytotoxicity were noted 2 evidenced objective tumor regression, 7 achieved stability or possible benefit in their clinical course, and only 3 were treatment

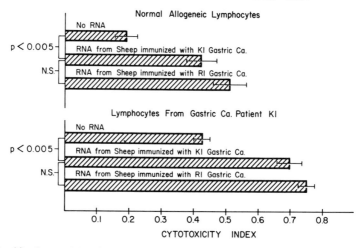

Fig. 13. *Cytotoxicity for gastric carcinoma (RI-H) target cells of normal allogeneic lymphocytes and of pretreatment lymphocytes from gastric carcinoma patient KI incubated (a) with no RNA, (b) with RNA from a sheep immunized with KI's own tumor cells and (c) with RNA from a sheep immunized with RI-H gastric carcinoma cells.*

failures. Of 8 patients in whom no change in lymphocyte-mediated cytotoxicity was noted, 4 evidenced stability or possible benefit while 4 failed treatment. In one patient, a decrease in cytotoxicity was noted and this patient was a treatment failure. There appeared to be a possible correlation between increases in lymphocyte-mediated cytotoxicity assessed *in vitro* and clinical response.

TABLE 5.
*Correlation Between Changes in Lymphocyte-Mediated Cytotoxic Responses **In Vitro** and Clinical Response*

Cytotoxicity	Improvement	Stability or Possible Benefit[1]	Failure	Total
Increase	2	7	3	12
Decrease	0	0	1	1
No Change	0	4[2]	4	8

[1] Patients with minimum residual disease and no evidence of recurrence are grouped in this category.

[2] One of these patients had a very high level of cytotoxicity prior to treatment which remained unchanged.

The results of experiments performed in 5 of these patients is presented in detail in Figures 14-18. Figures 14, 15 and 16 illustrate the results of testing lymphocytes from three different melanoma patients for cytotoxic activity on ED-H melanoma target cells. In the patients illustrated in Figures 14 and 15 significant increases in cytotoxicity were observed following the initiation of immune RNA therapy. Lymphocytes from the patient depicted in Figure 16 were highly cytotoxic for melanoma cells prior to the initiation of therapy and remained highly cytotoxic during therapy.

In Figure 17a the results of testing the cytotoxicity of lymphocytes from gastric carcinoma patient KI on RI-H gastric carcinoma target cells is presented, while Figure 17b depicts the results of testing aliquots of the same lymphocyte samples on ED-H melanoma target cells. It is evident that, while the cytotoxicity of KI's lymphocytes for RI-H cells increased substantially following the initiation of therapy with immune RNA, no significant cytotoxic activity was evidenced against ED-H melanoma target cells. In Figure 18, the results of testing lymphocytes from melanoma patient VE against autologous tumor cells and autologous skin fibroblasts is depicted. It is apparent that, while the cytotoxicity of VE's lymphocytes for autologous melanoma cells increased, no significant cytotoxicity was evidenced against autologous fibroblasts.

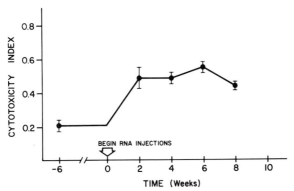

Fig. 14. *Cytotoxicity for ED-H melanoma cells of lymphocytes from melanoma patient DE prior to and following initiation of therapy with immune RNA extracted from the lymphoid organs of a sheep immunized with patient DE's autologous melanoma cells.*

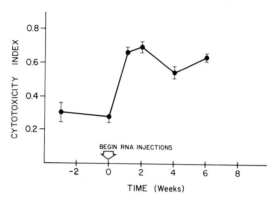

Fig. 15. *Cytotoxicity for ED-H melanoma cells of lymphocytes from melanoma patient MG prior to and following initiation of therapy with immune RNA extracted from the lymphoid organs of a sheep immunized with melanoma cells.*

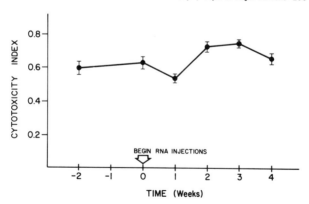

Fig. 16. *Cytotoxicity for ED-H melanoma cells of lymphocytes from melanoma patient HH prior to and following initiation of therapy with immune RNA extracted from the lymphoid organs of a sheep immunized with melanoma cells.*

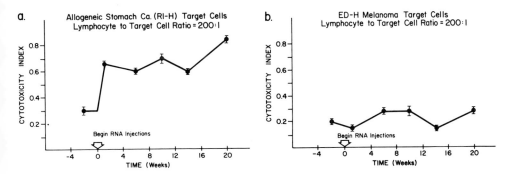

Fig. 17. *Cytotoxicity for RI-H gastric carcinoma target cells (Figure 17a) and for ED-H melanoma cells (Figure 17b) of lymphocytes from gastric carcinoma patient KI prior to and following the initiation of treatment with xenogeneic immune RNA extracted from the lymphoid organs of a sheep immunized with KI's own tumor tissue.*

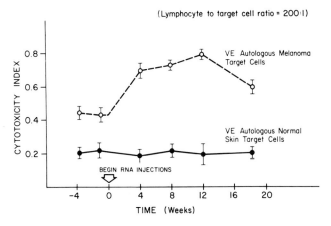

Fig. 18. *Cytotoxicity of lymphocytes from melanoma patient VE for VE's autologous melanoma cells and autologous skin fibroblasts prior to and following treatment with xenogeneic immune RNA extracted from the lymphoid organs of a sheep immunized against VE's autologous melanoma cells.*

These data suggest that treatment with I-RNA *does* effect at least certain host immunologic functions. This implies that these RNA preparations *are*, at least to some extent, immunologically active in man when administered in the dose and manner described above. These data also provide some evidence to suggest that the immune responses effected by immune RNA may, at least in part, be specific for the particular tumor type used to immunize the sheep from whose lymphoid tissues the I-RNA was extracted. When pretreatment lymphocytes were incubated with immune RNA, their cytotoxicity for tumor target cells was significantly increased. Moreover, in most of the patients tested, the cytotoxicity of peripheral blood lymphocytes for tumor target cells increased substantially following immune RNA therapy, and we have some evidence that this response is specific for the particular histologic type of tumor used to immunize the RNA donor. We do not know if the nature or magnitude of these effects are relevant to the clinical course of human cancer or if they can result in therapeutic benefit.

Finally, it is clear that sheep immune RNA, when prepared as described above and administered intradermally in the doses and schedules described, is completely free of significant local and systemic toxicity and is very well tolerated. Immune RNA offers several practical and theoretical advantages as a new modality for the immunotherapy of cancer. Much further study will be required to determine its potential immunotherapeutic efficacy.

REFERENCES

1. Pilch, Y. H. and K. P. Ramming (1970). Transfer of tumor immunity with ribonucleic acid. *Cancer 26*:630-637.

2. Ramming, K. P. and Y. H. Pilch (1970). Mediation of immunity to tumor isografts in mice by heterologous ribonucleic acid. *Science 168*:492-493.

3. Ramming, K. P. and Y. H. Pilch (1971). Transfer of tumor-specific immunity with RNA: Inhibition of growth of murine tumor isografts. *J. Natl. Cancer Inst. 46*:735-750.

4. Bach, M. K. (1964). The inhibition of deoxyribonucleotidyl tranferase RNAse and DNAse by sodium polyethene sulfonic acid. Effect of the molecular weight of the inhibitor. *Biochem. Biophys. Acta.* (Amst.) *91*:619-626.

5. Fellig, J. and C. E. Wiley (1959). The inhibition of pancreatic ribonuclease by aniomic polymers. *Arch. Biochem. Biophys. 85*:313-316.

6. Mora, P. T. and B. G. Young (1959). Reversible inhibition of enzymes by interaction with synthetic polysaccharide macroanions. *Arch. Biochem. Biophys. 82*:6-20.

7. Deckers, P. J. and Y. H. Pilch (1971). Transfer of immunity to tumor isografts by the systemic administration of xenogeneic "Immune" RNA. *Nature* (New Biology) *231*:181-183.

8. Pilch, Y. H, K. P. Ramming and P. J. Deckers (1971). Transfer of tumor immunity with RNA. *Israel J. Med. Sci. 7*:246-258.

9. Deckers, P. J. and Y. H. Pilch (1972). Mediation of immunity to tumor-specific transplantation antigens by RNA: Inhibition of isograft growth in rats. *Cancer Res. 32*:839-846.

10. Ramming, K. P. and Y. H. Pilch (1970). Transfer of tumor-specific immunity with RNA: Demonstration by immune cytolysis of tumor cells *in vitro. J. Natl. Cancer Inst. 45*:543-553.

11. Kern, D. H., C. R. Drogemuller and Y. H. Pilch (1974). Immune cytolysis of rat tumor cells mediated by syngeneic "Immune" RNA. *J. Natl. Cancer Inst. 52*:299-302.

12. Kern, D. H. and Y. H. Pilch (1974). Immune cytolysis of Murine tumor cells mediated by xenogeneic "Immune" RNA. *Int. J. Cancer 13*:679-688.

13. Veltman, L. L., D. H. Kern and Y. H. Pilch (1974). Immune cytolysis of human tumor cells mediated by xenogeneic "Immune" RNA. *Cellular Immunol. 13*:367-377.

14. Pilch, Y. H., L. L. Veltman and D. H. Kern (1974). Immune cytolysis of human tumor cells mediated by xenogeneic "Immune" RNA. *Arch. Surg. 109*:30-34.

15. Pilch, Y. H., L. L. Veltman and D. H. Kern (1974). Immune cytolysis of human tumor cells mediated by xenogeneic Immune RNA: Implications for immunotherapy. *Surgery 76*:23-43.

16. Pilch, Y. H., J. B. deKernion, D. G. Skinner, K. P. Ramming, P. M. Schick, D. Fritze, P. Brower and D. H. Kern. Immunotherapy of cancer with "Immune" RNA: A preliminary report. *Am. J. Surg.*, submitted for publication.

17. Pilch, Y. H., D. Fritze and D. H. Kern (1975). Immune cytolysis of human melanoma cells mediated by Immune RNA. *Behring Inst. Mitt. 56*:184-196.

18. Skinner, D. G., J. B. deKernion and Y. H. Pilch. Advanced renal cell carcinoma: Treatment with xenogeneic immune ribonucleic acid (RNA) and appropriate surgery. *J. Urol.*, in press.

19. deKernion, J. B., D. S. Skinner, K. P. Ramming and Y. H. Pilch (1976). The clinical experience in the treatment of renal adenocarcinoma with immune RNA. *Immune RNA in Neoplasia.* Mary A. Fink, Ed. Academic Press, New York.

MEDIATION OF ANTITUMOR IMMUNE RESPONSES WITH I-RNA[1]

David H. Kern and Yosef H. Pilch

Department of Surgery
Harbor General Hospital
1000 West Carson Street
Torrance, California 90509
and the UCLA School of Medicine

Our laboratories have previously reported the transfer of antitumor immune responses with immune RNA (I-RNA) in a number of model systems, both *in vivo* and *in vitro*. Xenogeneic antitumor I-RNA extracted from the lymphoid organs of guinea pigs immunized with a chemically-induced mouse sarcoma transferred specific antitumor immune responses to mice of the same strain in which the tumor originated. Significant inhibition of the development of tumor isografts was mediated by syngeneic spleen cells preincubated with xenogeneic I-RNA (1). Similar inhibition of tumor development was obtained in experiments in which xenogeneic I-RNA was administered systemically (2). In a totally syngeneic system it was shown that rats challenged with a chemically-induced sarcoma synchronous with the first of three daily intravenous injections of syngeneic spleen cells preincubated with syngeneic I-RNA extracted from the spleens of hyperimmunized rats evidenced a lower incidence of tumor isograft development than did rats in groups receiving spleen cells alone without incubation with I-RNA, or spleen cells incubated with control RNA preparations (3).

Utilizing a plaque assay, we demonstrated the immune cytolysis of strain 2 guinea pig tumor cells *in vitro* by syngeneic spleen cells preincubated with I-RNA extracted from the spleens of immunized guinea pigs (4). Later, we used a quantitative microcytotoxicity assay to show that normal mouse spleen cells became specifically cytotoxic to mouse tumor target cells after incubation with xenogeneic I-RNA extracted from the lymphoid organs of immunized guinea pigs (5). We also reported, within a completely syngeneic system, that normal nonimmune Fischer rat spleen cells, after incubation with syngeneic I-RNA extracted from spleens of tumor bearing rats, became specifically cytotoxic to rat tumor target cells *in vitro* (6). The mediation of cytotoxic immune reactions

---

[1] Supported in part by grants CA 14846 and CA 15372 from the National Institutes of Health

against human tumor cells by xenogeneic I-RNA also has been demonstrated in recent studies. Normal nonimmune human peripheral blood lymphocytes became cytotoxic to human tumor cells *in vitro* following incubation with I-RNA extracted from the lymphoid organs of sheep or guinea pigs which had previously been immunized with cells from the particular tumor being studied (7,8).

We are now investigating the mechanisms by which I-RNA mediates antitumor immune responses. In this report, we present the results of experiments which demonstrate that antitumor I-RNA mediates cytotoxic reactions *in vitro* which are directed specifically against tumor associated antigens of murine tumor cells. The evidence indicates that the active component(s) of our I-RNA preparations is one or more species of ribonucleic acid. The kinetics of synthesis of I-RNA in immunized animals has been studied, and the optimum time for harvesting lymphoid tissues for extraction of I-RNA determined. I-RNA was fractionated according to sedimentation velocity on preparative sucrose density gradients. Active antitumor I-RNA was found only in fractions with sedimentation values between 8 and 16S. These immunoreactive fractions comprised only a small portion of the total RNA extracted from the lymphoid cells. Finally, we have shown that the antitumor activity of I-RNA is localized in the cytoplasm rather than the nuclei of lymphoid cells from immunized animals.

MATERIALS AND METHODS

Tumors

Sarcomas were induced in female C3Hf/HeCr mice with either methylcholanthrene (MC-1) or 3,4-benz(a)pyrene (BP-1) and maintained by serial transplantation in female C3Hf/HeCr mice. The tumors were explanted from the second transplantation generation into plastic tissue culture flasks and grown as monolayers. The tissue culture medium used was RPMl 1640 supplemented with 20% fetal calf serum and 0.15% glutamine, and containing 100IU penicillin and 100 µg streptomycin per ml with 0.05% amphotericin B (all from Flow Laboratories, Los Angeles, California).

MC-1 tissue culture cells were used in the 10th to 20th passage. Transplantation of MC-1 cells from the 11th passage into normal C3H mice resulted in progressive tumor growth. The BP-1 tumor cells were used in their 25th to 40th tissue-culture passage. Transplantation of BP-1 cells from the 36th passage into normal C3H mice also resulted in progressive tumor growth.

A fibrosarcoma was induced in a female Fischer 344/N rat by methylcholanthrene and serially transplanted. The tumor, designated MC3-R, was explanted into plastic tissue culture flasks and grown as monolayers. It was used in passages 10-20. The transplantation of cells from the tenth tissue culture passage into normal Fischer rats resulted in progressive tumor growth. Tissue culture medium was the same as that described above.

Immunization of RNA Donors and Preparation of I-RNA

A 40% w/v suspension of tumor cells in Hank's balanced salt solution was mixed with an equal volume of complete Freund's adjuvant and 0.25 ml was injected into each of the four foot pads of adult Hartley guinea pigs. An intraperitoneal injection of 1.0 ml of the tumor cell suspensions without adjuvant was also given. Control guinea pigs received similar injections of normal tissue (a mixture of spleen, liver, lung and kidney cells) instead of tumor cells. The spleens and lymph nodes of immunized guinea pigs were excised 13-14 days after immunization, and xenogeneic I-RNA was extracted by the hot phenol method previously described (1). For studies on the kinetics of I-RNA synthesis, spleens and nodes were excised at 1,4,7,14 and 21 days following immunizations.

Syngeneic I-RNA was extracted by the same procedure from the spleens of Fischer rats bearing growing MC3-R tumor transplants approximately 2 cm in diameter. Tumors of this size usually arose between 2 and 3 weeks after inoculation with 10^6 viable MC3-R tumor cells. For kinetic studies, spleens were excised at 1,7,14,21 and 28 days following tumor transplantation. Control RNA preparations were also extracted from the spleens of Fischer rats bearing a different benzypyrene-induced tumor, BP1-R.

RNA concentration was measured by the Orcinol reaction (9). DNA was measured by the diphenylamine method (9), and protein was determined by the method of Lowry et al. (10). The RNA preparations were characterized with respect to their sedimentation properties by ultracentrifugation in continuous 5-20% sucrose density gradients. In some experiments, I-RNA was treated with enzymes as follows: (a) 1 mg/ml RNA preparation was treated with 50 μg/ml RNAse for one hour at 37°C (enzyme-to-substrate ratio of 1:20); (b) 1 mg/ml RNA preparation was treated with RNAse-free DNAse at an enzyme-to-substrate ratio of 1:20 (based on the actual amount of DNA in the sample) for one hour at 37°C; and (c) 1 mg/ml RNA, made 0.2m with respect to CaCl2, was treated with pronase at an enzyme-to-substrate ratio of 1:10 (based on the actual amount of protein in the sample) for one hour at 37°C.

Preparation of Nuclear and Cytoplasmic Fractions

For some experiments, I-RNA was extracted separately from the nuclear and cytoplasmic fractions of sensitized lymphoid cells. Spleens were removed from the donor animals, minced, and pressed through 40 and 80 mesh stainless-steel sieves. Spleen cells were mechanically lysed at 4°C in 0.25 M sucrose containing 0.006 M $MgCl_2$. Nuclei were sedimented by centrifugation at 1000 X g for 4 minutes at 4°C. The supernatant (cytoplasmic) fraction was removed and RNA precipitated by the addition of 3 volumes of cold ethanol and chilling at $-20°C$ for 2 hours. Nuclei were purified by ultracentrifugation through 2.2 M sucrose (containing 0.002 M $MgCl_2$ and 0.05 mM $CaCl_2$) at 100,000 × g for 30 minutes at 4°C. Electron microscopy revealed minimal cytoplasmic contamination of nuclei except for one preparation of nuclei described below. RNA was extracted from intact spleen cells (Total RNA), from nuclei fractions (Nuclear RNA) and from cytoplasmic fractions (Cytoplasmic RNA) by the hot phenol method. Electrophoretic migration patterns of all RNA preparations were determined on 2.5% polyacrylamide disc gels. All RNA specimens had profiles resembling that of a known, undegraded RNA standard. RNA, DNA and protein concentrations were measured as described above.

Fractionation of I-RNA

A 5-20% linear sucrose density gradient was prepared in buffer (0.01M tris, 0.1N NaCl, 0.2% SDS and 1mM EDTA, pH 7.0) in cellulose nitrate centrifuge tubes. 3mg of RNA was layered on top of the gradient; a total of 18mg of RNA was fractionated per run. The samples were centrifuged in a Beckman L2-65B ultracentrifuge at 40,000 rpm for 6 hours at +20°C. Fractions were collected on an ISCO Model 640 Density Gradient Fractionator at a flow rate of 3 ml/min. UV Absorbence at 254 mm was monitored by an ISCO Model UA-5 Absorbence Monitor. Appropriate fractions were pooled, made 0.1N with NaCl, and precipitated with 2.5 volumes of absolute ethanol overnight at $-20°C$. After precipitation, samples were centrifuged at 2800 rpm in an IEC-PR-6000 centrifuge for 30 minutes at 5°C. The pellet was resuspended in buffer (0.01M tris, 0.1N NaCl, 0.015M $MgCl_2 \cdot 6H_2O$, pH 7.5) and stored at $-40°C$ until used. Sucrose density gradients and polyacrylamide gel electrophoresis were used to characterize the fractions. The RNA, DNA and protein concentrations on the fractions were determined as described above.

Incubation of Spleen Cells with RNA

Spleens from normal C3H mice or from Fischer or Wistar Furth rats were removed aseptically, minced in Hank's balanced salt solution, and gently pressed through 40-mesh and 80-mesh stainless steel sieves in sequence. Red blood cells were lysed by a 5-minute exposure to 0.85% NH_4Cl followed by two washings with RPMI 1640. Viability was always greater than 95% as judged by trypsin blue exclusion. Cell concentration was adjusted to 3×10^7 cells/ml in RPMI. One ml of cells and the appropriate concentration of RNA were incubated together for 20 minutes at 37°C in a gyratory shaking water bath. Control spleen cell preparations were incubated without RNA, but otherwise were treated identically to the spleen cells incubated with RNA. Cells were washed twice with RPMI and suspended in RPMI with 20% fetal calf serum, glutamine and antibiotics. Cell viability was determined and was usually greater than 90%.

Spleen cell concentration was then adjusted as appropriate for the experiment and the cells were used immediately in the microcytotoxicity assay described below.

Microcytotoxicity Assay

10^6 tumor cells in 5 ml of medium containing 100 µc ^{125}I-Iododeoxyuridine were added to a 30 ml plastic tissue culture flask. After incubation for 24 hours at 37°C, the cells were removed with 0.25% trypsin. The cell concentration was adjusted to 10^4 cells/ml and 2000 cells in 0.25 ml were added to each well of a Falcon Microtest II plate (Falcon Plastics, Oxnard, California). The cells were incubated in a humid atmosphere of 95% air and 5% CO_2 for 4-18 hours at 37°C to allow the cells to adhere to the plastic surface. Spleen cells preincubated with one or another RNA preparation as described above, or incubated without RNA, were suspended in tissue culture medium. These spleen cells were then added to the target cells in varying concentrations in a volume of 0.2 ml. Six replicate wells were used for each variable of each experiment. The plates were incubated for 48 hours at 37°C. Medium containing nonadherent cells was then aspirated from the wells and the wells washed four times with medium to remove residual nonadherent cells. After drying in air, the plates were sprayed with Aeroplast (Park-Davis, Detroit, Michigan) and the individual wells were separated with a band saw. The residual radioactivity in each well was counted for 10 minutes with a gamma counter. Means and standard errors of means (S.E.M.) were calculated and groups were compared for significane using Students' t-test for unpaired data. Results were expressed in terms of cytotoxicity index (CI), calculated according to the formula:

$$CI = \frac{\text{cpm no RNA} - \text{cpm RNA}}{\text{cpm no RNA}}$$

where cpm no RNA = counts per minute remaining for tumor cells exposed to spleen cells incubated with no RNA, and cpm RNA = counts per minute remaining for tumor cells exposed to spleen cells incubated with RNA (I RNA for one of the control RNA preparations).

For the experiments depicted in Figures 5-7, a slightly different labeling method was used. The tumor target cells were prelabeled with 3H-thymidine as previously described (6). The remainder of the assay procedure was the same as that described above except that residual activity in each well was determined by digesting the adherent cells in each well with 0.1N NaOH and counting in a liquid scintillation counter.

RESULTS

Specificity of Antitumor Immune Reactions Mediated by Xenogeneic I-RNA

Xenogeneic immune RNA extracted from the lymphoid organs of guinea pigs immunized with MC-1 or BP-1 tumor cells mediated cytotoxic immune responses specific for that particular tumor used to immunize the RNA donor guinea pig (5). I-RNA was extracted from the lymphoid organs of adult female Hartley guinea pigs which had been immunized with either MC-1 or BP-1 tumor cells in complete Freund's adjuvant. Control guinea pigs were similarly immunized with a pool of normal C3H tissues. Spleen cells from normal nonimmune C3H mice were incubated with one or the other of the I-RNA preparations, or with control (normal tissue) RNA. The cytotoxicity of these spleen cells for either MC-1 or BP-1 tumor target cells following incubation with I-RNA was then measured.

In one experiment, representative of many performed in our laboratory and reported in detail elsewhere (5), normal nonimmune C3H mouse spleen cells preincubated with 1 mg of I-RNA extracted from the lymphoid organs of guinea pigs immunized with MC-1 tumor cells were found to be cytotoxic for MC-1 target cells with a cytotoxicity index (CI) of 0.45 ± 0.04 (Figure 1). Spleen cells preincubated with RNA from guinea pigs immunized with either normal C3H tissues or with BP-1 tumor cells evidenced no significant cytotoxicity for MC-1

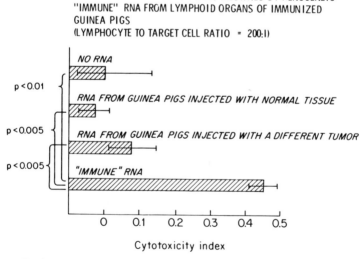

Fig. 1. *Immune cytolysis of MC-1 mouse tumor target cells mediated by xenogeneic immune RNA extracted from the lymphoid organs of guinea pigs immunized with MC-1 tumor cells. Controls include RNA from guinea pigs immunized with normal C3H mouse tissues or with a different syngeneic tumor (BP-1).*

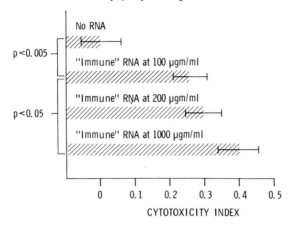

Fig. 2. *Immune cytolysis of MC-1 tumor cells mediated by xenogeneic I-RNA. Effects of varying I-RNA concentration.*

cells and were not statistically different from spleen cells incubated with no RNA. Similar results were obtained at lymphocyte-to-target ratios of 100:1 and 50:1.

While 1 mg/ml of RNA was the concentration of I-RNA employed in the above experiment, concentrations of RNA of 500, 200 and 100 µg/ml also were effective. In Figure 2, it can be seen that a concentration of I-RNA as low as 100 µg/ml was sufficient to mediate significant cytolysis of MC-1 target cells.

Since the I-RNA preparations contained small but finite amounts of DNA and protein, the effects of RNAse, DNAse and pronase treatment on the active anti-MC-1 I-RNA was evaluated. These results are shown in Figure 3. Normal C3H spleen cells preincubated with anti-MC-1 I-RNA at a concentration of 1 mg/ml yielded a cytotoxicity index of 0.32 ± 0.05. Anti-MC-1 I-RNA treated with RNAse had no activity. However, C3H spleen cells preincubated with anti-MC-1 I-RNA that had been treated with DNAse or pronase resulted in cytotoxicity indices of 0.36 ± 0.06 and 0.28 ± 0.03 respectively. There were no significant differences between the activity of anti-MC-1 I-RNA treated with either DNAse or pronase and untreated anti-MC-1 I-RNA. This suggests that the active component of these immune RNA preparations was one or more species of RNA and not contaminating DNA or protein.

Utilizing BP-1 tumor cells as target cells, experiments similar to those described above were performed. The purpose of these experiments was to determine whether or not anti-MC-1 I-RNA could convert normal C3H spleen

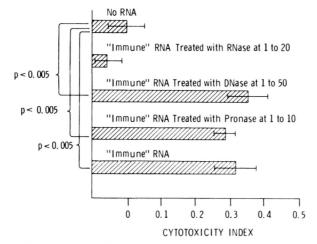

Fig. 3. *Effects of enzymes on the activity of anti-MC-1 immune RNA. RNAse was used at an enzyme to substrate ratio of 1 to 20. DNAse and pronase were used at enzyme to substrate ratios of 1 to 50 and 1 to 10 respectively based on the actual amount of DNA or protein in the I-RNA preparation.*

cells to become cytotoxic to the antigenically different BP-1 tumor cells and whether or not anti-BP-1 I-RNA, which was not active in the MC-1 system, could mediate immune responses against BP-1 target cells. In these experiments, it was shown that normal nonimmune C3H spleen cells preincubated with xenogeneic I-RNA from the lymphoid organs of guinea pigs immunized with BP-1 tumor cells became specifically cytotoxic for BP-1 target cells (Figure 4). However, C3H spleen cells preincubated with anti-MC-1 I-RNA were not cytotoxic for BP-1 target cells. Only RNA extracted from guinea pigs specifically immunized with BP-1 tumor cells mediated immune responses against BP-1 target cells. RNA extracted from guinea pigs immunized with normal C3H tissues (not shown) or with a different tumor (MC-1) did not mediate immune response against BP-1 target cells.

Specificity of Antitumor Immune Reactions Mediated by Syngeneic I-RNA

Using a somewhat different cell-mediated microcytotoxicity assay (in which target cells were prelabeled with ^{3}H-thymidine) in the MC3-R Fischer rat tumor host system, we found that syngeneic I-RNA extracted from the spleens of Fischer rats bearing growing MC3-R tumor transplants mediated immune cytolysis of MC3-R tumor target cells *in vitro* (6). The tumor used, the lymphoid

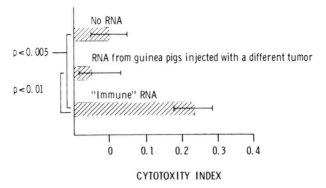

Fig. 4. *Immune cytolysis of BP-1 mouse tumor target cells mediated by xenogeneic immune RNA extracted from the lymphoid organs of guinea pigs immunized with BP-1 tumor cells. Control I-RNA was extracted from guinea pigs immunized with MC-1 tumor cells.*

tissues from which I-RNA was extracted, and the spleen cells incubated with the I-RNA were all obtained from members of a single inbred strain of rat. I-RNA was prepared either from spleens of Fischer rats bearing 1-2 cm tumors or from spleens of rats hyperimmunized to this tumor. Control I-RNA preparations were also extracted from the spleens of Fischer rats bearing a different chemically-induced syngeneic tumor, BP1-R [induced by benz(a)pyrene].

The results of these experiments indicated that normal nonimmune Fischer spleen cells preincubated with 500 μg of I-RNA extracted from the spleens of rats bearing growing MC3-R tumor transplants became significantly cytotoxic to MC3-R target cells (Figure 5). When normal spleen cells were incubated with I-RNA from the spleens of rats bearing growing BP1-R tumors, no significant cytotoxicity resulted. RNAse treatment abolished the activity of the I-RNA preparations, but DNAse or pronase treatment did not affect their activity (Figure 6). The concentration of RNA was apparently not critical within the limits tested (100-2500 μg) although increasing RNA concentrations resulted in moderate increases in cytotoxic responses (Figure 7).

Kinetics of Synthesis of I-RNA

We determined the optimum time following immunization of the I-RNA donor at which to extract antitumor I-RNA. The first system studied utilized syngeneic I-RNA prepared from the spleens of tumor bearing rats. Fischer rats were given a tumor cell innoculum of 1×10^6 viable MC3-R tumor cells.

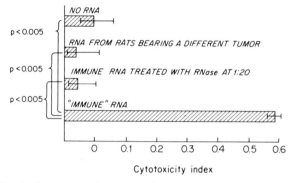

Fig. 5. *Immune cytolysis of MC3-R rat tumor cells mediated by syngeneic immune RNA extracted from the spleens of rats bearing growing MC3-R tumor transplants. Control RNAs included RNA from the spleens of rats bearing a different tumor (BP-1R) and active anti-MC-R Immune RNA which had been treated with RNAse at an enzyme to substrate ratio of 1 to 20.*

Groups of rats were then sacrificed 1, 7, 14, 21 and 28 days later. At the time of sacrifice, tumor size was noted and spleens frozen in dry ice for extraction of I-RNA. Normal rat spleen cells were incubated with these I-RNAs and tested for cell-mediated cytotoxicity against MC3-R target cells. Results of these kinetic studies are given in Figure 8. Immune RNA activity was low up

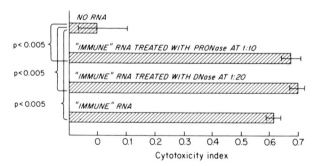

Fig. 6. *Effect of DNAse and pronase treatment on the activity of anti-MC3-R immune RNA. The I-RNA was treated with DNAse at an enzyme to substrate ratio of 1 to 20 and with pronase at an enzyme to substrate ratio of 1 to 10 based on the actual amount of DNA or protein in the I-RNA preparation.*

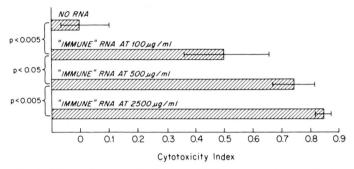

Fig. 7. *Effect of varying the immune RNA concentration on the immune cytolysis of MC3-R tumor cells.*

to seven days after tumor innoculation. However, between 21-28 days, the antitumor activity of the I-RNA was maximal and induced normal rat spleen cells to become cytotoxic to MC-R target cells with a cytotoxicity index approaching 0.50.

The kinetics of synthesis of xenogeneic I-RNA in immunized guinea pigs was then determined. Guinea pigs were immunized with MC-1 tumor cells in complete Freund's adjuvant. Groups of guinea pigs were sacrificed at 1,4,7,14 and 21 days following immunization. I-RNA was extracted from the lymphoid

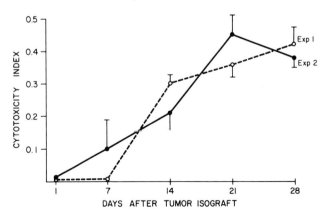

Fig. 8. *Kinetics of synthesis of syngeneic immune RNA by rats bearing growing transplants of MC3-R tumor.*

organs of each group of guinea pigs. Aliquots of nonimmune C3H spleen cells following incubation with each of these I-RNAs were tested for cytotoxic activity against MC-1 target cells. The results of these kinetic studies indicated that in as little as seven days I-RNA preparations were obtained that were active in converting normal nonimmune C3H spleen cells to antitumor immunoreactive status. Optimum activity was noted at 14-21 days. These data are illustrated in Figure 9.

Fractionation of Antitumor I-RNA

In order to identify the active component(s) of our I-RNA preparations, antitumor I-RNA was fractionated by sedimentation velocity through sucrose density gradients. Figure 10 is a typical sucrose density gradient profile of syngeneic antitumor I-RNA. The vertical dashed lines indicate the fractions of I-RNA which were collected. These fractions were then concentrated and purified by reprecipitating the RNA in three volumes of ethanol overnight at $-20°C$. The precipitate was washed, resuspended in RSB buffer and dialyzed against cold phosphate buffered saline, pH 7.4. The RNA fractions were assayed for antitumor cytotoxic activity against MC3-R target cells in the microcytotoxicity assay in which the MC3-R cells are prelabeled with 125 IUDR. Aliquots of spleen cells from normal nonimmune Fischer rats were incubated separately with 200 µg of each of the four fractions of antitumor I-RNA isolated from the total RNA extracted from the spleens of Fischer rats

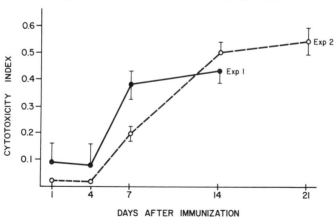

Fig. 9. *Kinetics of synthesis of xenogeneic immune RNA by guinea pigs immunized with MC-1 mouse tumor cells.*

bearing MC3-R tumors. I-RNA from the second fraction with a sedimentation value of 8-16S converted normal spleen cells to become cytotoxic to MC3-R

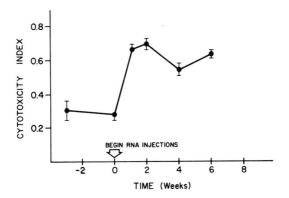

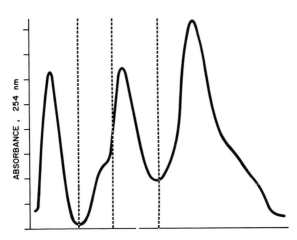

Fig. 10. *A sucrose density gradient profile of syngeneic immune RNA extracted from the spleens of rats bearing growing MC3-R tumor transplants. The low molecular weight (4S) RNA is to the left, and the higher molecular weight (28S) RNA is to the right. Vertical dashed lines indicate where the immune RNA was separated into four fractions.*

target cells with a cytotoxicity index 0.33 ± 0.06 (Figure 11). However, RNAs with sedimentation values of 4-8S, 16-20S or 20-35S, when incubated with normal spleen cells, effected no significant cytolysis of MC3-R target cells.

In another experiment, syngeneic I-RNA extracted from the spleens of rats bearing MC3-R tumors was fractionated into six components. Normal spleen cells were incubated separately with each of these fractions and tested for cytotoxicity to MC3-R target cells. As can be seen in Figure 12, the activity of antitumor I-RNA was predominately located in one region representing sedimentation values of 12-16S.

In similar experiments, xenogeneic I-RNA against the MC-1 sarcoma of C3H mice extracted from the lymphoid organs of immunized guinea pigs was fractionated. Fractions were incubated with normal C3H spleen cells and tested for cytotoxic activity against MC-1 target cells. The results are depicted in Figure 13. Here again, the activity was concentrated in the 12-16S fraction. Both the unfractionated I-RNA and the active fraction were electrophoresed on polyacrylamide gels. The resulting profiles are illustrated in Figure 14. It can be seen that the active fraction contains more than one molecular species of RNA.

Intracellular Localization of Antitumor I-RNA

Experiments were performed to determine the intracellular localization of immunologically active immune RNA. Syngeneic I-RNA was extracted from

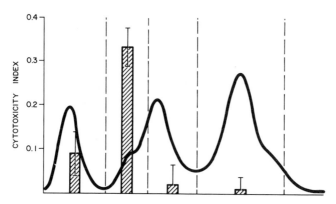

Fig. 11. *Localization of the active moiety of syngeneic immune RNA extracted from the spleens of rats bearing growing MC3-R tumors. Aliquots of normal rat spleen cells were incubated separately with each of the four fractions of anti-MC3-R immune RNA, and the cytotoxicity of these RNA-incubated spleen cells for MC3-R target cells was measured.*

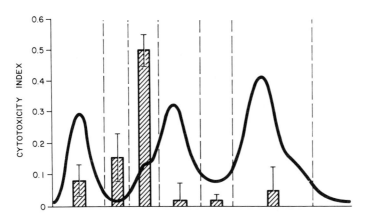

Fig. 12. *Localization of the active moiety of syngeneic immune RNA extracted from the spleens of rats bearing growing MC3-R tumors. Aliquots of normal rat spleen cells were incubated separately with each of the six fractions of anti-MC-3-R immune RNA, and the cytotoxicity of these spleen cells for MC3-R target cells was then measured.*

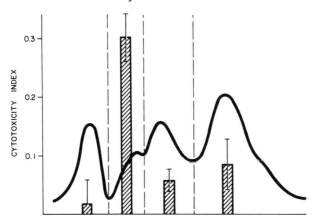

Fig. 13. *Localization of the active moiety of xenogeneic immune RNA extracted from the lymphoid organs of guinea pigs immunized with MC-1 mouse tumor cells. Aliquots of normal mouse spleen cells were incubated separately with each of the four fractions of anti-MC-1 immune RNA, and the cytotoxicity of these spleen cells for MC-1 target cells was then measured.*

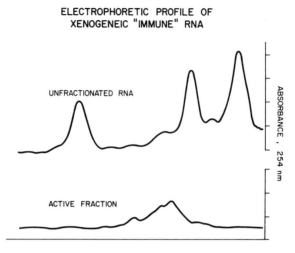

Fig. 14. *An electrophoretic profile of xenogeneic immune RNA extracted from the lymphoid organs of guinea pigs immunized with MC-1 mouse tumor cells. The low molecular weight RNA (higher mobility) is to the left of the top profile. The bottom profile is of the active fraction of I-RNA isolated by preparative sucrose density gradient centrifugation.*

intact spleens and separately from the nuclei and cytoplasm of spleen cells from rats bearing growing MC3-R tumor transplants. Also, xenogeneic I-RNA was extracted from the intact spleens and separately from the nuclei and cytoplasm of spleen cells from guinea pigs immunized with MC3-R tumor cells. Control xenogeneic I-RNA preparations were extracted from intact spleens and nuclei and cytoplasm of spleen cells of guinea pigs immunized with complete Freund's adjuvant. Control syngeneic I-RNA preparations were extracted from intact spleens and nuclei and cytoplasm of spleen cells from normal Fischer rats. Allogeneic spleen cells from normal nonimmune Wistar-Furth rats or syngeneic spleen cells from normal nonimmune Fischer rats were incubated separately with 500 μg each of these various I-RNA preparations and the cytotoxicity of these cells for MC3-R target cells was measured in the cell-mediated microcytotoxicity assay.

The cytotoxicity of the allogeneic effector cells to MC3-R tumor target cells following incubation with xenogeneic I-RNA fractions is depicted in Figure 15. Normal Wistar-Furth rat spleen cells, after incubation with xenogeneic I-RNA extracted from the cytoplasmic fraction of spleens of guinea pigs immunized with MC3-R tumor cells, became as cytotoxic to MC3-R target cells as spleen cells incubated with I-RNA from the intact spleens of immunized guinea pigs. Nuclear RNA from the spleens of guinea pigs immunized with MC3-R tumor cells was completely inactive. All control

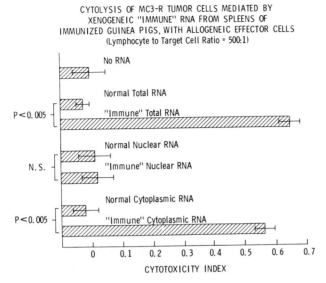

Fig. 15. *Immune cytolysis of MC3-R tumor target cells mediated by xenogeneic immune RNA extracted from intact spleen cells and from nuclear and cytoplasmic fractions of spleen cells from guinea pigs immunized with MC3-R tumor cells. The effector cells were spleen cells from normal allogeneic Wistar-Furth rats.*

RNAs from guinea pigs non-specifically immunized with complete Freund's adjuvant were also inactive.

Previous work from our laboratory had shown that when xenogeneic I-RNA is incubated with normal spleen cells that are syngeneic with respect to the immunizing tumor, the resultant cytotoxic immune responses (as determined on target cells of the same tumor) are directed specifically against tumor-associated antigens of that tumor (5). Therefore, the same RNAs used in the previous experiment were next incubated with normal nonimmune syngeneic Fischer rat spleen cells, and the resultant cytotoxicity for MC3-R target cells measured. I-RNA extracted either from whole spleen cells or from the cytoplasmic fraction of spleen cells from guinea pigs immunized with MC3-R tumor cells converted normal Fischer rat spleen cells to become cytotoxic to MC3-R target cells to about the same degree (Figure 16). RNA from the nuclear fraction was inactive. RNA extracted from intact spleens or the cytoplasmic or nuclear fraction of spleens from non-specifically immunized guinea pigs were also inactive.

In a totally syngeneic system, syngeneic I-RNA prepared from the cytoplasm of spleens of rats bearing MC3-R tumors was as active as RNA from whole intact spleens in converting normal syngeneic Fischer rat spleen cells to immunoreactive status against MC3-R target cells (Figure 17). In this

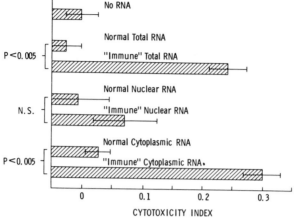

Fig. 16. *Immune cytolysis of MC3-R tumor target cells mediated by xenogeneic immune RNA extracted from intact spleen cells and from nuclear and cytoplasmic fractions of spleen cells from guinea pigs immunized with MC3-R tumor cells. The effector cells were spleen cells from normal syngeneic Fischer rats.*

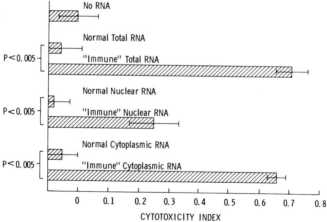

Fig. 17. *Immune cytolysis of MC3-R tumor target cells mediated by syngeneic immune RNA extracted from intact spleen cells and from nuclear and cytoplasmic fractions of spleen cells from Fischer rats bearing growing MC3-R tumor transplants. The effector cells were spleen cells from normal syngeneic Fischer rats.*

syngeneic system, these immune responses may be said to be directed against the tumor-associated antigens of the MC3-R target cells. RNA isolated from the nuclear fraction of spleen cells from Fischer rats bearing MC3-R tumors was considerably less active than either RNA from intact cells or the cytoplasmic fraction. The nuclei from which this RNA preparation was extracted evidenced some small amount of cytoplasmic contamination as indicated by electron micrography. However, the immunologic activity was confined mainly to the cytoplasmic RNA fraction. The partial activity of the nuclear RNA fraction may have been due to the noticeable amount of cytoplasmic contamination of the nuclei.

DISCUSSION

Specificity of Antitumor Immune Reactions Mediated by Xenogeneic I-RNA

These studies demonstrated that normal nonimmune murine spleen cells could be converted to effector cells ("killer cells") specifically cytotoxic to syngeneic murine tumor cells by incubation *in vitro* with xenogeneic I-RNA extracted from the lymphoid organs of specifically sensitized guinea pigs. This response was shown to be specific for the particular tumor used to immunize the RNA donor animals. Normal mouse spleen cells incubated with I-RNA extracted from the lymphoid organs of guinea pigs immunized with MC-1 tumor cells were cytotoxic only to MC-1 target cells and not to target cells of the syngeneic but antigenically different tumor, BP-1. Spleen cells incubated with I-RNA from guinea pigs immunized with BP-1 tumor cells were cytotoxic to BP-1 target cells but not to MC-1 target cells.

The guinea pigs immunized with mouse tumor cells were undoubtedly sensitized to a broad spectrum of murine transplantation antigens as well as to the tumor-associated antigens present on the tumor cells. Yet, immunization with tumor cells appeared to be necessary for I-RNA to mediate an antitumor immune response and these responses were specific for the particular tumor used to immunize the RNA donor. RNA extracted from the lymphoid organs of guinea pigs that had been immunized with normal C3H mouse tissues failed to induce C3H spleen cells to mediate an immune response against C3H tumor cells. Certainly, our anti-MC-1 I-RNA and anti-BP-1 I-RNA preparations must have contained I-RNAs directed against C3H transplantation antigens as well as to MC-1 or BP-1 tumor-associated antigens. Why, then, were the immune responses tumor-specific? These I-RNA preparations were incubated with lymphoid cells which were syngeneic with respect to the tumor target cells. We postulate that, at least within our system, recognition of self occurred at the level of interaction between the I-RNA and the lymphoid cells. I-RNAs to C3H transplantation antigens were recognized as self by the C3H lymphoid cells, and, consequently, no immune responses to C3H histocompatibility antigens were initiated. However, I-RNAs against tumor-associated antigens were not recognized as self by C3H spleen cells and specific antitumor immune responses to these antigens were elicited.

Specificity of Antitumor Immune Reactions Mediated by Syngeneic I-RNA

Within a totally syngeneic system, it was shown that normal nonimmune rat spleen cells became specifically cytotoxic for MC3-R tumor target cells following incubation *in vitro* with syngeneic I-RNA extracted from spleens of either tumor bearing or hyperimmunized rats. The degree of cytotoxicity obtained following incubation with I-RNA was found to be comparable to the cytotoxicity of spleen cells from rats hyperimmunized with the MC3-R tumor. Since the MC3-R and BP1-R tumors, the lymphoid tissues from which the I-RNA was extracted, and the spleen cells incubated with the I-RNA were all obtained from members of a single inbred strain, the immune responses elicited upon incubation of syngeneic spleen cells with syngeneic I-RNA could be directed only against the tumor-associated antigens of the MC3-R tumor and were presumedly mediated by an I-RNA specific for the tumor-associated antigens of the MC3-R sarcoma. This response was not due to nonspecific effects of rat lymphoid RNA, since spleen cells incubated with I-RNA from the spleens of rats bearing growing BP1-R tumors were not cytotoxic for MC3-R tumor cells. Again, these immune responses were abrogated when I-RNA was treated with RNAse before incubation with spleen cells; but treatment with DNAse or pronase did not affect the activity.

Our I-RNA preparations contained small but significant amounts of protein, part of which might consist of tumor and/or transplantation antigens. However, treatment of the active I-RNA preparations with pronase did not remove their activity, suggesting that the sensitization of the spleen cells was not affected by simple passive transfer of protein antigens. This possibility cannot be ruled out, however, since protein antigens conjugated with RNA may be more resistant to pronase degradation than proteins unassociated with RNA. However, the time period between incubation of spleen cells with I-RNA and completion of the cytotoxic reaction was probably too brief to allow for the *in vitro* sensitization of nonimmune lymphoid cells by tumor antigens which might be present as contaminants within the I-RNA preparations. The finding that treatment of the active I-RNA preparations with ribonuclease inactivated them, whereas treatment with pronase or deoxyribonuclease did not, indicates that one or more species of RNA is the active moiety in our I-RNA preparations.

Kinetics of Synthesis of I-RNA

The activity of syngeneic I-RNA obtained at periodic intervals following innoculation of rats with MC3-R tumor cells was determined. No appreciable cytotoxic activity was apparent up to 7 days. However, significant cytotoxic activity was mediated by I-RNA extracted from spleens of tumor bearing rats 14 days after tumor isograft. Maximum activity was reached at 21 days, and the activity remained high at day 28.

By contrast, xenogeneic I-RNA extracted from the lymphoid organs of immunized guinea pigs reached maximum activity approximately one week earlier. Significant cytotoxic immune reactivity was evidenced within 7 days after immunization of the guinea pigs with tumor cells. Maximum immune reactivity of I-RNA was reached by day 14 rather than 21. While this may have been due to differences in the immune responses of the two different species of I-RNA donor animals, it is more likely a function of the different immunization methods used. The rats received 10^6 syngeneic tumor cells without adjuvant, and approximately 7 days was required for tumor growth before palpable tumors could be detected. It is likely that there was a significant latency period between the time of tumor innoculum and the presence of sufficient tumor antigen in the lymphoid system to result in significant synthesis of I-RNA. The guinea pigs, on the other hand, were immunized with much larger doses of xenogeneic tumor cells (approximately 10 times greater than that given the rats). Guinea pigs also received Freund's adjuvant at the time of immunization. The fact that the guinea pigs received xenogeneic cells and a larger tumor innoculum in the presence of adjuvant probably accounted for the more rapid synthesis of significant quantities of I-RNA.

Fractionation of Antitumor I-RNA

Antitumor I-RNA preparations were fractionated by sedimentation velocity through sucrose density gradients, and the active component was found principally in the 12-16S region. No significant antitumor immunoreactivity was found in fractions below 8S or above 16S. A quantitative determination of the relative amounts of I-RNA in each fraction indicated that the active component comprised 5 to 7% of the total RNA extracted from the lymphoid tissues. Thus, in our system, active I-RNA was present only in small quantities relative to the total RNA extracted from the lymphoid cells of immunized animals. The techniques used to separate the RNA fractions were taxed to the limit. By sucrose gradient centrifugation, we were able to separate the RNA into four major components. However, more sophisticated techniques will have to be used for further resolution of the active I-RNA fraction. It is possible that active antitumor I-RNA represented only a small portion of the RNA contained in the 12-16S fractions.

Sedimentation values of 12-16S are compatible with molecular weights between 100,000 and 200,000. This is approximately in the range of most mammalian messenger RNA species. This, plus the fact that our I-RNA preparations were inactivated by ribonuclease but not by deoxyribonuclease or pronase, suggests that the immune responses observed may have been mediated by informational RNA molecules. We are presently attempting to further characterize these I-RNA preparations to determine if antitumor immunologic reactivity can be shown to reside in RNA moieties with characteristics of "messenger-type" RNA molecules.

It is not clear whether or not the cytotoxic immune reactions observed in our studies were due to direct lymphocyte/tumor-cell interactions (T-cell killing). It is possible that antibodies may have been elaborated from the spleen cells after incubation with I-RNA. Such antibodies may have played a role in mediating cytotoxic immune reactions. I-RNA no doubt is involved in a number of humoral as well as cellular immune responses. We have measured only one type of immune reaction mediated by I-RNA. While we have isolated the active fraction of I-RNA responsible for cell-mediated cytotoxic antitumor reactions, it is possible that other types of I-RNA are present in sensitized lymphoid cells which mediate other types of immune reactions (e.g., antibody production). These may be detectable by techniques other than the cell-mediated microcytotoxicity assay we employed in our studies.

Intracellular Localization of Antitumor I-RNA

We obtained RNA rich extracts from homogenates of whole intact spleen cells, from nuclei alone, and from nuclei-free cytoplasmic fractions. The immunologically active components were found to be confined mainly to RNA from cytoplasmic fractions. In only one experiment was there some activity in a nuclear fraction. This may have been due to a small but noticeable amount of cytoplasmic contamination of the nuclei as detected by electron microscopy. Our cytoplasmic fractions were entirely free of intact nuclei, but we cannot entirely rule out the possibility that some nuclear material from nuclei which may have been disrupted during homogenization of the whole cells may have contaminated our cytoplasmic fractions. The cytoplasmic fractions of both syngeneic and xenogeneic I-RNA were as active on a weight per weight basis as their respective total RNA counterparts.

It is possible that the appearance of I-RNA in the nucleus is an early event which was not detected in our experiments. We may have observed the later appearance of active I-RNA in the cytoplasm following translocation from the nucleus. However, in our experiments in which I-RNA was extracted from lymphoid tissues two weeks or longer following the immunization, the immunologically active fraction of syngeneic and xenogeneic antitumor I-RNA appeared to be localized in the cytoplasm of sensitized lymphoid cells.

SUMMARY

Using a quantitative microcytotoxicity assay, we have shown that normal nonimmune C3H mouse spleen cells were converted to effector cells specifically cytotoxic to chemically induced syngeneic C3H tumor cells by incubation with xenogeneic I-RNA extracted from the lymphoid organs of specifically immunized guinea pigs. This response was specific for the tumor used to immunize the I-RNA donor. In a totally syngeneic system, we showed that syngeneic I-RNA extracted from the spleens of tumor bearing rats mediated cytotoxic immune reactions which were directed specifically against the tumor-associated antigens of syngeneic rat tumor target cells. Active antitumor I-RNA synthesis in the lymphoid organs of I-RNA donor animals reached a maximum between 14 and 21 days, depending on the route of administration and the nature of the immunizing tumor. Active I-RNA preparations were insensitive to treatment with deoxyribonuclease or pronase, but were inactivated by ribonuclease treatment, thereby indicating that the active moiety was one or more species of RNA. The active fractions of the I-RNA preparations had sedimentation values in sucrose density gradients of between 12-16S, and comprised only a small fraction of the total RNA present in the lymphoid cells. Active antitumor I-RNA appeared to be localized in the cytoplasm of sensitized lymphoid cells, rather than in the nucleus. Our data are consistent with but do not prove the hypothesis that I-RNA is an information-containing ribonucleic acid molecule capable of mediating immune reactions *in vitro* which are specific for the tumor associated antigens of the tumor used to immunize the I-RNA donor.

REFERENCES

1. Ramming, K. P. and Y. H. Pilch (1971). Transfer of tumor specific immunity with RNA: Inhibition of growth of murine tumor isografts. *J. Nat. Cancer Inst.* 46:735-750.
2. Deckers, P. J. and Y. H. Pilch (1971). Transfer of immunity to tumor isografts by the systemic administration of xenogeneic "immune" RNA. *Nature (New Biology)* 231:181-183.
3. Deckers, P. J. and Y. H. Pilch (1972). Mediation of immunity to tumor specific transplantation antigens by RNA inhibition of isograft growth in rats. *Cancer Res.* 32:839-846.
4. Ramming, K. P. and Y. H. Pilch (1970). Transfer of tumor specific immunity with RNA: Demonstration by immune cytolysis of tumor cells *in vitro*. *J. Nat. Cancer Res.* 45:543-553.
5. Kern, D. H. and Y. H. Pilch (1974). Immune Cytolysis of Murine Tumor Cells mediated by xenogeneic "immune" RNA. *Internat. J. Cancer.* 13:679-688.
6. Kern, D. H., C. R. Drogemuller and Y. H. Pilch (1974). Immune cytolysis of rat tumor cells mediated by syngeneic "immune" RNA. *J. Nat. Cancer Inst.* 52:299-302.
7. Veltman, L. L., D. H. Kern and Y. H. Pilch (1974). Immune cytolysis of human tumor cells mediated by xenogeneic "immune" RNA. *Cellular Immunol.* 13:367-377.
8. Pilch, Y. H., L. L. Veltman and D. H. Kern (1974). Immune cytolysis of human tumor cells mediated by xenogeneic "immune" RNA: Implications for immunotherapy. *Surgery* 76:23-34.
9. Dische, Z. (1955). *The Nucleic Acids.* E. Chargaff and J. N. Davidson, Eds. Academic Press, New York.

10. Lowry, O. H., N. J. Rosebrough, A. L. Farr, and R. J. Randall (1951). Protein measurement with the folin phenol reagent. *J. Biol. Chem. 193*:265-268.

AUGMENTATION OF THE EFFERENT ARC OF A TUMOR SPECIFIC IMMUNE RESPONSE[1]

Peter J. Deckers, M.D., Bosco S. Wang, M.S.
and John A. Mannick, M.D.

Laboratory of Tumor Immunology
Department of Surgery
Boston University School of Medicine
80 East Concord Street
Boston, Massachusetts 02118

It has been conclusively demonstrated that the lymphoid cells of the spleen, regional lymph nodes, thoracic duct lymph and peritoneal cavity of animals immunized to tumor-specific antigens can adoptively transfer tumor-specific immunity to inbred normal animals (1-5). When these treated animals are subsequently inoculated with the specific syngeneic tumor, inhibition of tumor development occurs. However, the adoptive transfer of tumor immunity by allogeneic or xenogeneic leukocytes has had minimal success in man and in experimental animals (6-12). Immune elimination soon inactivates the transferred cells and attempts at immunosuppression of the host to protect these cells may result in lethal graft versus host disease and have been uniformly unsuccessful (13-15).

Failure of allogeneic immune lymphocytes to prevent tumor development has led to an interest in the potential role of subcellular components including the nucleic acids of these lymphocytes in transferring immune responses. The model for the immunotherapy of malignant disease with immune RNA was first suggested by Mannick and Egdahl (16). They demonstrated that allograft immunity could be transferred to previously untreated rabbits by the administration of autologous lymphoid cells which had been preincubated with RNA extracted from the lymphoid organs of specifically immunized donor rabbits. Rabbits immunized by the adoptive transfer of RNA-incubated lymphocytes rejected specific skin grafts in an accelerated fashion while indifferent third strain skin allografts were rejected in the normal first set fashion (17). Ramming and Pilch adopted this model to the study of tumor immunity and

---

[1] Supported by NIH Grants CA 15848 and AM 10824, NIH Contract AO 143949, and by NIH GRS RR 05487 General Research Support Grant.

demonstrated that immunity to murine tumor isografts could be transferred to previously untreated mice by the intraperitoneal administration of syngeneic spleen cells preincubated with RNA extracted from the lymph nodes and spleens of guinea pigs immunized with the specific mouse tumor being treated. Even though the guinea pigs were immunized to xenogeneic antigens as well as tumor antigens, the response was – indirectly at least – shown to be tumor-specific (18,19). In all of their experiments, test animals received a carefully quantitated number of viable tumor cells synchronously with the intraperitoneal injection of RNA-incubated spleen cells; and, therefore, transplantation resistance rather than immunotherapy was evaluated in each instance.

Deckers and Pilch then obtained direct evidence that immune RNA could mediate a specific immune response to tumor antigens alone by utilizing a system in which the tumor-cell donor, the RNA donor, the spleen cell donor and the recipient animal were all members of a single inbred strain (20). This eliminated the possibility that the observed tumor isograft rejection might, in part, be due to an immune response directed against histocompatibility antigens, and established conclusively the ability of spleen cells incubated with immune RNA to mediate destruction of tumor cells *in vivo*.

It had, therefore, been established that normal syngeneic spleen cells incubated with xenogeneic or syngeneic RNA extracted from the lymphoid organs of animals immunized with the tumor to be treated could be converted to a tumor-specific immunologic reactivity. Deckers and Pilch have argued that immune RNA may be an ideal form of immunotherapy for animal and human neoplasms by suggesting that the immunoreactivity of lymphocytes from patients with tumor may be augmented by incubation with RNA extracted from the lymphoid cells of animals immunized to these antigens and, thereby, might mediate a specific cytolytic effect on human tumor cells *in vivo* (21). However, no conclusive evidence of the regressions of established murine or human tumors by treatment with immune RNA has been reported. It remained to demonstrate that immune RNA could be effective treatment of established viable growing tumor isografts which have been inoculated several days prior to the initiation of immunotherapy or which have become palpable tumors. In addition, the practical application of the immunotherapy model of Deckers and Pilch was delayed by a failure to develop a quantitative assay to document the *in vitro* effectiveness of immune RNA prior to its *in vivo* application.

Likewise, in all previous *in vivo* and *in vitro* studies, normal lymphoid cells were converted to an immunoreactive state by incubation with immune RNA preparations. Animals with growing tumors, however, ordinarily demonstrate a detectable immunologic reaction against the tumor-specific antigens of these tumors (22). It remained to demonstrate that the immunoreactivity of lymphoid cells from animals that have been immunized by progressive growth of a tumor isograft could be augmented by incubation of these already committed lymphocytes with RNA extracted from the lymphoid organs of other animals immunized with the particular tumor being treated.

Experiments were therefore designed to:

1. develop and test reliable *in vitro* quantitative assays of the transfer of tumor-specific immunity in mice with RNA.

2. determine whether RNA extracted from the lymphoid organs of animals immunized with murine tumors could augment the response of lymphocytes from mice immunized to the tumor-specific antigens of these syngeneic tumors.

3. develop a model for the immunotherapy of rodent tumors with immune RNA.

4. determine whether xenogeneic RNA can augment the cytotoxicity of lymphocytes from patients with tumors *in vitro* and, thereby, establish a rational basis for the immunotherapy of cancer in man.

MATERIALS AND METHODS

In the first part of these studies, two antigenically-distinct murine tumors, designated S-17 and S-18, were induced by the injection of 0.1 ml of a 1% solution of 20-methylcholanthrene in sesame oil into adult female C57BL/6 mice. The immunogenicity and antigenic distinctiveness of these two tumors was demonstrated by the transplantation resistance techniques of Prehn and Main. Animals immunized by amputation of growing S-17 isografts rejected a second inoculation with S-17 tumor cells, but inoculation of S-18 cells produced viable growing tumors in each instance. Similarly, animals amputated of S-18 tumors rejected a second S-18 inoculum whereas an S-17 isograft grew progressively.

Hartley guinea pigs were immunized by an injection of a mixture of tumor cells and Freund's complete adjuvant as previously described (19). The animals were sacrificed 10-14 days later and immune RNA was extracted from their lymphoid organs by a modification of the procedure of Scherrer and Darnell (24). All RNA preparations were analyzed by determining their sedimentation characteristics in sucrose density gradients. RNA preparations usually have three peaks corresponding to the sedimentation characteristics of 4-8S, 16-18S and 23-28S, respectively. If the slower sedimenting 4-8S peak was higher than the 23-28S peak, the RNA was considered degraded and was not used in this study.

Immune lymphocytes were obtained as previously reported (22) from C57BL/6 mice that had been immunized by amputation of growing S-17 or S-18 isografts. These lymphocytes were incubated in immune RNA (1.0-1.5 mg/10^7 cells/ml) in Hanks' balanced salt solution (HBSS) at 37°C for 30 minutes. As controls, lymphocytes were also incubated with (a) HBSS without

RNA; (b) immune RNA that had been degraded by RNase (15 μg/ml RNA); and (c) RNA extracted from the lymphoid tissues of nonimmunized guinea pigs.

After RNA incubation, lymphocytes were washed twice with HBSS and then suspended in medium RPMI 1640 containing 10% heat-inactivated human serum with 1% L-glutamine, 100 units/ml of penicillin, and 100 μg/ml of streptomycin. The lymphocyte concentration was adjusted to 4 $\times$ 10^6 viable cells/ml of culture. Single tumor-cell suspensions were prepared by a modification of the method of Hammond *et al* (25). Fifty μg of mitomycin C was then added for every 10^7 viable tumor cells, and these cells were incubated at 37°C for 60 min. These cells were then washed twice with HBSS and resuspended in complete medium as described above. Various concentrations of mitomycin C-treated tumor cells (0-2 $\times$ 10^5/ml) were then added to the lymphocytes, and these mixed lymphocyte-tumor cell (MLTC) cultures were incubated at 37°C in a 5% CO_2 water-saturated atmosphere. Tests were always performed in triplicate. Culture tubes were maintained in stationary vertical position for the first 24 hrs and then were put on a roller for another 24 hrs. Two μCi of ^{3}H-thymidine (specific activity = 20 Ci/mmole) was then added, and incubation was continued for an additional 16-18 hr. The cells were then washed twice with 2 ml of 0.85% NaCl and disrupted by 0.01N NaOH. Each sample was added to 5 ml of cocktail D, the mixture of 100 g of naphthalene, 5 g of PPO, and 1 liter of 1,4-dioxane, and counted in a liquid scintillation counter.

The means and standard deviation of the counts per minute (cpm) were obtained from each triplicate cell culture. Students' t tests were calculated to determine P values between experimental and control groups. To eliminate possible mistakes that might be made by inaccurate cell counting in the different groups, the percentage of increase in cpm (stimulations index, S.I.) was also calculated by the following formula:

$$\text{S.I. (\%)} = \frac{\text{cpm of lymphocytes with tumor cells} - \text{cpm of lymphocytes without tumor cells}}{\text{cpm of lymphocytes without tumor cells}} \times 100$$

For other experiments, inbred male and female C3H/HeN mice were used. The tumors employed were the antigenically distinct sarcomas, BP-8 and BP-9, and were induced in adult female C3H/HeN mice by the subcutaneous injection of 0.1 ml of a 1% solution of 3,4 benz(a)pyrene in sesame oil. The BP-8 and BP-9 sarcomas were immunogeneic within their murine strain of origin and were maintained by serial transplantation in adult female C3H/HeN mice. They were used in their third through tenth transplant generations.

Single tumor cell suspensions were prepared. Again, Hartley guinea pigs were immunized to either one of these two tumors. Control guinea pigs received similar injections of mouse normal tissues (a mixture of lung, liver, kidney and spleen cells) in Freund's complete adjuvant. After ten to fourteen days, the spleens and axillary, popliteal and inguinal lymph nodes from all groups of guinea pigs were excised, frozen in dry ice and the RNA extracted.

For these *in vitro* studies, C3H/HeN mice were immunized by progressive tumor growth. We have previously demonstrated that mice with growing tumors develop a tumor-specific immunologic reaction against these tumors and that tumor-specific transplantation immunity can be transferred to previously normal untreated mice by intraperitoneal inoculation of lymphoid cells from tumor bearing mice (22). The adoptive transfer of tumor immunity in this fashion is fully developed when lymphoid cells are removed from tumor bearing mice 14 days after inoculation of the viable tumor isografts and is lost at some indefinite period when the tumors of the lymphoid cell donor become excessive in size. This phenomenon is referred to as concomitant tumor immunity, and demonstrates that animals with growing tumors are immunized to the tumor-specific transplantation antigens of these tumors and that this immunity resides in the lymphoid cells of these animals.

Spleens were excised from mice immunized by progressive growth of a tumor isograft. In addition, spleens were also removed from normal C3H/HeN mice. These spleens were individually minced with scissors in cold HBSS and single cell suspensions prepared. For incubation with RNA, the cell button was resuspended in the RNA solution at a concentration of 10^7 to 10^8 cells per ml. RNA concentrations were always 500-1000 μg/ml. The mixtures of RNA and spleen cells were then incubated at 37°C for 20 minutes in a gyratory shaking water bath, washed once with cold HBSS, counted and suspended in RPMI 1640 with 20% fetal calf serum at a concentration of 4×10^7 viable lymphoid cells/ml. Viability was determined by trypan blue exclusion. In addition, for the *in vitro* studies, purification of the lymphoid cell populations was achieved by the Ficoll-Hypaque technique of Perper (26) prior to incubation with RNA. An examination of the recovered mononuclear cell layer after centrifugation revealed a population of greater than 95% lymphocytes with a viability of greater than 98% as judged by trypan blue exclusion. These purified lymphocytes were used in the *in vitro* cytotoxicity assay to be described. Prior to incubation with immune RNA they were washed at least three times with tissue culture medium.

A modification of the cell-mediated cytotoxicity assay using ^{125}I-iododeoxyuridine (^{125}I-UdR)-labeled target cells, as originally described by Cohen (27), was used for these studies. Single cell BP-8 tumor cell suspensions were prepared and approximately 2,000 BP-8 cells in 0.2ml of RPMI 1640 containing 1% L-glutamine, 100 μ/ml of penicillin, 100 μg/ml of streptomycin and 10^{-6} M fluorodeoxyuridine/ml were then added to each well of a Microtest II (Falcon Plastics) culture plate. One μCi of ^{125}I-UdR was also added to each well and the cultures were incubated at 37°C in a 10% CO_2 water-saturated atmosphere for 24 hours. Excess ^{125}I-UdR was then aspirated and each well was washed once with RPMI 1640. Lymphocytes in 0.2 ml of RPMI 1640 were then added. In some experiments, these lymphocytes were normal previously untreated lymphocytes removed from mice never immunized to the BP-8 tumor or these were normal lymphocytes which had been incubated with various RNA preparations. In still other experiments, these were immune lymphocytes removed from animals immunized by progressive growth of a tumor isograft or were similar immune lymphocytes which had been preincubated with RNA extracted from the lymphoid organs of xenogeneic animals immunized to the BP-8 tumor.

After 48 hours of incubation, lymphocytes and killed target cells were removed by washing and the remaining radioactive adherent target cells were sprayed with aeroplast. Each well was then cut with a band saw and radioactivity was measured in a gamma counter. Five to eight replicate wells were tested in each group and a mean cpm together with the standard deviation of each mean was calculated for each group.

For the immunotherapy studies *in vivo*, Hartley guinea pigs were again immunized with either the BP-8 or BP-9 tumor, with Freund's adjuvant alone or with normal mouse tissues and used as donors of immune RNA. In some experiments, BP-8 or BP-9 immune RNA was degraded with RNase prior to incubation with normal mouse spleen cells. Four groups of C3H/HeN mice were inoculated on Day 1 with 10^3 viable BP-8 tumor cells. Five days later these groups of mice were treated respectively with normal syngeneic spleen cells which had been incubated with either BP-8 immune RNA, Freund's adjuvant RNA, or with HBSS alone. Each mouse received 10^7 to 10^8 lymphocytes, treated as described above, intraperitoneally five, seven, nine, and eleven days after the initial tumor isograft. Mice in all groups were then observed for tumor development. Two times each week, tumor size for all the mice in each group was calculated and a mean tumor volume for all the animals so studied was then calculated by the formula of Attia and Weiss (28) where volume = 0.4 ab^2. In a second similar experiment, four groups of syngeneic C3H/HeN mice were inoculated on Day 1 with 10^3 BP-8 tumor cells. Seven days later these animals were treated by the intraperitoneal inoculation of 10^7 to 10^8 viable syngeneic lymphocytes which had been preincubated with either BP-8 immune RNA, BP-8 immune RNA previously degraded with RNase (15 μg/ml of RNA), or BP-9 immune RNA. Again all animals were followed for tumor development. The size of each tumor was calculated as described previously.

In still other immunotherapy trials, tumor bearing mice received treatment with RNA-incubated spleen cells 5,7,9,11 and 17,19,21 and 23 days after initial tumor inoculation to determine whether a second treatment course would further inhibit tumor development. In other studies, mediation of immune cytolysis of human tumor cells *in vitro* by xenogeneic immune RNA was shown utilizing the same ^{125}I-UdR cytotoxicity assay.

Tumor removed surgically from patients with disseminated malignant melanoma and aliquots mixed with complete Freund's adjuvant were used to immunize Hartley guinea pigs. Ten days later these pigs were sacrificed and immune RNA extracted from their lymphoid organs. At this point, further tumor was excised from these patients and monolayers of these tumor cells were established in each well of a Microtest II culture plate as described above. These cells were labeled with one μCi of ^{125}I-UdR and incubated at 37°C in a 10% CO_2 water-saturated atmosphere for 24 hours. Excess ^{125}I-UdR was then aspirated and each well was washed once with RPMI 1640. Lymphocytes from these patients purified by centrifugation in Ficoll-Hypaque gradients in 0.2 ml of RPMI 1640 were then added. In some experiments, these lymphocytes were putatively normal lymphocytes from patients never known to have had, or been exposed to, malignant melanoma. The lymphocytes from patients with melanoma and from normal donors were added to the wells either without or after incubation with xenogeneic immune RNA extracted from the lymphoid organs of guinea pigs immunized with the melanoma tumor as described above. In some experiments, the immune RNA was degraded with RNase (15 μg/ml at 37°C for 30 min) prior to incubation with the patients or normal lymphocytes. Again, after 48 hours, lymphocytes and killed target cells were removed by washing and the radioactivity remaining was determined as an expression of adherent, presumably still viable, tumor cells.

RESULTS

As shown in Table 1, lymphocytes from mice immunized by amputation of growing S-17 isografts proliferated significantly *in vitro* when co-cultured with syngeneic mitomycin C-treated S-17 cells. After incubation of these lymphocytes with xenogeneic S-17 immune RNA, they responded, in the presence of S-17 cells, with a considerably augmented incorporation of ^{3}H-thymidine. RNase treatment abrogated the response, and incubation of these lymphocytes with normal RNA had no significant effects.

Partial specificity for the immunizing tumor is demonstrated in Table 2. In this experiment two groups of guinea pigs were immunized independently with S-17 and S-18 tumor cells. Lymphoid RNAs were then extracted from these two groups and designated as S-17 RNA and S-18 RNA respectively. Lymphocytes from S-17 immune mice were then incubated with each RNA and cultured separately with mitomycin C-treated S-17 cells. Again, lymphocytes from mice immunized by amputation of S-17 isografts proliferated in the presence of S-17 cells. This proliferation was significantly increased when these lymphocytes were preincubated with S-17 immune RNA but not with S-18 immune RNA.

TABLE 1.
Results

	Concentration of tumor cells (S-17)		
	0	2 × 10³	2 × 10⁴
Lymphocytes*	2352.0 ± 518.3[+++]	2591.5 ± 601.7	4338.3 ± 326.8**
S.I. (%)	0	10.2	84.4
Lymphocytes* S-17 RNA	1972.6 ± 484.5	5227.3 ± 593.8**	6294.0 ± 113.1***
S.I. (%)	0	165.0	219.1
Lymphocytes* + S-17 RNA-RNase[+]	2885.0 ± 340.8	3654.0 ± 608.1	3965.5 ± 524.0
S.I. (%)	0	26.7	37.5
Lymphocytes* + N-RNA[++]	2812.0 ± 490.0	3352.0 ± 157.7	4380.3 ± 126.8
S.I. (%)	0	19.2	55.8
No lymphocytes	95.5 ± 38.8	100.6 ± 41.6	173.0 ± 73.6

*Lymphocytes (4 × 10⁶/ml/culture) from C57BL/6 mice immunized by amputation of growing S-17 tumors

[+]Immune RNA preincubated with ribonuclease (15 μg/ml)

[++]RNA extracted from normal guinea pigs

[+++]Mean cpm ± SD

**P < 0.01, compared with the control (lymphocytes without tumor cells) of same group

***P < 0.001, compared with the control of same group

TABLE 2.

		Concentration of tumor cells		
	0	S-17 2 × 10³	S-17 2 × 10⁴	S-17 2 × 10⁵
Lymphocytes*	1367.6 ± 110.5**	1337.0 ± 654.8	2165.5 ± 150.6	2160.0 ± 390.3
S.I. (%)	0	−2.2	58.3	57.9
Lymphocytes* + S-17 RNA	1679.0 ± 165.0	2836.6 ± 962.8	2924.6 ± 789.2	4286.0 ± 4.2***
S.I. (%)	0	68.9	74.2	155.3
Lympohcytes* + S-18 RNA	1931.0 ± 385.5	2113.0 ± 629.9	1450.6 ± 323.0	2063.0 ± 193.7
S.I. (%)	0	9.5	−24.9	6.8
No lymphocytes	116.6 ± 41.7	161.6 ± 52.2	312.6 ± 114.7	758.5 ± 86.9

*Lymphocytes (4 × 10⁶/ml/culture) from S-17 immune C57BL/6 mice

**Mean cpm ± SD

***P < 0.001, compared with the control (lymphocytes without tumor cells) of the same group

The MLTC assay had, therefore, been useful in demonstrating that immune RNA could increase the immunoreactivity of lymphocytes from animals immunized by amputation of growing tumor isografts, thereby confirming our belief that immune RNA could augment the efferent arc of tumor specific immune response. In our hands, however, the MLTC was not consistently adaptable to several different tumor-host systems or to man so the ^{125}I-UdR assay was utilized to determine if the immunoreactivity of lymphocytes from animals and patients with growing tumors could be augmented with immune RNA.

Our initial *in vitro* experiments were designed to test whether normal nonimmune syngeneic C3H/HeN mouse spleen cells when preincubated *in vitro* with xenogeneic immune RNA extracted from the lymphoid organs of guinea pigs immunized with BP-8 tumor cells could be converted into cells specifically cytotoxic to BP-8 tumor cells *in vitro*. The results of these experiments are given in Figure 1. At a lymphoid cell to target cell ratio of 150 to 1, normal C3H/HeN spleen cells preincubated with BP-8 immune RNA caused cytolysis of BP-8 target cells. Statistically significant cytotoxicity was

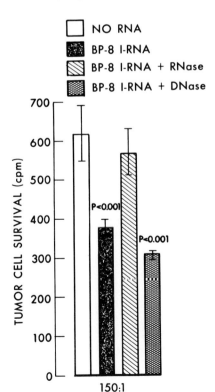

Fig. 1. Cytolysis of BP-8 tumor cells by normal syngeneic lymphocytes preincubated with RNA from the lymphoid organs of immunized guinea pigs.

observed at all lymphoid cell to target cell ratios from 50 to 500 to one, above which nonspecific killing was observed. In addition, no significant cytolysis of BP-8 tumor cells by normal lymphocytes was ever demonstrated. Furthermore, the pretreatment of immune RNA with RNase resulted in destruction of the RNA and complete abrogation of the immune cytolysis observed. This fact indicates the necessity for intact immune RNA in these experiments. However, treatment with DNase resulted in no change in the immune cytolysis observed, indicating that DNA is not an integral factor in this immune reaction.

In all previous experiments, lymphoid cells were removed from normal mice and then preincubated with appropriate RNA solutions. In a separate experiment, mice were inoculated with BP-8 tumor on Day 0; and then 7,14 and 21 days later these mice were sacrificed, and their spleen cells removed. These lymphoid cells were either directly tested *in vitro* against BP-8 tumor cells or were first preincubated with xenogeneic BP-8 immune RNA extracted from the lymphoid organs of guinea pigs immunized with BP-8 tumor.

It is demonstrated in Figure 2 that on Day 0 lymphoid cells from BP-8 tumor bearing animals demonstrated very little cytotoxicity, but that preincubation of these lymphoid cells with BP-8 immune RNA resulted in a very significant cytotoxicity compared to control animals. At 7, 14 and 21 days after the initial tumor inoculation, the lymphoid cells of tumor bearing mice were specifically cytotoxic to BP-8 tumor cells *in vitro* independent of any

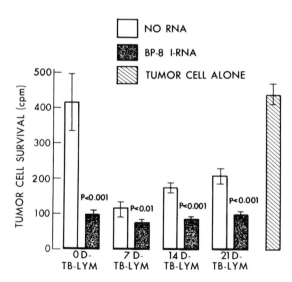

Fig. 2. *Cytolysis of BP-8 tumor cells by lymphocytes from mice with growing BP-8 tumors prior to and after incubation of these lymphocytes with BP-8 immune RNA.*

preincubation with immune RNA. However, incubation of these tumor-specific effector lymphoid cells with BP-8 immune RNA extracted from the lymphoid organs of guinea pigs immunized with the BP-8 tumor resulted in a significant augmentation of this immunoreactivity relative to control groups. This data suggests that the cytolytic capacity of the immune response is not quantitatively maximal in the tumor bearing animal and that this capacity can be augmented by preincubation with immune RNA.

In the experiment shown in Figure 3, mice were inoculated with BP-8 tumor cells on Day 0 and tumors were allowed to develop through five weeks of normal tumor growth. Tumor bearing mice were sacrificed on Day 0, Day 7, Day 21, and Day 35 of this experiment and lymphoid cells from these mice were tested at a lymphoid cell to target cell ratio of 125 to 1 against BP-8 tumor cells *in vitro*.

Lymphoid cells from these mice were specifically immunoreactive against BP-8 tumor cells 7 and 21 days after inoculation of the BP-8 tumor, but by 35 days after inoculation this immunoreactivity had been lost. However, on Day 0, 7, 21, and 35 days after inoculation with the BP-8 tumor cells, lymphoid cells from these mice when mixed with BP-8 immune RNA were cytotoxic against BP-8 tumor cells *in vitro*. This data again indicates that concomitant tumor immunity can be augmented by incubation of tumor-specific immune lymphocytes with specific immune RNA. In addition, this

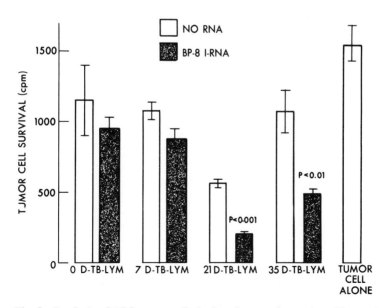

Fig. 3. *Cytolysis of BP-8 tumor cells by lymphocytes from mice with growing BP-8 tumors prior to and after incubation of these lymphocytes with BP-8 immune RNA.*

data confirms that at large tumor volumes the immunoreactivity of the host against the growing neoplasm may be lost; but, significantly, the data suggests that this immunoreactivity may be restored by preincubation of these lymphocytes with specifically immune RNA. Immune RNA, therefore, cannot only augment the immunoreactivity of lymphocytes from animals with growing tumors but may be able to restore a tumor specific efferent arc weakened or destroyed by progressive growth of the antigenic tumor mass.

As mentioned above, for the *in vitro* immunotherapy study, groups of Hartley guinea pigs were immunized with Freund's adjuvant alone, Freund's adjuvant mixed with BP-8 tumor cells, Freund's adjuvant mixed with normal mouse tissue, and Freund's adjuvant mixed with antigenically distinct BP-9 tumor cells. Ten to fourteen days after this immunization all groups of animals were sacrificed. The lymphoid RNA was extracted as described previously. Again, all experimental mice were inoculated with 10^3 BP-8 tumor cells on Day 1. Five, 7, 9 and 11 days later these animals received intraperitoneal inoculations of previously untreated syngeneic C3H/HeN lymphocytes which had been preincubated with either BP-8 immune RNA, BP-9 immune RNA, normal tissue RNA, Freund's adjuvant RNA, or BP-8 immune RNA which had been degraded by incubation with RNase. In other words, immunotherapy was begun five days after the injection of the tumor isograft and at a time when the tumor isograft was, presumably, growing steadily. Other control groups included animals which received tumor cells alone and animals which received only normal spleen cells as a method of immunotherapy.

It can be seen from Figure 4 that in the first experiment there was some decrease in the incidence of tumor development early in the experiment and that this delay in tumor appearance was initially statistically significant. However, eventually the incidence of tumor development equalized in all groups. Although initially there was a striking decrease in the size of the palpable tumors in the animals treated with BP-8 immune RNA, eventually the difference in tumor size between these groups was no longer demonstrable. In a second experiment, the tumor specificity of these observations as well as the necessity for intact immune RNA was also demonstrated. Again, animals were treated on Day 5,7,9, and 11 and observed for tumor development. As can be seen from Figure 5, the results were similar to those observed in the first immunotherapy experiment with the BP-8 tumor and the C3H/HeN mice. Mice which received treatment with BP-8 immune RNA-incubated lymphoid cells had an early statistically significant decrease in tumor development. At this early point in the experiment, these observations were tumor-specific. Eventually, however, all animals again developed tumors and the growth rate of the tumors equalized in all of these experiments.

Attempts to increase this observed response by treating these mice with a second course of RNA-incubated spleen cells 17, 19, 21 and 23 days after initial tumor inoculum were not successful. In each instance, results similar to

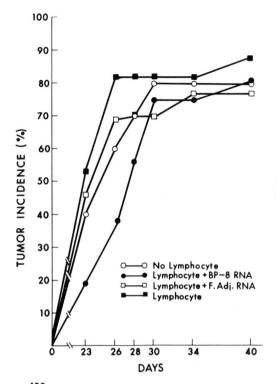

Fig. 4. *Immunotherapy of a murine fibrosarcoma with immune RNA.*

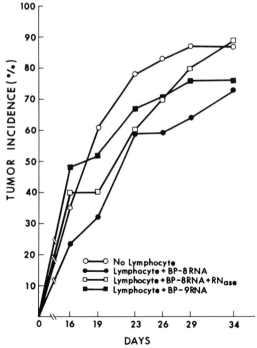

Fig. 5. *Immunotherapy of a murine fibrosarcoma with immune RNA.*

those in Figure 4 were observed. These attempts to retard the growth of the tumor when it is palpable and growing logarithmically were not rewarding and suggests that if immunotherapy with RNA-incubated cells is to be successful, it may need to be used on subclinical foci of residual tumor or as an adjuvant against disseminated disease after more conventional local measures have controlled the primary tumor.

The ^{125}I-UdR assay was also effective in measuring the immunoreactivity of lymphocytes from patients with cancer against the antigens of autologus tumor cells. In Figure 6, the immunoactivity of lymphocytes from a patient with disseminated malignant melanoma (stage 3) was studied. Several weeks earlier, prior to widespread dissemination of this disease, the patient's lymphocytes were significantly cytotoxic against monolayers of autologous melanoma cells. At this large tumor burden, however, no cytotoxicity was demonstrated.

However, it these same non-reactive lymphocytes from the tumor bearing patient were first mediated with xenogeneic RNA extracted from the lymphoid organs of guinea pigs immunized 14 days earlier with the melanoma being studied, significant cytotoxicity was observed indicating that immune RNA could not only augment the efferent arc of a tumor-specific immune response in humans but also could overcome the immunologic exhaustion

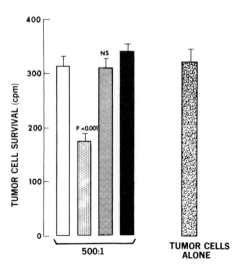

Fig. 6. *Cytolysis of autologous malignant melanoma cells by lymphocytes preincubated with xenogeneic immune RNA.*

witnessed at large tumor volumes. RNase treatment abrogated this effect, thereby establishing this necessity for intact RNA in this observation.

In Figure 7, the cytotoxicity of lymphocytes from a patient with stage 2 malignant melanoma is seen. At this lower total body tumor burden, the patient's lymphocytes are cytotoxic against autologous tumor cells at two different lymphocyte to tumor cell ratios independent of any incubation with immune RNA.

However, incubation of these lymphocytes with RNA extracted from the lymphoid tissues of guinea pigs immunized with either the patient's tumor or that of another patient with malignant melanoma significantly increased the cytotoxicity observed. Again non-degraded immune RNA was necessary for this effect. Incubation of these lymphocytes with RNA extracted from the lymphoid tissues of guinea pigs immunized with Freund's adjuvant only or with another histologically different tumor did not augment the cytotoxicity of the patients' lymphocytes alone indicating the need for RNA specifically directed against an antigen peculiar to malignant melanoma alone.

DISCUSSION

Tumor-specific transplantation immunity is usually demonstrated by the rejection of tumor isografts in animals immunized by excision of growing transplants of the tumor under investigation. The lymphoid cells of animals so

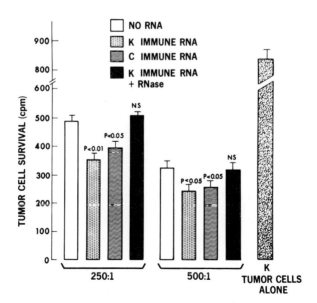

Fig. 7. *Cytolysis of autologous malignant melanoma cells by lymphocytes before and after incubation with RNA extracted from the lymphoid organs of guinea pigs immunized with two different melanoma cell lines.*

immunized can adoptively transfer tumor-specific immunity against subsequent tumor isografts to normal animals.

Much recent evidence suggests that a tumor-specific immunological response can be observed in most animal tumor-host systems despite the relentless growth of the immunizing tumor isograft. Indeed, the progressive growth of a primary tumor isograft despite an immune response that prevents a reinoculation of the same tumor from growing elsewhere in the tumor bearing host has been repeatedly demonstrated (29). We have demonstrated, in a syngeneic murine tumor-host system, that the adoptive transfer of spleen cells from mice with growing tumors to previously untreated syngeneic mice will prevent the growth of an isograft of the same tumor in the spleen cell recipients. In addition, we and other investigators have confirmed *in vitro* tests that progressive growth of a primary tumor does not necessarily indicate the absence of an antitumor response (30). Rosenau and Morton reported that lymphoid cells from mice bearing transplants of a MCA-induced sarcoma suppressed the growth of explants of the immunizing tumor *in vitro* (31). Barski and Youn, similarly, noted that peritoneal exudate cells that had been removed from BALB/c mice bearing tumors induced by an attenuated Rauscher virus inhibited colony formation of these tumor cells *in vitro* (32).

Recently, the existence of tumor-specific antigens in a variety of human tumors has been confirmed (33-39), and evidence exists to suggest that patients with cancer mount a significant immune response to these antigens despite the persistent growth of the primary or metastatic tumor (33,38-40). These immune responses directed against tumor-associated antigens and detected in a host bearing a progressively growing tumor are referred to as concomitant tumor immunity. The effective immunotherapy of cancer may depend upon understanding the paradox of progressive growth of antigenic tumor in an immunologically competent host despite the concomitant existence of immune responses against the tumor-specific antigens of that tumor.

The existence of concomitant tumor immunity has interesting implications for those who advocate the adoptive transfer of immune cells or subcellular fractions as a method of passive immunotherapy. Much experimental evidence implies that the growth of the primary tumor may represent a relative inadequacy of the cell-mediated immune response. The proliferative capacity of the viable tumor cells may exceed the tumor-specific cytolytic ability of specifically sensitized lymphocytes. This viewpoint accepts the belief that the problem of continued tumor growth despite the presence of a strong antitumor immune response is one primarily of either logistics (inadequate numbers of sensitized cells), or of a quantitative inefficiency of adequate numbers of sensitized cells, or both. Passive immunotherapy attempts to either increase the numbers of sensitized cells, augment their killing efficiency, or both. It is toward the possibility of augmenting the killing efficiency of lymphocytes from immunized animals and man with ribonucleic acid that this report was directed.

In the first series of experiments, we studied the reliability of the MLTC in a syngeneic murine tumor-host system and used this assay to determine whether RNA from animals immunized with this tumor could augment the immunoreactivity of mice immunized by amputation of growing tumor isografts. If the incorporation of ^{3}H-thymidine by lymphocytes in response to the presence of a specific antigen is an accurate *in vitro* correlate of *in vivo* cellular immune reactivity, and if *in vivo* tumor destruction is mediated by lymphocytes in a cellular immune response, then this assay may help in predicting which RNAs may have potential for immunotherapeutic use. Increased lymphocyte stimulation was repeatedly observed whenever lymphocytes from S-17 immune mice were incubated with S-17 immune RNA and co-cultured with S-17 cells. The increased proliferative response was abrogated when immune RNA was treated with RNase before incubation with spleen cells, suggesting that one or more species of RNA was the active moiety. Moreover, the response was not due to nonspecific effects of guinea pig lymphoid RNA on murine spleen cells since spleen cells incubated with RNA from nonimmunized guinea pigs were not effective. In addition, partial specificity for the immunizing tumor was demonstrated. S-17 immune lymphocytes, preincubated with S-17 immune RNA, proliferated significantly only in the presence of S-17 cells. Conversely, S-17 immune lymphocytes preincubated with S-18 immune RNA showed no increase activity in the presence of S-17 cells.

As mentioned above, the proliferative response of lymphocytes from S-17 immune mice was significantly greater when these lymphocytes were preincubated with S-17 immune RNA than when they were cultured with S-17 cells without RNA preincubation. Since in all previous experiments involving the transfer of tumor immunity with RNA, normal untreated lymphocytes were the cells converted to a specific immunoreactive state, we believe these experiments demonstrate specific augmentation of the immunoreactivity of lymphocytes already reactive against the tumor specific antigens of a murine tumor. This data implies that immune RNA can specifically augment the efferent arc of an immune response. If it is accepted that the lymphocytes of tumor bearing animals and patients often manifest a detectable degree of specific immune reactivity against the TSA of the tumor in question, then this type of augmentation may be required for the successful immunotherapy of cancer. However, in these initial experiments, lymphocytes were removed from animals immunized by amputation of growing tumors; and this somewhat artificial situation may not correlate with the clinical situation of progressive growth of the immunizing tumor. It remained to demonstrate that RNA could augment the cytotoxicity of lymphocytes from animals and man bearing progressively increasing tumor burdens.

The ^{125}I-iododeoxyuridine microcytotoxicity assay proved more reliable in this regard and was more readily adaptable to human tumor studies. In these experiments, we again demonstrated that normal nonimmune murine lymphoid cells can be converted to effector cells specifically cytotoxic to syngeneic murine tumor cells by incubation *in vitro* with xenogeneic immune RNA extracted from the lymphoid tissues of specifically immunized guinea pigs. We have repeatedly observed that treatment of the active immune RNA preparations with ribonuclease inactivates them, whereas treatment with deoxyribonuclease does not, indicating that one or more species of RNA is the active moiety in our immunoreactive nucleic acid preparations. Utilizing an inbred syngeneic murine tumor host system, we have also demonstrated that the ^{125}I-UdR cytotoxicity assay is useful in determining quantitatively the immunoreactivity of lymphocytes from animals with growing tumors against monolayers of that same tumor. Using this system, we have demonstrated that preincubation of lymphoid cells from animals with growing tumors with xenogeneic immune RNA extracted from the lymphoid organs of animals immunized with the murine tumor being studied is capable of significantly augmenting the immunoreactivity of these lymphoid cells against specific tumor cells *in vitro*. In other words, concomitant tumor immunity can be increased by incubation of already committed specifically sensitized lymphoid cells with tumor-specific immune RNA. This finding, as discussed previously, is clearly important if RNA is to be employed clinically as a means of immunotherapy. Obviously, animals with growing tumors and patients with growing tumors may have intact immune responses against these particular tumors and one problem may well be a quantitative inadequacy of their lymphocyte-mediated tumor-specific immune response. These data confirm the fact that immune RNA can augment the efferent arc of a tumor-specific immune response. Moreover, evidence was presented which confirms the fact that at some large, but presumably variable tumor volume, tumor specific immunity is lost. Significantly, this data suggests that not only can immune RNA augment the cytotoxicity of lymphocytes from tumor bearing animals, but RNA can restore a tumor-specific response to lymphocytes "weakened," "exhausted" or "paralyzed" by excessive tumor burden. These data, therefore, made it logical to proceed with this modality as a method of passive immunotherapy of established rodent tumors.

We have also presented our initial experiments in the immunotherapy of rodent tumors with immune RNA. In these early experiments, a statistically significant delay in tumor development was demonstrated. This event was shown to be specific for the immunizing tumor and to be independent of any nonspecific effect of murine lymphoid cells themselves. In addition, it was shown that intact immune RNA was necessary for this result to occur.

However, in both of these early immunotherapy experiments, despite treatment beginning only five days after inoculation of the viable tumor isograft and proceeding every other day through the next week, no statistically significant decrease in final tumor incidence or in the growth of these tumors was ever demonstrated. Moreover, multiple treatment courses with immune RNA-incubated lymphoid cells did not increase this response. Nevertheless, these preliminary experiments suggest that immune RNA may be a successful method for treatment of established tumors. Experiments are in progress to determine the optimal time of RNA harvest, the cell or tissue productive of the most reliable and efficacious RNA, and to test the possibility that direct injections of immune RNA with or without an RNase inhibitor may be effective in eradicating established rodent tumors.

Evidence to support use of this more direct and simpler approach was recently presented by Pilch who has demonstrated the successful immunotherapy of a transplantable, spontaneously metastasizing, chemically-induced mammary adenocarcinoma in female Fischer 344/N rats with direct injections of xenogeneic immune RNA mixed with the RNase inhibitor, sodium dextran sulfate. He has initiated Phase I trials of the immunotherapy of cancer in man with direct injections of immune RNA (41). Much work still remains in determining the optimal concentration of immune RNA, the optimal RNase inhibitor, the best treatment schedule and the potential toxicity of this treatment.

Furthermore, the *in vitro* augmentation by immune RNA of the cytotoxicity of human lymphocytes against autologous tumor cells is reported. This confirms and extends the work of Pilch (42). Patients with malignant melanoma have a tumor-specific, lymphocyte-mediated immunological reaction against their tumor cells which is easily quantitated by the ^{125}I-UdR assay. At a variable, but large tumor volume, this immunoreactivity was lost; but, at all times, during progressive tumor growth and metastasis, RNA extracted from the lymphoid organs of animals immunized with this human tumor tissue could augment the cytotoxicity of these lymphocytes against specific tumor cells. This data implies that an insufficient or weakened tumor-specific efferent arc of an immune response can be significantly strengthened or restored, and this data makes clinical trials with immune RNA reasonable. Much work remains to be done both experimentally and clinically, but it appears that immune RNA may have a role to play in the adoptive immunotherapy of human cancer.

REFERENCES

1. Bard, D. S. and Y. H. Pilch (1969). The role of the spleen in the immunity to a chemically-induced sarcoma in C3H mice. *Cancer Res.* 29:1125-1131.

2. Old, L. J., E. A. Boyse, D. A. Clarke and E. A. Carswell (1962). Antigenic properties of chemically induced tumors. *Ann N.Y. Acad. Sci. 101*:80-106.

3. Bard, D. S., W. G. Hammond and Y. H. Pilch (1969). The role of the regional lymph nodes in the immunity to a chemically-induced sarcoma in C3H mice. *Cancer Res. 29*:1379-1384.

4. Delorme, E. J. and P. Alexander (1964). Treatment of primary fibrosarcoma in the rat with immune lymphocytes. *Lancet 2*:117-120.

5. Wepsic, H. T., B. M. Zbar, H. R. Rapp and T. Borsos (1970). Systemic transfer of tumor immunity: Delayed hypersensitivity and suppression of tumor growth. *J. Nat. Can. Inst. 44*:955-963.

6. Alexander, P., E. J. Delorme and J. G. Hall (1966). The effect of lymphoid cells from the lymph of specifically immunized sheep on the growth of primary sarcomata in rats. *Lancet 1*:1186-1189.

7. Balme, R. H., M. D. Dockerty, J. H. Grindlay and T. J. Litzow (1969). The use of immune lymph in the treatment of mouse cancer. *Minn. Med. 45*:892-899.

8. Woodruff, M. F. A. and B. Nolan (1963). Preliminary observations on treatment of advanced cancer by injection of allogeneic spleen cells. *Lancet 2*:426-429.

9. Woodruff, M. F. A., M. O. Symes and A. E. Stuart (1963). The effect of rat spleen cells on two transplanted mouse tumors. *Brit. J. Cancer 17*:320-327.

10. Nadler, S. H. and G. E. Moore (1966). Clinical immunologic study of malignant disease: Response to tumor transplants and transfer of leucocytes. *Ann. Surg. 164*:482-490.

11. Nadler, S. H. and G. E. Moore (1969). Immunotherapy of malignant disease. *Arch. Surg.* (Chicago) *99*:376-381.

12. Fisher, B. and E. R. Fisher (1971). Studies concerning the regional lymph node in cancer. I. Initiation of immunity. *Cancer 27*:1001-1004.

13. Fefer, A. (1970). Immunotherapy of primary maloney sarcoma-virus-induced tumors. *Int. J. Cancer 5*:327-337.

14. Woodruff, M. F. A. and J. L. Boak (1965). Inhibitory effect of pre-immunized CBA spleen cells on transplants of A strain mouse mammary carcinoma in (CBA × A)F_1 hybrid recipients. *Brit. J. Cancer 19*:411-417.

15. Borberg, H., H. F. Oettgen, K. Choudry and E. J. Beattie, Jr. (1972). Inhibition of established transplants of chemically induced sarcomas in syngeneic mice by lymphocytes from immunized donors. *Int. J. Cancer 10*:539-547.

16. Mannick, J. A. and R. H. Egdahl (1964). Transfer of heightened immunity to skin homografts by lymphoid RNA. *J. Clin. Invest. 43*:2166-2177.

17. Mannick, J. A. and R. H. Egdahl (1971). Ribonucleic acid in "transformation" of lymphoid cells. *Science 137*:976-977.

18. Ramming, K. P. and Y. H. Pilch (1970). Mediation of immunity to tumor isografts in mice by heterologous ribonucleic acid. *Science 168*:492-493.

19. Ramming, K. P. and Y. H. Pilch (1971). Transfer of tumor specific immunity with RNA: Inhibition of growth of murine tumor isografts. *J. Nat. Cancer Inst. 46*:735-750.

20. Deckers, P. J. and Y. H. Pilch (1972). Mediation of immunity to tumor-specific transplantation antigens by RNA: Inhibition of isograft growth in rats. *Cancer Research 32*:839-846.

21. Deckers, P. J. and Y. H. Pilch (1971). RNA-mediated transfer of tumor immunity — a new model for the immunotherapy of cancer. *Cancer 28*:1219-1228.

22. Deckers, P. J., B. W. Edgerton, B. S. Thomas and Y. H. Pilch (1971). The adoptive transfer of concomitant immunity to murine tumor isografts with spleen cells from tumor-bearing animals. *Cancer Res. 31*:734-742.

23. Prehn, R. T. and J. M. Main (1957). Immunity to methylcholanthrene-induced sarcomas. *J. Nat. Cancer Inst. 18*:769-778.

24. Scherrer, K. and J. E. Darnell (1962). Sedimentation characteristics of rapidly labelled RNA from Hela cells. *Biochem. Biophys. Res. Commun. 7*:486-490.

25. Hammond, W. G., J. C. Fisher and R. T. Rolley (1967). Tumor-specific transplantation immunity to spontaneous mouse tumors. *Surgery 62*:124-133.

26. Perper, T., T. W. Zee, and M. Mickelson (1968). Purification of lymphocytes and platelets by gradient centrifugation. *J. Lab. Clin. Med. 72*:842-848.

27. Cohen, A. M., J. F. Burdick and A. S. Ketcham (1971). Cell-mediated cytotoxicity: An assay using ^{125}I-Iododeoxyuridine-labelled target cells. *J. Immunol. 107*:895-898.

28. Attia, M. A. M. and D. W. Weiss (1966). Immunology of spontaneous mammary carcinomas in mice. V. Acquired tumor resistance and enhancement in strain A mice infected with mammary tumor virus. *Cancer Res. 26*:1787-1800.

29. Deckers, P. J., B. W. Edgerton, B. S. Thomas and Y. H. Pilch (1971). Tumor immunity in inbred mice with progressively growing syngeneic tumors. A study of concomitant immunity. *Current Topics Surg. Res. 3*:371-384.

30. Deckers, P. J., R. C. Davis, G. H. Parker and J. A. Mannick (1973). The effect of tumor size on concomitant tumor immunity. *Cancer Res. 33*:33-39.

31. Rosenau, W. and D. L. Morton (1966). Tumor-specific inhibition of growth of methylcholanthrene-induced sarcomas *in vivo* and *in vitro* by sensitized isologous lymphoid cells. *J. Nat. Cancer Inst. 36*:825-834.

32. Barski, G. and J. K. Youn (1969). Evolution of cell-mediated immunity in mice bearing an antigenic tumor. Influence of tumor growth and surgical removal. *J. Nat. Cancer Inst. 43*:111-121.

33. Eilber, F. R. and D. L. Morton (1970). Immunologic studies of human sarcomas: Additional evidence suggesting an associated sarcoma virus. *Cancer 26*:558-596.

34. Gold, P., M. Gold and S. O. Friedman (1968). Cellular location of carcinoembryonic antigens of the human digestive system. *Cancer Res. 28*:1331-1334.

35. Hellstrom, I., K. E. Hellstrom, G. E. Pierce and J. P. S. Yang (1968). Cellular and humoral immunity to different types of human neoplasms. *Nature 220*:1352-1354.

36. Klein, G., P. Clifford, E. Kelin, R. T. Smith, J. Minowada, R. M. Kourilsky and J. H. Burchenal (1967). Membrane immunoflourescence reactions of Burkitt's lymphoma cells from biopsy specimens and tissue cultures. *J. Nat. Cancer Inst. 39*:1027-1044.

37. Lewis, M. G., R. L. Ikonopisov, R. D. Nairn, T. M. Phillips, G. H. Fairley, D. C. Bodenham and P. Alexander (1969). Tumor-specific antibodies in human malignant melanoma and their relationship to the extent of the disease. *Brit. Med. J. 3*:547-552.

38. Morton, D. L., R. A. Malmgren, E. C. Holmes and A. S. Ketcham (1968). Demonstration of antibodies against human malignant melanoma by immunoflourescence. *Surgery 64*:233-240.

39. Oettgen, H. F., T. Aoki, L. J. Old, E. A. Boyse, E. DeHarven and G. M. Mills (1968). Suspension culture of a pigment-producing cell line derived from a human malignant melanoma. *J. Nat. Cancer Inst. 41*:827-843.

40. Hellstrom, I., K. E. Hellstrom, G. E. Pierce and A. H. Bill (1968). Demonstration of cell-bound and humoral immunity against neuroblastoma cells. *Proc. Nat. Acad. Sci. U.S. 60*:1231-1238.

41. Pilch, Y. H., D. Fritze, D. H. Kern and J. B. DeKernion (1975). Immunotherapy of cancer with "immune" RNA. *Proc. Amer. Assoc. Cancer Res. 16*:258.

42. Pilch, Y. H., L. L. Veltman and D. H. Kern (1974). Immune cytolysis of human tumor cells mediated by xenogeneic "immune" RNA: Implication for immunotherapy. *Surgery 76*:23-34.

THE RELATIONSHIP OF MYELOMA "RNA" TO THE IMMUNE RESPONSE[1]

Paul Heller, N. Bhoopalam,
Y. Chen and V. Yakulis

*Veterans Administration West Side Hospital
and
Departments of Medicine and Microbiology
University of Illinois at the Medical Center
Chicago, Illinois*

The relationship of oncogenic viruses and of the malignant tumor itself to immunosuppression is one of the most fundamental and extensively explored problems of tumor immunology (1,2), but the pathologic basis of the immunosuppressive effect of malignant cells is still incompletely understood. Whatever the general mechanism of this puzzling phenomenon, the immunologic subversion exerted by the malignant proliferation of cells of the immune system itself, such as multiple myeloma in man or murine plasmacytoma, has its own specific peculiarities which are well described but also are not fully clarified. The features of the immunologic deficiency associated with these tumors are reduced plasma levels and decreased synthesis of normal immunoglobulins and diminished primary responsiveness to antigens (3,4,5).

It is generally assumed that B lymphocytes are essential for the primary immune response and that the idiotypic portion of the Fab fragment of the surface immunoglobulin (SIg) is involved in antigen reception (6,7,8). Previous studies of the relationship of lymphocyte SIg and the immunologic deficiency in BALB/c mouse plasmacytoma (9,10,11) and in human myeloma (12,13,14) led us to the hypothesis that a RNA containing factor is released from the tumor cells that changes normal SIg of a large population of B lymphocytes into SIg with the immunochemical characteristics of the respective myeloma globulin by a mechanism still to be clarified. This conversion of SIg to a uniform idiotype ("cell conversion") could be responsible for the diminution of the primary response to antigens characteristics of these two diseases.

The evidence that led to this hypothesis consists of the following:

a. The relative proportion of lymphocytes with normal SIg begins to decrease within a few days after the implantation of BALB/c plasmacytoma

---

[1] Supported by Veterans Administration Research Funds and a grant from the Leukemia Research Foundation.

cells (9). This decrease is accompanied by an increase of lymphocytes that have plasmacytoma specific SIg demonstrable with antiidiotypic antibody. The decrease of lymphocytes with normal SIg was shown by immunocytoadhesion and immunofluorescence and the increase of cells with idiotypic SIg by these technics and more recently by a radioimmunoassay using the cells as immunoabsorbent.

b. Experiments in which normal lymphocytes in a Millipore chamber with nitrocellulose membranes of 0.1μ pore size (VCWP 1300, Millipore Corp., Bedford, Mass.) were implanted into tumor bearing animals and tumor cells in such a chamber into normal animals indicated that in both cases the SIg of the normal cells assumed the immunochemical properties of the myeloma globulin (9). Absorption of myeloma globulin to the lymphocytes was excluded by extensive washing and by regeneration of SIg after trypsinization (11,12,13).

c. An RNA rich extract from the plasmacytoma and also from the plasma of mouse and man with myeloma induced the change of SIg *in vitro* (10,11), and the injection of such an extract into normal mice brought about the same change in circulating and splenic lymphocytes (10,11,15). When such cells were cultured, the proportion of lymphocytes with idiotypic SIg increased without increase in the total number of cells, suggesting persistence and possibly replication of the RNA after its incorporation into the cell (16).

d. The injection of RNA followed 72 hours later by the injection of bovine serum albumin or sheep erythrocytes diminished the primary responsiveness to these antigens (12,13,15), but already established secondary responses were not affected. The effectiveness of the RNA preparation depended on intact RNA molecules.

e. The physico-chemical properties of the RNA, effective in cell conversion, were studied by sucrose gradient ultracentrifugation and oligo (dT) cellulose chromatography (11,17). The effective RNA was found among the 14-18S fractions. A thermolabile 40-50S fraction was also found to be effective (17). The effectivenss of the RNA preparations was abolished by RNase and inhibited by puromycin and cycloheximide, but not by actinomycin (11,12,13,15,17). The RNA molecules with immunosuppressive activity have similar physico-chemical properties. Active RNA contained poly A (Giacomoni and Katzmann, this volume).

This combination of the effects of RNA on the immunochemical characteristics of SIg and on the primary immune response suggested, but did not prove, that the suppressive action of the injected plasmacytoma RNA on the immune response occurred because of the resulting uniformity of the Fab of SIg of B lymphocytes and the incapacity of these lymphocytes to react with the multiplicity of antigens. This suggestion would be strengthened if B lymphocytes of animals with a plasmacytoma that produces an Ig with specific antibody activity to an antigen were found to continue to react with this antigen and remain capable of an immune response to it while becoming unresponsive to others. The possibility that the RNA extracted from such a tumor would reproduce this phenomenon became more attractive because *in vivo* testing of this hypothesis in myeloma bearing animals is prevented by the high levels of the myeloma globulin in the plasma.

We chose for this study MOPC 104E producing an IgM-λ that is an antibody to B1355 dextran S with α-1,3 linked oligosaccharides extractable from *Leuconostoc mesenteroides* (18). The anitbody response to this antigen has been found to be T cell independent and the antibodies elicited in normal BALB/c mice are heterogeneous containing at least two idiotypes one of which is identical with the IgM secreted by MOPC 104E (19,20).

MATERIALS AND METHODS

RNA was prepared from MOPC 104E, LPC-1 and normal mouse spleen by the usual phenol extraction procedure according to the method of Scherrer and Darnell (21) with some modifications (22). Sucrose gradient fractionation and agarose gel electrophoresis of these extracts showed the normal pattern of 4, 18 and 28 S peaks. In the immunocytoadhesion reaction (9,10,11), the MOPC 104E RNA and LPC-1 RNA preparations were found to be effective in inducing on normal lymphocytes the appearance of SIg with the immuno-chemical specificity of the idiotype of MOPC 104E IgM or LPC-1 IgG respectively, both *in vitro* and *in vivo*. The quantities of RNA used were determined spectrophotometrically at 254 nm, 1 O.D. unit corresponding to 50μg RNA.

In all experiments 200μg of the respective RNA preparations in 0.2 ml of barbital buffered saline, pH 7.8 was injected intraperitoneally into several groups of mice. In some of the control experiments MOPC 104E RNA was treated with RNase (Pancreatic, Calbiochem, La Jolla, CA.), DNase (Worthington Biochem. Corp., Freehold, N.J.) and Pronase (Calbiochem, La Jolla, CA.) as described previously (11). Seventy two hours later, an interval that previously had been found to be optimal for the demonstration of the immunosuppressive effect of plasmacytoma RNA (15), the animals were immunized with dextran, lipopolysaccharide (LPS) or DNP_{39} $AECM_{83}$ Ficoll, all T cell independent antigens; or with sheep red blood cells (SRBC), bovine serum albumin (BSA) or saline. The immune response in all animals was measured by passive hemagglutination and the Jerne plaque technique (23). The interval between antigen injection and these measurements was 5 days for the T independent antigens and 7 days for SRBC and BSA. The test antigens were sheep erythrocytes coated with the hapten-albumin conjugate (24,25) by the chromic chloride technique (26) except for LPS which was attached by the method of Britton (27). The dose response curve for dextran was determined by injection intraperitoneally 0.25μg to 15μg of the antigen into 60 normal CF_1 mice.

RESULTS AND DISCUSSION

Effect of MOPC 104E RNA on surface immunoglobulins

The peripheral blood lymphocytes of 8 BALB/c mice with MOPC 104E were tested for the immunochemical characteristics of SIg at weekly intervals after the implantation of the tumor. In agreement with previous findings (9), the percentages of lymphocytes with normal SIg gradually diminished and those with idiotypic SIg increased (Fig. 1). When erythrocytes coated with dextran-albumin were used for the demonstration of the dextran reactive SIg, similar results were obtained.

RNA extracts from the tumor induced 104E IgM SIg on normal lymphocytes *in vitro*. 10^6 normal peripheral blood lymphocytes were incubated with 200μg of MOPC 104E RNA for 30 minutes and washed three times and the immunocytoadhesion reaction was performed. Surface Ig of the 104E IgM idiotype was found on 4% (±2.5) of lymphocytes before and on 32% (±5) after incubation with MOPC 104E RNA. In the *in vivo* experiments, 72 hours after the injection of 200μg of MOPC 104E RNA, the proportion of peripheral lymphocytes with 104E IgM SIg was maximal and decreased

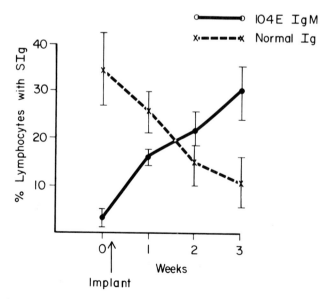

Fig. 1. *Percentage of peripheral lymphocytes with normal and plasmacytoma specific surface immunoglobulin after implantation of MOPC 104E plasmacytoma (n:8).*

thereafter (Fig. 2). RNA treated with RNase was ineffective in SIg conversion while DNase and pronase treated RNA retained this ability.

Effect of RNA on the immune response

The normal dose-response curve for intraperitoneally injected dextran-S is shown in Fig. 3 (open bars). Because only a small increase of antibody titers and plaque forming units per 10^6 spleen cells (PFU) over background occurred when 0.5μg of this antigen was injected, 1μg was considered to be the minimally effective dose. The maximal response occurred with 8μg of dextran-S. In animals, however, which were injected with MOPC 104E-RNA 72 hours prior to the antigen, 1μg resulted in a response of approximately the same magnitude as 8μg in the animals which had not been injected with RNA (Fig. 3, solid bars).

The interval of 72 hours between the injection of RNA and the antigen was chosen as standard because during this period not only the percentage of lymphocytes with 104E IgM idiotype was maximal but also this specific immune response to dextran (Fig. 4). Both the antibody titer and the number of PFU diminished again with the increase of this interval.

The response to 1μg of dextran in animals injected with RNA degraded by RNase was comparable to the response obtained in control mice without

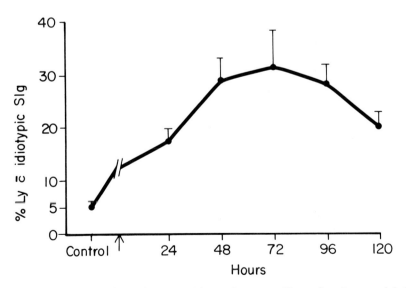

Fig. 2. *Increase of lymphocytes with myeloma specific surface immunoglobulin following injection of MOPC 104E RNA (n:48; arrow indicates injection of 200μg RNA).*

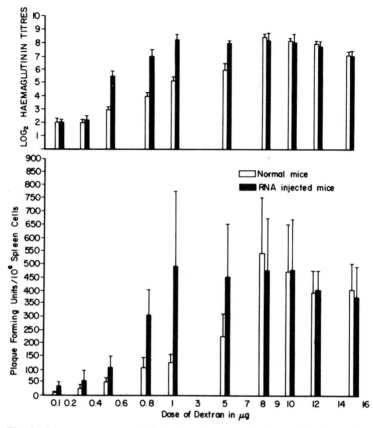

Fig. 3. Dose response to α-1,3 dextran in normal mice and in mice injected with MOPC 104E RNA.

prior RNA injection (Fig. 5). The effect of RNA was not significantly altered by treatment of the RNA preparation with DNase or pronase. Although MOPC 104E RNA increased the immune response to 1μg of dextran to the same level as 8μg in control mice, amounts of dextran larger than 1μg did not further increase it (Fig. 3).

When MOPC 104E RNA was given 72 hours prior to the injection of other antigens such as SRBC, BSA, DNP-Ficoll and LPS, a profound immunosuppressive effect was observed (Fig. 6) in agreement with previous observations of the effects of other plasmacytoma RNAs. RNA from the LPC-1 tumor decreased the immune response to dextran-S, and RNA from normal mouse spleens had no effect (Table 1).

The results of these studies are illuminating in several respects. They confirm the previous observations that RNA extracted from the plasmacytoma

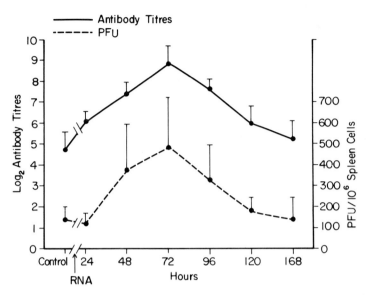

Fig. 4. *Immune response to 1µg dextran at various intervals after injection of MOPC 104E RNA (n:42).*

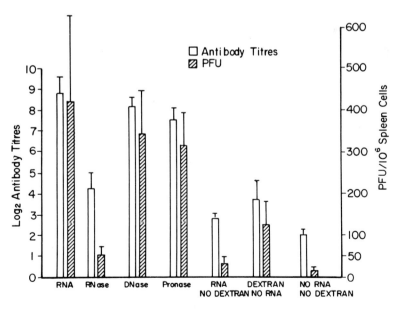

Fig. 5. *Influence of various enzymes on the effect of MOPC 104E RNA on the immune response to dextran (explanation in text).*

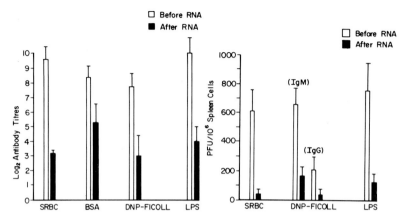

Fig. 6. *Effect of MOPC 104E RNA on immune response to antigens other than dextran.*

induces on normal B lymphocytes the synthesis of surface immunoglobulins with the immunochemical characteristics of the respective plasmacytoma globulin, thus reproducing the phenomenon of SIg conversion that has been observed during the natural history of murine plasmacytoma and of human myeloma by several investigators including ourselves (9,28,29,30). This reproduction of a natural pathobiological phenomenon points toward the possibility that the tumor releases a RNA containing factor which alters the surface characteristics of B lymphocytes and disturbs their functional competence in the encounter with antigens. There is indeed highly suggestive evidence for the existence of such a factor in the plasma of tumor carrying mice and of patients with myeloma (12,13,14). A humoral factor that induces diminished

TABLE 1.
Effect of Injection of RNA from Various Sources on the Immune Response to 1µg and 8µg of dextran (HA: Hemagglutinins)

Source of RNA	1µg		8µg	
	HA	PFU	HA	PFU
Saline	4.9 ± 0.8	126 ± 54	8.3 ± 0.2	548 ± 229
Normal Spleen	5.0 ± 1.0	104 ± 14	8.5 ± 0.5	416 ± 190
	($p<.001$)	($p<0.025$)		
MOPC 104E	8.8 ± 0.9	494 ± 311	8.8 ± 0.4	461 ± 148
	($p<0.001$)	($p<0.01$)	($p<0.001$)	($p<0.01$)
LPC-1	4.1 ± 1.2	56 ±12	3.3 ± 0.5	108 ± 52

reactivity to sheep red cells has also been demonstrated in plasmacytomatous mice by other investigators (31,32). It should be emphasized that these factors are specific for myeloma and have no physico-chemical resemblance to other agents of small molecular size (e.g., prostaglandin) that has been postulated to be responsible for the diminished immunologic capacity of tumor carrying animals (33). The nature of the effective RNA fraction is still under investigation in Giacomoni's laboratory, but the evidence strongly suggests messenger characteristics (Giacomoni and Katzmann, this volume). Another aspect of the mechanism of the immune response is being stressed by the obtained results, namely the importance of the role of the surface immunoglobulins as antigen receptors in the initiation of a specific immune response, a function that has recently been doubted (34). When the induced monoclonal surface immunoglobulin is capable of combining with the antigen, the immune response to this particular antigen is preserved; but to others, both T cell dependent and independent ones, is suppressed. The B cell dysfunction in myeloma appears to be attributable to their monoclonal surface immunoglobulin that has no affinity to common antigens.

The question must be raised whether the plasmacytoma RNA has any effect on T cell function. We have determined the proportion of circulating T cells in tumor carrying animals by fluorescein labeled anti-theta antisera. The proportion of T cells remained normal at approximately 45% throughout the 3-4 week life span of the tumor carrying animal (10-15% are null cells). We also found that the incubation of isolated T cells with plasmacytoma RNA does not induce the synthesis of surface immunoglobulin. The mitogenic responses of lymphocytes from mice that had been injected with plasmacytoma RNA to PHA, Concanavalin A and Pokeweed were normal (unpublished observations). These findings do not exclude the possibility that the injection of plasmacytoma RNA affects suppressor or helper function of T cells. This question needs to be further investigated.

ACKNOWLEDGEMENTS

We thank Drs. Dray, D. Giacomoni and J. Katzmann for helpful discussion and advice; Mr. L. Hall for skillful technical assistance, and Drs. A. Jeanes and D. E. Mosier for generous supplies of α-1,3 dextran and DNP Ficoll, respectively.

REFERENCES

1. Dent, P. B. (1972). Immunodepression by oncogenic viruses. *Prog. Med. Virol.* *14*:1-35.

2. Kersey, J. H., B. D. Spector and R. A. Good (1973). Primary immunodeficiency and cancer. *Adv. Cancer Res.* *18*:211-230.

3. Fahey, J. L., R. Scoggins, J. P. Utz and C. F. Swed (1963). Infection, antibody response, and gamma globulin components in multiple myeloma and macroglobulinemia. *Amer. J. Med. 35*:698-707.

4. Cone, L. and J. W. Uhr (1964). Immunological deficiency disorders associated with chronic lymphocytic leukemia and multiple myeloma. *J. Clin. Inves. 42*:2241-2248.

5. Cwynarski, M. T. and S. Cohen (1971). Polyclonal immunoglobulin deficiency in myelomatosis and macroglobulinemia. *Clin. Exp. Immunol. 8*:237-248.

6. Makela, O. (1970). Analogies between lymphocyte receptors and the resulting humoral antibodies. *Transplantation Reviews 5*:3-18.

7. Walters, C. S. and H. Wigzell (1970). Demonstration of heavy and light chain antigenic determinants on the cell-bound receptor for antigen. Similarities between membrane attached and humoral antibodies produced by the same cell. *J. Exp. Med. 132*:1233-1249.

8. Wigzell, H. (1970). Specific fractionation of immunocompetent cells. *Transplantation Reviews 5*:76-104.

9. Yakulis, V., N. Bhoopalam, S. Schade and P. Heller (1972). Surface immunoglobulins of circulating lymphocytes in mouse plasmacytoma. I. Characteristics of lymphocyte surface immunoglobulins. *Blood 39*:453-464.

10. Bhoopalam, N., V. Yakulis, N. Costea and P. Heller (1972). Surface immunoglobulins of circulating lymphocytes in mouse plasmacytoma. II. The influence of plasmacytoma RNA on surface immunoglobulins of lymphocytes. *Blood 39*:465-471.

11. Giacomoni, D., V. Yakulis, S. R. Wang, A. Cooke, S. Dray and P. Heller (1974). *In vitro* conversion of normal mouse lymphocytes by plasmacytoma RNA to express idiotypic specificities on their surface characteristic of the plasmacytoma immunoglobulin. *Cell. Immunol. 11*:389-400.

12. Heller, P., V. Yakulis, N. Bhoopalam, N. Costea, V. Cabana and R. D. Nathan (1972). Surface immunoglobulins on circulating lymphocytes in mouse and human plasmacytoma. *Trans. Assoc. Am. Physicians 85*:192-202.

13. Heller, P., N. Bhoopalam, V. Cabana, N. Costea and V. Yakulis (1973). The role of RNA in the immunological deficiency of plasmacytoma. *Ann. N. Y. Acad. Sci. 207*:468-480.

14. Chen, Y., Bhoopalam, V. Yakulis and P. Heller. Changes in lymphocyte surface immunoglobulins in myeloma and the effect of an RNA-containing plasma factor. *Ann. Int. Med.*, in press.

15. Yakulis, V., V. Cabana, D. Giacomoni and P. Heller (1973). Surface immunoglobulins of circulating lymphocytes in mouse plasmacytoma. III. The effect of plasmacytoma RNA on the immune response. *Immunol. Commun. 2*:129-139.

16. Bhoopalam, N., V. Yakulis, D. Giacomoni and P. Heller. Surface immunoglobulins of lymphocytes in mouse plasmacytoma IV. Evidence for the persistence of the effect of plasmacytoma-RNA on the surface immunoglobulins of normal lymphocytes *in vivo* and *in vitro*. *Clin. Exp. Immuno.*, in press.

17. Katzmann, J., D. Giacomoni, V. Yakulis and P. Heller (1975). Characterization of two plasmacytoma fractions and their RNA capable of changing lymphocyte surface immunoglobulins (cell conversion). *Cell. Immunol. 18*:98-109.

18. Leon, M. A., N. M. Young and K. R. McIntire (1970). Immunochemical studies of the reaction between a mouse myeloma macroglobulin and dextrans. *Biochemistry 9*:1023-1030.

19. Young, N. M., I. B. Jocius and M. A. Leon (1971). Binding properties of a mouse immunoglobulin M myeloma protein with carbohydrate specificity. *Biochemistry 10*:3457-3460.

20. Carson, D. and M. Weigert (1973). Immunochemical analysis of the crossreacting idiotypes of mouse myeloma proteins with anti-dextran activity and normal anti-dextran antibody. *Proc. Nat. Acad. Sci.* (U.S.A.) *70*:235-239.

21. Scherrer, K. and J. E. Darnell (1962). Sedimentation characteristics of rapidly labeled RNA from HeLa cells. *Biochem. Biophys. Res. Commun.* *7*:486-490.

22. Bhoopalam, N., Yakulis and P. Heller (1974). Monoclonal IgM surface immunoglobulin on lymphocytes of aging NZB mice and its induction in young mice by RNA. *Clin. Exp. Immunol.* *16*:243-258.

23. Jerne, N. K., A. A. Nordin and C. Henry (1963). The agar plaque technique for recognizing antibody-producing cells. *Cell-bound Antibodies*. B. Amos and H. Koprowski, Eds. The Wistar Institute Press, Philadelphia.

24. Sanderson, C. J. and D. V. Wilson (1971). Methods for coupling protein or polysaccharide to red cells by periodate oxidation. *Immunochemistry* *8*:163-168.

25. Eisen, H. N., S. Belman and M. E. Carsten (1953). The reaction of 2,4-dinitrobenzene sulfuric acid with free amino groups of proteins. *J. Amer. Chem. Soc.* *75*:4583-4585.

26. Gold, E. R. and H. H. Fudenberg (1967). Chromic chloride: A coupling reagent for passive hemagglutination reactions. *J. Immunol.* *99*:859-866.

27. Britton, S. (1969). Regulation of antibody synthesis against Escherichia Coli endotoxin, II. Specificity, dose requirements and duration of paralysis induced in adult mice. *Immunology* *16*:513-526.

28. Mellstedt, H., S. Hammarstrom and G. Holm (1962). Monoclonal lymphocyte population in human plasma cell myeloma. *Clin. Exp. Immunol.* *17*:371-384.

29. Lindstrom, F. D., W. R. Hardy, B. J. Eberle and R. C. Williams, Jr. (1973). Multiple myeloma and benign monoclonal gammopathy: Differentiation by immunofluroescence of lymphocytes. *Ann. Int. Med.* *78*:837-844.

30. Abdou, N. I. and N. L. Abdou (1975). The monoclonal nature of lymphocytes in multiple myeloma. Effect of therapy. *Ann. Int. Med.* *83*:42-45.

31. Zolla, S., D. Naor and P. Tanapatchaiyapong (1974). Humoral immunosuppressive substance in mice bearing plasmacytomas. *J. Immunol.* *112*:2068-2076.

32. Tanapatchaiyapong, P. and S. Zolla (1974). Humoral immunosuppressive substance in mice bearing plasmacytomas. *Science* *186*:748-750.

33. Plescia, O. J., A. H. Smith and K. Grinwich (1975). Subversion of immune system by tumor cells and role of prostaglandins. *Proc. Nat. Acad. Sci.* (U.S.A) *72*:1848-1851.

34. Coutinho, A. and G. Moller (1973). B cell mitogenic properties of thymus-independent antigens. *Nature* (New Biology) *245*:12-14.

AN *IN VITRO* MODEL FOR TRANSFER OF TUMOR SPECIFIC SENSITIVITY WITH "TUMOR IMMUNE" RNA EXTRACTS AND LOCALIZATION OF IMMUNOLOGICALLY ACTIVE RNA FRACTIONS

Ronald E. Paque

Department of Microbiology
The University of Texas
Health Science Center at San Antonio
San Antonio, Texas

INTRODUCTION

Earlier studies from our laboratory have demonstrated that cell-mediated hypersensitivity can be transferred to guinea pig peritoneal exudate cells (GP-PEC) with RNA extracts prepared from lymphoid tissues of sensitized outbred guinea pigs (1). Transfer of sensitivity was assessed by the cell-migration-inhibition correlate of delayed hypersensitivity. Other studies have demonstrated that transfer of sensitivity occurs across species barriers (i.e., monkey-to-guinea pig and monkey-to-man) with RNA extracts prepared from lymphoid tissues of sensitized rhesus monkeys (2,3). Transfers of specific sensitivity within species and interspecies to nonsensitized GP-PEC with RNA have, for the most part, involved non-tumor antigens such as purified protein derivative (PPD), keyhole limpet hemocyanin (KLH) and coccidioidin (1-3).

We have extended these RNA transfer studies to include a syngeneic strain-2 guinea pig tumor system. Line-10 and line-1 transplantable tumors are diethylnitrosomine-induced hepatomas possessing antigenically distinct tumor specific transplantation antigens (4). The RNA transfer studies in the line-10-line-1 guinea pig tumor system offer an *in vitro* model to study the molecular mechanisms responsible for the induction and expression of cell-mediated immunity to tumor specific antigens. This report is a summary of our studies in the development of an *in vitro* model for the transfer of tumor-specific sensitivity with RNA extracts and of our experiments that were designed to isolate and localize the immunobiologically active fraction contained in the RNA extracts.

MATERIALS AND METHODS

The methods utilized in these studies have been described in detail elsewhere (1-5). The immunization and skin testing of guinea pigs were done according to protocols developed by Rapp et al. (4-6). The extraction of soluble line-1 or tumor antigens was done according to the methods of Reisfeld et al. and Meltzer et al. (7-8).

RESULTS

Specific Transfer of Line-10 or Line-1 Sensitivity to Nonsensitized Guinea Pig Peritoneal Exudate Cells (GP-PEC).

The experimental plan for transfer of line-10 sensitivity involves collection of nonsensitized syngeneic strain-2 peritoneal exudate cells. The GP-PEC are incubated at 37°C with phenol-extracted RNA from the spleens or lymph nodes of strain-2 guinea pigs immunized against viable line-10 hepatoma cells + BCG or viable line-1 tumor cells alone. After incubation with RNA, the GP-PEC are packed into capillary tubes and assayed for inhibition of migration in the presence of line-10 or line-1 tumor antigens. Keyhole limpet hemocyanin (KLH) or histoplasmin are included as unrelated antigens. The results of a typical experiment in which line-10 tumor-specific sensitivity was transferred to "nonimmune" PEC is shown in Figure 1. The RNA-converted PEC were specifically inhibited in their migration in the presence of soluble, salt-extracted line-10 antigen or PPD; the mean migration indices (M.M.I.'s) were 58±6 and 54±4 respectively (Figure 1). On the other hand, the RNA-treated strain-2 PEC were not inhibited when an unrelated antigen, KLH, was employed; the M.M.I. was 94±8 (Figure 1). The RNA extracts incubated with GP-PEC alone were unable to inhibit the migration of the PEC; the mean M.I. was 90±7 (Figure 1). To be certain the GP-PEC used in our experiments were nonsensitized, the GP-PEC were incubated in the presence of each antigen and assessed for inhibition of migration. As shown in Figure 1, line-10 solubilized antigen, KLH, or PPD failed to inhibit the migration of the PEC; the M.M.I.'s ranged from 97±9 to 112±10 (Figure 1). Identical experiments were performed utilizing histoplasmin as the unrelated antigen with essentially similar results. In a total of 12 similar experiments, utilizing RNA from animals skin test sensitive to line-10, GP-PEC were specifically inhibited in the presence of the line-10 antigen; the M.I.'s ranged from 36±3 to 64±3 (5). In five individual experiments, in which RNA was extracted from animals immunized against line-1 tumors, similar results were obtained. GP-PEC incubated with line-1 "immune" RNA were specifically inhibited in their migration in the presence of line-1 antigen – the mean M.I.'s ranged from 39±6 to 65±4

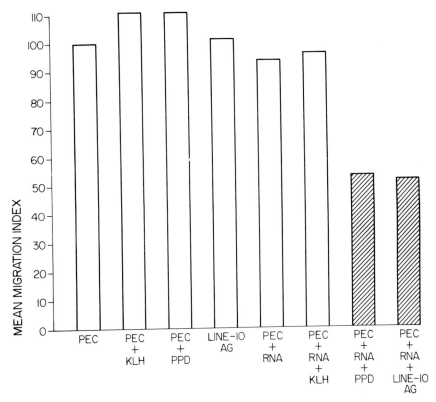

Fig. 1. *A typical experiment comparing the mean migration indices of strain-2 GP-PEC incubated with and without RNA extracts from strain-2 animals immunized against line-10 tumor cells and BCG. The shaded bars indicate cell-migration-inhibition in the experimental system as compared to the controls (clear bars).*

(5) – but not with line-10 antigen – the mean M.I.'s ranged from 87±5 to 95±7 (5).

To further assess the specificity of transfer of line-1 or line-10 tumor specificity via RNA to a single population of GP-PEC, an experimental design similar to that described above was utilized. A single pool of strain-2 PEC was divided into three equal portions of packed cells. The first portion was incubated with RNA extracted from line-10 immunized animals, the second portion was incubated in the RNA extracted from line-1 immunized guinea pigs, and the third portion was incubated without RNA. The results of a typical experiment are shown in Figure 2. When GP-PEC were incubated with RNA extracted from animals immunized against line-10 tumor cells, the GP-PEC were inhibited in their migration; the M.M.I. was 33±4 (Figure 2). On the other hand, incubation of these same PEC with the line-1 soluble antigen

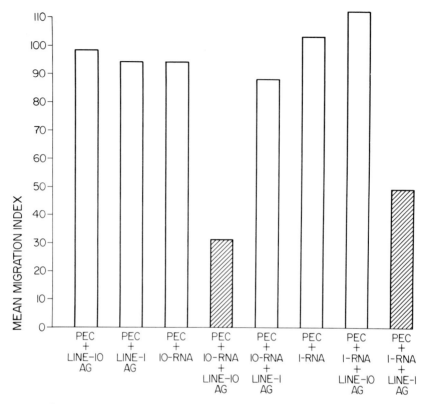

Fig. 2. Mean migration indices of a single population of strain-2 GP-PEC incubated with either line-10 "immune" or line-1 "immune" RNA and line-10 or line-1 solubilized antigen. The shaded bars indicate the mean migration index (M.M.I.) of the experimental system; clear bars indicate the M.M.I. of the controls.

failed to inhibit their migration; the mean M.I. was 89±9 (Figure 2). Similarly, the second portion of GP-PEC incubated with RNA extracted from line-1 immunized animals was inhibited in its migration with the line-1 antigen (M.M.I. = 46±6), but was not inhibited in the presence of the line-10 antigen; the M.M.I. was 110±12 (Figure 2). The third portion of PEC incubated alone was not inhibited by either line-10 or line-1 solubilized tumor antigens; the M.M.I.s were 99±10 and 95±7 respectively (Figure 2). Again, RNA prepared from either line-10 or line-1 immunized animals in the absence of solubilized tumor specific antigens failed to inhibit the migration of the GP-PEC; M.M.I.s were 95±8 and 106±10 respectively (Figure 2).

For controls, transfer of specific line-10 sensitivity was attempted utilizing RNA extracts prepared from the kidney, muscle, and liver of strain-2 guinea pigs immunized against line-10 tumor cells and RNA extracts prepared from the spleens of non-immunized animals. In a total of six experiments, line-10 or line-1 sensitivity was not transferred to GP-PEC utilizing RNA extracts prepared from nonlymphoid or nonimmune tissues (5).

Fractionation and Localization of RNA Fractions Able to Transfer Line-10 or Line-1 Tumor Specific Sensitivity.

To isolate and localize the RNA moiety responsible for transferring line-10 or line-1 tumor specific sensitivity to GP-PEC, approximately 1.0 mg of RNA from either line-10 or line-1 immunized animals was layered on individual linear sucrose density gradients. The gradients were centrifuged for 18 hrs at 5°C in a Spinco SW-41 rotor in a Beckman L3-50 centrifuge. After centrifugation, the gradients were fractionated on an ISCO Model 640 density gradient fractionator. The RNA fractions were cut, collected, and assessed for their ability to transfer line-10 or line-1 tumor specific sensitivity to "nonsensitive" GP-PEC. The position of each fraction cut from the characteristic RNA profile consisting of 4S, 18S, and 28S peaks is shown in Figure 3. A total of five fractions – designated as A, B, C, D, and E – were each tested for transfer of line-1 or line-10 tumor specific sensitivity in an experimental design similar to that described above. The results of two typical experiments in which individual fractions prepared from anti-line-10 RNA or anti-line-1 RNA were tested is shown in Figure 4. In the experiment utilizing fractionated anti-line-10 RNA (clear bars, Figure 4), strain-2 GP-PEC incubated with fraction "B" and corresponding to about 5-12S were specifically inhibited in

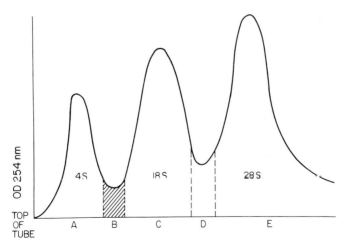

Fig. 3. *Representative cuts of a typical sucrose density gradient containing 1.0mg of either line-10 or line-1 "immune" RNA after centrifugation at 28,000 rpm for 18 hr at 5°C. The shaded portion of the RNA profile represents the "B" fraction.*

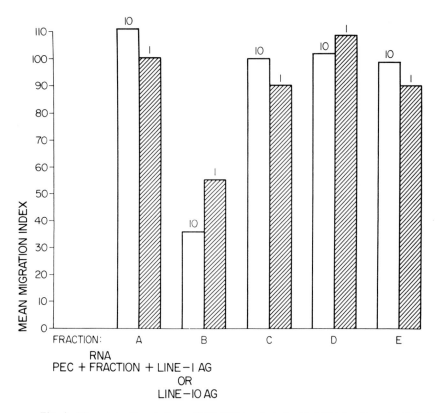

Fig. 4. Mean migration indices of GP-PEC incubated with RNA fractions A, B, C, D or E and either line-10 or line-1 solubilized tumor antigens. The shaded bars indicate fractions prepared from line-1 immunized animals; the clear bars represent fractions prepared from line-10 immunized animals.

their migration in the presence of line-10 tumor antigen; the M.M.I. was 39±5 (Figure 3). On the other hand, fractions A, C, D and E roughly corresponding to 4S, 18S, 22S and 28S failed to transfer line-10 tumor specific sensitivity to GP-PEC; M.M.I.'s were 112±10, 99±9, 100±8 and 95±6 respectively (Figure 4). Similarly, in another experiment (shaded bars, Figure 4) utilizing anti-line-1 RNA, the GP-PEC were specifically inhibited in their migration when incubated with fraction "B" and line-1 tumor antigen. Fractions A, C, D and E were unable to transfer line-1 sensitivity to PEC; the M.M.I.s were 105±5, 86±4, 106±6, 90±8 (Figure 4).

DISCUSSION

The data presented in this report indicate that tumor-specific sensitivity for line-10 or line-1 hepatoma tumor cell antigens can be transferred with

RNA extracts to "nonsensitized" GP-PEC as assessed by the cell-migration-inhibition correlate of delayed hypersensitivity. Specificity of transfer was demonstrated by the fact that a single pool of GP-PEC was inhibited with line-10 or line-1 solubilized antigen after incubation with RNA extracts from line-10 or line-1 immunized animals. The GP-PEC incubated with anti-line-10 RNA were not inhibited by the line-1 antigen; similarly, the GP-PEC treated with anti-line-1 RNA were not inhibited with the line-10 antigen. Unrelated antigens such as KLH or histoplasmin failed to inhibit the RNA-treated GP-PEC. Transfer of tumor-specific sensitivity failed to occur when the RNA extracts were prepared from liver, muscle or kidney of animals immunized against line-10 or line-1 tumor cells or from the spleens of unimmunized strain-2 guinea pigs.

Our *in vitro* studies in the line-1-line-10 tumor system extend the results of Jureziz *et al.* (1) who demonstrated transfer of PPD and coccidioidin sensitivity to GP-PEC with RNA extracts from animals sensitized to each antigen. Other data in the guinea pig system demonstrating *in vitro* transfer of specific sensitivity to non-tumor antigens with RNA extracts have been reported. For example, Schlager *et al.* (11) transferred specific sensitivity for mono-(p-azobenzenarsonate)-N-chloroacetyl-L-tyrosine (ARS-NAT), a low molecular weight antigen, to nonsensitized GP-PEC with RNA extracts from animals sensitized to ARS-NAT. Paque *et al.* have transferred specific sensitivity for DNP-oligolysines to "nonresponder" strain-13 GP-PEC utilizing RNA extracted from lymphoid tissues of "responder" guinea pigs sensitized to the DNP-oligolysines (12). The *in vitro* studies transferring sensitivity with RNA extracts for chemically defined antigens are in agreement with those reported here for the line-1-line-10 tumor system and those using non-chemically defined antigens.

We have attempted to localize and isolate the immunologically active fraction from the RNA extracts responsible for transferring tumor-specific sensitivity to GP-PEC. The results indicate that a fraction located between the 4S and 18S peaks of a sucrose density gradient is able to transfer line-10 or line-1 sensitivity to GP-PEC. This data is similar to that reported by Thor and Dray (13) who fractionated allogeneic RNA-rich extracts obtained from human lymphoid tissue, and those of Paque and Dray (14) who fractionated xenogeneic RNA-rich extracts prepared from sensitized rhesus monkeys. In both instances, the biologically active moiety responsible for transferring specific sensitivity to human cultured lymphoid cells or peripheral blood leucocytes respectively was found between the 4S and 18S peaks. Our data in the line-10-line-1 tumor system are consistent and in agreement with RNA fractionation experiments reported for xenogeneic and allogeneic RNA extracts utilizing non-tumor antigens.

The RNA-conversion systems previously reported and the results reported here all depend on the presence of high-molecular-weight RNA in the RNA extracts. Even when RNA extracts are incubated with RNase, they have failed to transfer line-1 or line-10 tumor specific sensitivity. The presence of 4S material in degraded RNA extracts also fails to transfer line-1 or line-10 tumor specific sensitivity. Indeed, in our fractionation studies, the presence of greater quantities of 4S material would suggest degradation of the molecular species found between the 4S and 18S peaks, thus destroying the immunological activity contained in the RNA extracts. The substantial quantities of lysosomes present in GP-PEC that contain high levels of RNase could account for degradation of the immunobiologically active "tumor immune" RNA species in our system. Approximately 30% of our RNA preparations prepared from sensitized lymphoid tissues of guinea pigs are broken down when assessed by sucrose density gradient centrifugation (5).

While we have found that at least 500 µg of unfractionated phenol-extracted RNA is required for transfer of specific sensitivity in our *in vitro* systems, only 30 to 150 µg is necessary when one utilizes fraction "B". These data suggest that a dose-response relationship and the relative purity of each RNA preparation are essential technical parameters necessary for transferring tumor specific sensitivity in this system and perhaps other non-tumor systems.

Though the isolated "B" fraction in our experiments is found well within the molecular sizes reported for messenger RNA in rabbit and chick globin synthesizing cell-free systems (15), it has been argued that the immuno-biological activity of the RNA is due to RNA-antigen complexes acting as "super" immunogens and thus capable of directly sensitizing the immuno-competent GP-PEC *in vitro*. There are data available that argue against this idea. For example, a direct analysis of RNA extracts able to transfer specific sensitivity for ARS-NAT antigen utilizing atomic absorption spectroscopy for assessment of arsenic as a chemical marker in the antigen indicated that ARS-NAT antigen could be present in 500 µg of RNA in an amount of no more than 0.0000065% (11). Furthermore, we have transferred specific sensitivity for DNP-oligolysines to genetic "nonresponder" strain-13 GP-PEC with RNA extracts prepared from outbred and strain-2 "responder" animals. It is difficult to reconcile the idea of a "super" immunogen directly sensitizing *in vitro* GP-PEC which are genetically incapable of responding to the DNP-oligolysines *in vivo*.

The other mechanism frequently suggested for transfer of specific sensitivity is that the RNA could function as an informational molecule. In this role, the RNA could utilize the biosynthetic mechanisms of the recipient cells, interact with ribosomal components, and thus code for the synthesis of immunoreactive determinants, perhaps receptors, capable of reacting with the solubilized line-1 or line-10 tumor antigens. Indeed, in a mouse plasmocytoma system, Bhoopalam *et al.* have demonstrated the appearance of specific myeloma proteins on the surface of normal mouse lymphoid cells after incubation with RNA extracted from the mouse plasmacytoma tumor cells (16). Because the question of the mechanism is still in doubt, it becomes imperative to develop experiments designed to biochemically isolate and characterize the RNA moiety and the "B" fraction responsible for transferring tumor specific sensitivity in this system. In addition, further investigation in assessing the functions of these RNA moieties in a cell-free system could help resolve the question as to the mechanisms whereby RNA extracts transfer tumor-specific sensitivity to GP-PEC.

REFERENCES

1. Jureziz, R. E., D. E. Thor and S. Dray (1968). Transfer with RNA extracts of the cell migration inhibition correlate of delayed hypersensitivity in the guinea pig. *J. Immunol. 101*:823-829.

2. Paque, R. E. and S. Dray (1970). Interspecies transfer of delayed hypersensitivity *in vitro* with RNA extracts. *J. Immunol. 105*:1334-1338;

3. Paque, R. E. and S. Dray (1972). Monkey to human transfer of delayed hypersensitivity *in vitro* with RNA extracts. *Cell Immunol. 5*:30-41.

4. Rapp, H. J., W. H. Churchill, B. S. Kronman, R. T. Rolley, W. G. Hammond and T. Borsos (1968). Antigenicity of a new diethylnitrosamine-induced transplantable guinea pig hepatoma: Pathology and formation of ascites variant. *J. Natl. Cancer Inst. 41*:1-11.

5. Paque, R. E., M. S. Meltzer, B. Zbar, H. J. Rapp and S. Dray (1973). Transfer of tumor-specific delayed hypersensitivity *in vitro* to normal guinea pig peritoneal exudate cells using RNA extracts from sensitized lymphoid tissues. *Cancer Res. 33*:3165-3171.

6. Zbar, B., I. Bernstein, I. Tanaka and H. J. Rapp (1970). Tumor immunity produced by the intradermal inoculation of living tumor cells and living *Mycobacterium bovis* (Strain BCG). *Science 170*:1217-1218.

7. Meltzer, M. S., E. J. Leonard, H. J. Rapp and T. Borsos (1971). Tumor-specific antigen solubilized by hypertonic potassium chloride. *J. Natl. Cancer Inst. 47*:703-709.

8. Reisfeld, R. A., M. A. Pellegrino and B. D. Kahan (1971). Salt extraction of soluble HLA antigens. *Science 172*:1134-1136.

9. Scherrer, K. and J. E. Darnell (1962). Sedimentation characteristics of rapidly labeled RNA from HeLa cells. *Biochem. Biophys. Res. Commun. 7*:486.

10. Paque, R. E., P. J. Kniskern, S. Dray and P. Baram (1969). *In vitro* studies with "transfer factor": Transfer of the cell-migration inhibition correlate of delayed hypersensitivity in humans with cell lysates from humans sensitized to histoplasmin. Coccidioidin, or PPD. *J. Immunol. 103*:1014-1021.

11. Schlager, S. I., S. Dray and R. E. Paque (1974). Atomic spectroscopic evidence for the absence of a low-molecular weight (486) antigen in RNA extracts shown to transfer delayed-type hypersensitivity *in vitro*. *Cell. Immunol. 14*:104-122.

12. Paque, R. E., M. Ali and S. Dray (1975). RNA extracts of lymphoid cells sensitized to DNP-oligolysines convert nonresponder lymphoid cells to responder cells which release migration inhibition factor. *Cell. Immunol. 16*:261-268.

13. Thor, D. E. and S. Dray (1968). The cell-migration-inhibition correlate of delayed hypersensitivity. Conversion of human nonsensitive lymph node cells to sensitive cells with an RNA extract. *J. Immunol. 101*:469-480.

14. Paque, R. E. and S. Dray (1974). Transfer of delayed hypersensitivity to nonsensitive human leukocytes with rhesus-monkey lymphoid RNA extracts. *Transplantation Proc.* 6:203-207.

15. Williamson, R., M. Morrison, G. Lanyon, R. Eason and J. Paul (1971). Properties of mouse globin messenger ribonucleic acid and its preparation in milligram quantities. *Biochemistry 10*:3014.

16. Bhoopalam, N., V. Yakulis, N. Costea and P. Heller (1972). Surface immunoglobulins of circulating lymphocytes in mouse plasmacytoma. II. The influence of plasmacytoma RNA on surface immunoglobulins of lymphocytes. *Blood 39*:465-471.

IMMUNOTHERAPY OF A GUINEA PIG HEPATOMA WITH TUMOR-IMMUNE RNA[1]

Seymour I. Schlager and Sheldon Dray

Department of Microbiology
University of Illinois at the Medical Center

INTRODUCTION

Our model of immunotherapy is based on the use of syngeneic or xenogeneic RNA-rich extracts obtained from the lymphoid tissues of tumor-immunized animals (1,2). The tumor system employed is the transplantable line-10 ascites variant of a diethylnitrosamine-induced hepatoma which can grow i.d. in strain 2 inbred guinea pigs. The line-10 tumor was selected for investigation because it offers certain advantages over other tumor systems: the ascites form of the tumor is readily transplantable and can be used to immunize syngeneic and xenogeneic animals; soluble tumor-specific antigen can be readily extracted from these cells; 1×10^6 cells growing i.d. are uniformly lethal within 60-90 days; regional metastases are known to occur within 6 days after i.d. injection of 1×10^6 line-10 cells; other lines of hepatomas with different tumor antigens have been characterized; and considerable information is available concerning the biology of these tumors, their prevention by immunization, and their therapy with BCG or its derivatives (3). For example, guinea pigs injected i.d. with 1×10^6 tumor cells admixed with 6×10^6 living BCG organisms never develop an established tumor and are resistant to subsequent challenge with the line-10 tumor (1,3,4). When the BCG is injected intratumorally 6 or 12 days after the injection of 1×10^6 cells, complete tumor regression is observed in approximately 85% or 20% of the animals, respectively (5-10).

RNA-rich extracts prepared by the phenol method (11-13) from lymphoid tissues of animals immunized to a specific antigen have been shown to confer specific immunological sensitivity for that antigen to lymphoid cells of nonsensitized animals (11,14-21). This principle has been applied to several models of tumor prophylaxis and immunotherapy. For example, RNA-treated

---
[1] Supported by contract no. CP 23205 and grant no. CA 05291 from the National Cancer Institute.

lymphoid cells or RNA extracts alone (i.e., without lymphoid cells) have been injected i.p. or s.c. to prevent tumor induction in an animal which was later challenged with the tumor (17,22-25); generally, tumor induction was thereby prevented in approximately 25-60% of the animals treated and occasional inhibition of tumor growth was observed (17,22-25). In other experiments, the RNA-treated lymphoid cells were injected after the tumor cells as immunotherapeutic agents; in almost all of these, incomplete, transient, or no tumor regression was observed (26,29). Ramming and Pilch (30) demonstrated that incubation of normal mouse spleen cells with RNA from guinea pigs immunized to the mouse tumor caused the conversion of these cells to a state of specific immunological reactivity against the tumor since subsequent adoptive transfer of these RNA-treated cells resulted in significant growth inhibition of the challenge tumor isograft.

The effect of RNA extracts in tumor systems has also been assayed *in vitro*. Paque *et al.* (31) reported that tumor-specific delayed hypersensitivity *in vitro* was transferred to peritoneal exudate cells (PEC) obtained from unimmunized strain 2 guinea pigs when the PEC were preincubated with RNA-rich extracts from lymphoid tissues of syngeneic guinea pigs immune to either the line-1 or line-10 tumor. In these experiments, tumor-specific immunological reactivity was demonstrated by the inhibition of migration (from capillary tubes) of the RNA-treated PEC in the presence of the soluble tumor-specific antigen (TSAg) (31). In an attempt to transfer antitumor immunological reactivity *in vitro* with xenogeneic RNA, a cytotoxicity assay was used by Veltman *et al.* (32), but they could not eliminate the possibility that the RNA might be transferring immunological reactivity against histocompatibility antigens as well. Thus, both syngeneic and xenogeneic RNA extracts have been used for *in vivo* and *in vitro* studies on the transfer of specific immunological reactivity for a TSAg (30,31).

In this paper, we present a unique utilization of syngeneic and xenogeneic RNA extracts in an *in vivo* immunotherapy system involving the injection of lymphoid cells, RNA and TSAg directly into a host already bearing a lethal dose of tumor cells (1). Also, we have used the RNA regimen to monitor the effects of treating one line-10 tumor on the development of an identical tumor 10 cm away (2).

MATERIALS AND METHODS

Preparation of Tumor-Immune RNA

Strain 2 inbred guinea pigs (450-500 gm) were immunized to the line-10 or line-1 tumor and Rhesus monkeys, *Macacca mulatta*, were immunized to the line-10 tumor as previously described (1,31). RNA was extracted from the spleens and inguinal and axillary lymph nodes of the immunized guinea pigs or Rhesus monkeys. The method of RNA extraction was the hot-cold phenol procedure previously described (1). As a control, bacterial RNA was extracted from a stock broth culture of *Escherichia coli* as described (1). All RNA extracts were analyzed by sucrose density gradient centrifugation to assess their physical integrity (1). All of these RNA extracts revealed a characteristic distribution among 4S, 18S, and 28S (4S, 16S, and 23S for the *E. coli* RNA) RNA moieties indicating that they had not been significantly degraded.

Tumor Specific Antigen

Water-soluble tumor specific antigen (TSAg) extracts used in the *in vivo* and *in vitro* experiments were prepared by the method of Meltzer *et al.* (33). When tumor-immunized guinea pigs were skin tested with the TSAg, specific delayed-type hypersensitivity reactions were observed (33-36).

In Vitro Testing of RNA Activity

Before using the guinea pig or Rhesus monkey line-1 or line-10-immune RNA extracts *in vivo*, the RNA was tested for biological activity *in vitro* in the direct cell-migration-inhibition assay with guinea pig nonsensitive peritoneal exudate cells (NS-PEC) as previously described (1,21,31). Two pools each of RNA obtained from strain 2 guinea pigs immunized to the line-10 or line-1 tumor or from Rhesus monkeys immunized to the line-10 tumor were tested for their ability to transfer tumor-specific immunological reactivity *in vitro* of NS-PEC in the cell-migration-inhibition assay. The NS-PEC used were obtained from inbred strain 2 guinea pigs and cell migration areas were measured 36 hours after the RNA-treated or untreated PEC were incubated alone or with TSAg (1,21). A Mean Migration Index (MMI) was calculated for each group of cells under the various experimental conditions as previously described (21).

EXPERIMENTAL DESIGN

Immunotherapy with Immune RNA — A Single Tumor Site

Strain 2 male guinea pigs (350-400 gm) were injected i.d. with 1×10^6 line-10 tumor cells on the right flank. In one experiment, 28 animals were injected with 10th ascites passage line-10 cells at the same time and randomly distributed into 7 groups of 4 animals per group. In a second experiment, 60 animals were injected with 12th ascites passage line-10 cells at the same time and randomly distributed into 6 groups of 10 animals per group. Five days later, each animal was injected s.c. under the tumor bleb with one of the therapeutic regimens outlined in Table 1. In a third experiment, 40 strain 2 guinea pigs were injected with 15th ascites passage line-10 cells at the same time and randomly distributed into 5 groups of 8 animals per group. Seven, 12, 17 or 21 days later, these animals were injected s.c. under the tumor bleb with the therapeutic regimens outlined in Table 1. As another control, 6 animals each were injected with line-1 or line-10 tumor-specific antigen alone. All animals were monitored for tumor growth and development for several months.

TABLE 1.
Tumor Regression in Strain 2 Guinea Pigs Injected I.D. with a Uniformly Lethal Dose of Line-10 Tumor Cells (10^6) and Subsequently Treated with Immune RNA.[a]

Treatment[b]	Days Between Tumor Injection and Treatment	Number of Guinea Pigs in Group	Final Tumor Incidence[c]	Mean Survival (Days + S.D.)
NS-PEC	5	14	14/14	73 ± 9
NS-PEC + Line-10 TSAg	5	14	14/14	73 ± 23
Syngeneic Line-10-Immune RNA	5	4	4/4	94 ± 6
NS-PEC + E. Coli RNA	5	4	4/4	72 ± 9
NS-PEC + Syngeneic Line-1-Immune RNA + Line-1 TSAg[d]	5	4	4/4	100 ± 2
Line-10 TSAg	7, 11, 16 & 23	6	6/6	63 ± 3
Line-1 TSAg	7, 11, 16 & 23	6	6/6	46 ± 14
NS-PEC + Syngeneic Line-10-Immune RNA + Line-10 TSAg	5	14	0/14	>300[c]
NS-PEC + Xenogeneic Line-10-Immune RNA + Line-10 TSAg[d]	5	10	0/10	>180[e]
Saline controls	5	14	14/14	65 ± 9

[a]Summary of 2 experiments. All animals in each experiment were injected with the identical passage of line-10 cells (10th or 12th ascites passage) and randomly distributed.

[b]All treatments given s.c. under the tumor papule. 10^7 nonsensitive guinea pig peritoneal exudate cells (NS-PEC), 2.5 mg syngeneic line-10 or line-1 immune RNA (from strain 2 guinea pigs), 2.5 mg xenogeneic line-10 immune RNA (from Rhesus monkeys), 2.0 mg *E. coli* RNA, and 1.0 mg line-1 or line-10 tumor specific antigen (TSAg) were used.

[c]Cumulative incidence of tumor development expressed as number of guinea pigs with tumors/number of guinea pigs in group.

[d]10^7 NS-PEC injected first, followed by 2.5 mg RNA 1-2 min later, and 1.0 mg TSAg 1 hr later.

[e]Still living.

Effect of Immunotherapy of One Tumor Site with RNA on a Separate Untreated Tumor Site

Twenty strain 2 guinea pigs (350-400 gm) were injected i.d. with 1×10^6 line-10 tumor cells (14th ascites passage) at each of two sites 10 cm apart on the right flank and randomly distributed into 2 groups of 10 animals per group. One group of 10 animals was injected with the RNA therapeutic regimen and the second group was injected with 1 ml of saline s.c. under the tumor bleb as shown in Table 3. All animals were monitored for tumor growth and development at both tumor sites for several months.

RESULTS

RNA Activity *In Vitro*

Strain 2 NS-PEC were treated with guinea pig (syngeneic) line-10 or line-1-immune RNA or with Rhesus monkey (xenogeneic) line-10-immune RNA. The PEC were then incubated in the cell-migration-inhibition assay, either alone or in the presence of line-1 or line-10 TSAg; 36 hours later, the migration areas were measured and the MMIs were calculated. None of the 3 RNA extracts alone was inhibitory to the strain 2 NS-PEC at the concentrations used; the MMIs ranged from 89 to 112% (Fig. 1: A, B and C). Also, both the strain 2 guinea pig line-10-immune RNA and the Rhesus monkey line-10-immune RNA were effective in transferring sensitivity for the line-10 TSAg to the NS-PEC; the

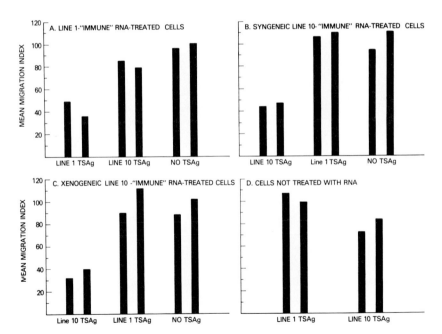

Fig. 1. *Mean migration indices of strain 2 guinea pig peritoneal exudate cells incubated with RNA extracted from strain 2 guinea pigs that were skin test sensitive to line-1 cells (line-1-"immune" RNA), or from strain 2 guinea pigs or Rhesus monkeys that were skin test sensitive to line-10 cells (syngeneic or xenogeneic line-10-"immune" RNA, respectively). The RNA-treated (or untreated) cells were incubated in the cell-migration-inhibition assay either alone or with line-10 or line-1 tumor antigens.*

MMIs ranged from 32 to 47% (Fig. 1: B and C). RNA obtained from line-1-immune strain 2 guinea pigs transferred sensitivity for the line-1 TSAg; the MMIs were 36 and 49% (Fig. 1A). However, when line-10-immune RNA-treated cells were incubated with line-1 TSAg or when line-1-immune RNA-treated cells were incubated with line-10 TSAg, no inhibition of migration was observed. The MMIs ranged from 90 to 112% (Fig. 1: A, B and C). Finally, the line-1 and line-10 TSAg extracts alone were not inhibitory for the PEC at the levels used; the MMIs for the PEC not treated with RNA and incubated in the presence of line-1 or line-10 TSAg ranged from 80 to 107% (Fig. 1D).

In Vivo Immunotherapy with Immune-RNA

Strain 2 guinea pigs were injected i.d. with 1×10^6 line-10 tumor cells and 5 days later, various combinations of NS-PEC, RNA extracts and TSAg were injected s.c. under the developing tumor blebs. Injection of 1×10^7 NS-PEC, 2.5 mg RNA from line-10-immune strain 2 guinea pigs 1-2 min later, and 1.0 mg line-10 TSAg 1 hr later was effective for all 14 animals in causing complete local tumor regression within 60 days after the injection of the tumor cells (Table 1). These animals were still alive 300 days since the tumor inoculations (Table 1). Almost identical results were obtained in the 10 animals receiving RNA from line-10-immune Rhesus monkeys instead of the syngeneic RNA in the same therapeutic regimen (Table 1). Injection of any of the following under the tumor bleb 5 days after the tumor inoculations had no effect on the development of the tumor or on the survival time of the animals: NS-PEC alone; NS-PEC and line-10 TSAg; RNA from line-10-immune strain 2 guinea pigs alone; NS-PEC and RNA from *E. coli*; NS-PEC followed by RNA from line-1-immune strain 2 guinea pigs and line-1 TSAg 1 hr later; or line-10 or line-1 TSAg injected alone on days 7, 11, 16 and 23 (Table 1). All these animals exhibited progressive tumor growth and died within 60-90 days in the same manner as the tumor control animals which had been treated only with saline (Table 1). A significant decrease in tumor growth was noted when NS-PEC and RNA from line-10-immune strain 2 guinea pigs was injected without TSAg 5 days after the tumor inoculations and these animals exhibited a somewhat prolonged survival (100 ± 2 days) (Table 1).

When 8 strain 2 guinea pigs injected with line-10 tumor cells were treated 7 days later (instead of 5 days later) with NS-PEC, RNA from line-10-immune strain 2 guinea pigs and line-10 TSAg, all the animals still exhibited complete local tumor regression – but not until 80 days instead of 60. These animals were still alive and tumor-free 180 days after the tumor inoculations (Table 2). When the animals were treated with the RNA regimen 12 days after the tumor inoculations, 7 of the 8 animals developed progressively growing tumors and died within 86 days (Table 2). One animal in this group was alive at 180 days. Delaying the administration of the RNA therapeutic regimen until 17 days after

TABLE 2.
Response of a Uniformly Lethal Dose (10^6 Cells) of Line-10 Tumor (15th Ascites Passage) to Delayed Immune RNA Therapy.

Treatment[a]	Days Between Tumor Injection and Treatment	Number of Guinea Pigs in Group	Final Tumor Incidence[b]	Mean Survival (Mean + S.D.)
NS-PEC + Syngeneic Line-10-immune RNA + Line-10 TSAg[c]	7	8	0/8	>180[d]
	12	8	7/8	81 ± 5
	17	8	8/8	79 ± 22
	21	8	8/8	68 ± 8
Saline controls	5	8	8/8	79 ± 2

[a] All treatments given s.c. under the tumor papule. 10^7 nonsensitive guinea pig peritoneal exudate cells (NS-PEC), 2.5 mg syngeneic line-10-immune RNA (from strain 2 guinea pigs), and 1.0 mg line-10 tumor specific antigen (TSAg) were used.

[b] Cumulative incidence of tumor development expressed as number of guinea pigs with tumors/number of guinea pigs in group.

[c] 10^7 NS-PEC injected first, followed by 2.5 mg RNA 1-2 min later, and 1.0 mg TSAg 1 hr later.

[d] Still living.

tumor injection resulted in a slight depression of tumor growth in 2 of the 8 animals in the group and a prolongation of their survival until 102 days (Table 2). The other 6 animals treated on day 17, the animals in the group treated on day 21, and the control group treated with saline all exhibited progressive tumor growth and died within 86, 75 and 82 days, respectively (Table 2).

Effect of Immunotherapy with RNA on an Untreated Tumor Site

When strain 2 guinea pigs were injected simultaneously with 1×10^6 line-10 tumor cells at each of two sites 10 cm apart and 5 days later received injections of 1×10^7 NS-PEC followed by 2.5 mg RNA from line-10-immune strain 2 guinea pigs or line-10-immune Rhesus monkeys 1-2 min later and 1.0 mg line-10 TSAg s.c. under one tumor bleb, complete tumor regression was observed at both tumor sites in all 10 animals (Table 3). Control animals injected with saline s.c. under one tumor exhibited progressive tumor growth at both tumor sites and the group's mean survival was 44 days (Table 3).

DISCUSSION

We have presented an *in vivo* model of immunotherapy involving direct injection into the tumor of RNA-rich extracts obtained from animals immunized to the tumor as part of a total regimen which includes nonsensitive peritoneal exudate cells (NS-PEC) and tumor specific antigen (TSAg). In all 32 animals

TABLE 3.
Effects on Two Simultaneously Injected Line-10 Tumors (14th Ascites Passage) During Immune RNA Therapy at Only One Tumor Site.

Treatment[a]	Days Between Tumor Injection and Treatment	Number of Guinea Pigs in Group	Final Tumor Incidence[b]		Mean Survival (Mean + S.D.)
			Treated Site	Untreated Site	
NS-PEC + Syngeneic or Xenogeneic Line-10- immune RNA + Line-10 TSAg[c]	5	10	0/10	0/10	>180[d]
Saline controls	5	10	10/10	10/10	44 ± 7

[a] All treatments were given s.c. under one tumor papule. 10^7 nonsensitive guinea pig peritoneal exudate cells (NS-PEC), 2.5 mg syngeneic (from strain 2 guinea pigs) or xenogeneic (from Rhesus monkeys) line-10-immune RNA, and 1.0 mg line-10 tumor specific antigen (TSAg) were used.

[b] Cumulative incidence of tumor development expressed as number of injection sites with developing tumors/number of sites injected with tumor cells.

[c] 10^7 NS-PEC injected first, followed by 2.5 mg RNA 1-2 min later, and 1.0 mg line-10 TSAg 1 hr later.

[d] Still living.

injected i.d. with a uniformly lethal dose of line-10 cells and treated 5 to 7 days later with NS-PEC, syngeneic or xenogeneic line-10-immune RNA, and a line-10 TSAg preparation 1 hr later, complete local tumor regression was observed and all animals survived with no palpable tumors. Little or no tumor protection was observed if either the NS-PEC, RNA, or TSAg were omitted from the therapy; if bacterial RNA or RNA from animals sensitized to a different tumor (line-1) were used; if line-10 or line-1 TSAg were given alone; or if the administration of the RNA therapeutic regimen was delayed until days 12, 17, or 21.

We have also assessed systemic effects due to the RNA therapy on an identical untreated tumor 10 cm away from the treated tumor. Thus, a uniformly lethal dose of line-10 cells was injected i.d. simultaneously at the two sites. In 10 animals, when one site was treated 5 days later with syngeneic or xenogeneic tumor-immune RNA in a regimen including syngeneic NS-PEC and TSAg, complete and apparently specific regression of the tumors occurred at both the treated and untreated sites in all animals.

The antitumor effect of the RNA immunotherapeutic regimen is presumably a consequence of generating relatively large numbers of immunocompetent tumor-immune cells at the tumor site. Bernstein et al. (37) have reported that tumor bearing animals have an immunologically impaired lymphoid cell population as measured by impaired delayed cutaneous hypersensitivity to specific antigens. To circumvent the possible deficiency or insufficient numbers of immunocompetent host lymphoid cells at the tumor area, normal syngeneic NS-PEC were first injected under the growing tumor. The RNA was then injected to "convert" the NS-PEC to a specific state of immunological sensitivity to the line-10 TSAg. Furthermore, to attempt an amplification effect (i.e., increased blast transformation and cell proliferation) of lymphoid cells that might have been "converted" by the RNA, line-10 TSAg was injected later as a specific immunological stimulant (38). Injection of NS-PEC and RNA without subsequent TSAg injection did, in fact, result in a significant depression of tumor growth locally; but the antitumor effect was incomplete and transient. Thus, the inclusion of TSAg was critical to achieve complete tumor regression whereas TSAg alone was ineffective. The NS-PEC, RNA, and TSAg were injected s.c. under the growing tumor based on observations by den Otter et al. (39) and Borberg et al. (40) that s.c. immunization (e.g., to SL2 and TLC5 mouse lymphomas) results in a stronger and more generalized immunity than i.p. or i.d. injections. This fact is based on the greatly increased subdermal vascularization compared to dermal or peritoneal areas (39,40).

Our specificity controls in the immune RNA model were limited to the use of bacterial RNA or line-1-immune RNA to treat the line-10 tumor. Although the combination of NS-PEC, line-1-immune RNA and line-1 TSAg was shown to elaborate migration inhibitory factor (MIF) on *in vitro* assay, this regimen used *in vivo* demonstrated no therapeutic effect against the line-10 tumor. This suggests that MIF production alone is not sufficient for the regression of a growing neoplasm in our system; it is possible that an antigen-specific cytotoxic factor might be necessary to effect an antitumor response. In fact, Veltman et al. (32) demonstrated that xenogeneic RNA obtained from guinea pigs or sheep immunized to a human tumor and incubated with normal human lymphocytes *in vitro* could transfer antitumor cytolytic activity to the lymphocytes. They suggest that the RNA acted to "convert" the normal human lymphocytes into "killer cells" and presented no evidence that soluble factors were essential in the cytolysis of the tumor cells (32). Although the use of xenogeneic RNA makes the determination of specificity of the antitumor response more difficult, Ramming and Pilch (30) provided evidence that when xenogeneic tumor-immune RNA is incubated with lymphoid cells that are syngeneic with the tumor cells, tumor-specific but not histocompatibility immune responses were transferred. We could not test the effects of line-10-immune RNA in treating a line-1 tumor to serve as the reciprocal control for the experiment where the effects of line-1-immune RNA were tested against the line-10 tumor since the line-1 tumor regresses spontaneously. However, by treating another antigenically different tumor, which is uniformly lethal, with line-10 immune RNA and *vice versa*, it would be possible to firmly establish the specificity of the RNA as an antitumor agent.

The mechanism involved in the transfer of immunological sensitivity by RNA extracts has not been completely elucidated but an RNA moiety has been implicated to be the active component of these extracts since they are inactivated by RNase but not by DNase or trypsin (14,31,41-43). Furthermore, in contrast to transfer factor, the active component of RNA has been identified as an 8-12S fraction in human RNA (43) monkey RNA (44) and guinea pig RNA (see Paque, this volume). Our RNA preparations were assessed for molecular size by sucrose density gradient analyses and found to contain 4S, 18S, and 28S RNA moieties in typical proportions indicating minimal degradation of the RNA. The intact RNA may function in an informational role (e.g., by coding for cell surface receptors) as has been suggested for numerous systems (11,13,14,16,20,21,31,41-43) rather than as an RNA-antigen complex (or "super-antigen"). Further support for an information mechanism in the action of RNA has recently accumulated in two systems: Schlager et al. (21) demonstrated that no trace of antigen ($<6.5 \times 10^{-6}\%$) could be found in RNA extracts capable of transferring MIF reactivity *in vitro* for a chemically-defined, low MW antigen; also, Bhoopalam et al. (15) demonstrated that the *in vitro* incubation of RNA-extracts from a mouse plasmacytoma with normal mouse lymphocytes can convert the normal lymphocytes to express the specific plasmacytoma Ig on their surface. Isolation and characterization of the active RNA component in our *in vivo* immunotherapy system, however, requires further explicit investigation.

The i.d. injection of 1×10^6 line-10 cells was not only uniformly lethal (as reported by others), but this dose is also known to result in a high incidence of metastases within 6 days (5). Since the RNA therapy was not given until 5 or 7 days after the tumor cells were injected, the long term tumor-free survival of all treated animals indicates that a systemic immunity was induced which halted the early metastatic progression of the tumor. Animals that had been treated with the complete RNA therapeutic regimen were also resistant to a subsequent challenge with a lethal dose of line-10 tumor cells (1) indicating that systemic immunity against the line-10 tumor was induced by the complete regression of the local neoplasm. Systemic immunity was explicitly demonstrated in the RNA immunotherapy system when one of two tumors injected 10 cm apart was treated as described above. The mechanism by which the RNA therapy led to systemic antitumor immunity can be visualized as follows: the RNA regimen involved supplying a large number of additional immunocompetent lymphoid cells to the tumor site. These were then presumably converted by the RNA to a state of specific immunity for the tumor (31). Due to stimulation by the added TSAg, these converted cells were then presumably further activated to undergo proliferation (1,38). The effective systemic immunity observed after RNA treatment, therefore, could be due to the presence of a large number of lymphoid cells specifically sensitized to the tumor which could enter the circulation and reach the second untreated site.

Numerous other reports of immunotherapy models have been based on the passive transfer of sensitized lymphoid cells from an immunized donor to a tumor bearing host (37,45) or through the use of RNA extracts either injected directly (25-27) or incubated with lymphoid cells which were then injected into the tumor bearing recipient (22-25,28,29). For example, Bernstein et al. (37) passively transferred systemic tumor immunity with peritoneal exudate cells from immunized guinea pigs to unimmunized syngeneic recipients; however, this has restricted potential applicability to human cancer. Katz et al. (45) attempted a similar transfer of immunity to a mouse leukemia with allogeneic lymphoid cells; the results, however, were obscured due to the occurrence of a graft-versus-host reaction. Rigby (29), Pilch et al. (25), Ohno et al. (24) and Deckers et al. (22) reported varying degrees of success in treating established tumors or in attempting to prevent the incidence of tumors by injecting syngeneic lymphoid cells that had been incubated with syngeneic RNA from tumor-immunized animals. The direct injection of xenogeneic RNA extracts as immunotherapy agents was attempted by Alexander et al. (26,27). They found growth retardation and occasional temporary regression of benzpyrene-induced sarcomas in rats treated with footpad injections of RNA extracted from sheep immunized against the rat tumors. Londner et al. (28) observed similar results when the rats were given i.p. injections of syngeneic RNA. Pilch et al. (25) used direct s.c. injections of xenogeneic RNA prophylactically to provide protection against a BP-4 mouse tumor; they reported the prevention of detectable tumors in up to 60% of the animals treated. All these results differ from our system in that we incubate our lymphoid cells with the RNA directly in the tumor bearing recipient, we include TSAg in the therapeutic regimen, we inject all agents s.c. under the growing tumor, and we report complete local regression of established uniformly lethal tumors in 100% of the treated animals using either syngeneic or xenogeneic RNA. Furthermore, we have included specificity controls in our model by using bacterial RNA or immune RNA to an antigenically different tumor (line-1) in treating the line-10 tumor. Thus, our RNA model involves the specific stimulation of lymphoid cells against a particular tumor antigen associated with the neoplasm being treated.

If the RNA immunotherapy model is to be extended to man, we need to consider at least the following problems: a source of nonsensitive histocompatible immunocompetent lymphoid cells is not readily available but could possibly come from the tumor bearing individual; xenogeneic RNA would probably be preferable over allogeneic RNA since animals can be readily immunized to most tumors; the active component of the RNA should be isolated since "whole" xenogeneic RNA could give rise to complications in man; and the availability of TSAg would depend on the size of the primary tumor. In selected favorable circumstances, therefore, immune RNA therapy may become feasible. In our guinea pig tumor immunotherapy model with RNA, we have as yet not had the opportunity to develop optimal conditions for therapy due to limitations in the availability of strain 2 guinea pigs. However, the adaptation of our immunotherapy model to other tumors and in other species should be helpful in evolving more effective tumor therapy regimens.

REFERENCES

1. Schlager, S. I., R. E. Paque and S. Dray (1975). Complete and apparently specific local tumor regression using syngeneic or xenogeneic "tumor-immune" RNA extracts. *Cancer Research 35*:1907-1914.
2. Schlager, S. I. and S. Dray. Tumor regression at an untreated site during immunotherapy of an identical distant tumor. *Proc. Nat. Acad. Sci. USA 72*, in press.
3. Rapp, H. J. (1973). A guinea pig model for tumor immunology. *Israel J. Med. Sci. 9*:366-374.
4. Bast, R. C., B. Zbar, T. Borsos and H. J. Rapp (1974). BCG and cancer. *New Engl. J. Med. 290*:1413-1420 and 1458-1469.
5. Zbar, B., I. D. Bernstein, G. L. Bartlett, M. G. Hanna and H. J. Rapp (1972). Immunotherapy of cancer: regression of intradermal tumors and prevention of growth of lymph node metastases after intralesional injection of *Mycobacterium bovis*. *J. Natl. Cancer Inst. 49*:119-130.
6. Zbar, B., I. D. Bernstein and H. J. Rapp (1971). Suppression of tumor growth at the site of infection with living bacillus calmette-guerin. *J. Natl. Cancer Inst. 46*:831-839.
7. Zbar, B., I. D. Bernstein, T. Tanaka and H. J. Rapp (1970). Tumor immunity produced by the intradermal inoculation of living tumor cells and living *Mycobacterium bovis* (Strain BCG). *Science 170*:1217-1218.
8. Zbar, B., E. Ribi, T. Meyer, I. Azuma and H. J. Rapp (1974). Immunotherapy of cancer: regression of established intradermal tumors after intralesional injection of mycobacterial cell walls attached to oil droplets. *J. Natl. Cancer Inst. 52*:1571-1577.
9. Zbar, B. and T. Tanaka (1971). Immunotherapy of cancer: regression of tumors after intralesional injection of *Mycobacterium bovis*. *Science 172*:271-273.
10. Zbar, B., H. T. Wepsic, T. Borsos and H. J. Rapp (1970). Tumor-graft rejection in syngeneic guinea pigs: evidence for a two-step mechanism. *J. Natl. Cancer Inst. 44*:473-481.
11. Paque, R. E. and S. Dray (1972). Monkey to human transfer of delayed hypersensitivity *in vitro* with RNA extracts. *Cell. Immunol. 5*:30-41.
12. Scherrer, K. and J. E. Darnell (1962). Sedimentation characteristics of rapidly labeled RNA from HeLa cells. *Biochem. Biophys. Res. Commun. 7*:486.
13. Thor, D. E. and S. Dray (1968). The cell-migration-inhibition correlate of delayed hypersensitivity. Conversion of human non-sensitive lymph node cells to sensitive cells with an RNA extracts. *J. Immunol. 101*:469-480.
14. Bell, C. and S. Dray (1969). Conversion of non-immune spleen cells by ribonucleic acid from an immunized rabbit to produce γM antibody of foreign light chain allotype. *J. Immunol. 103*:1196-1211.
15. Bhoopalam, N., V. Yakulis, N. Costea and P. Heller (1972). Surface immunoglobulins of circulating lymphocytes in mouse plasmacytoma. II. The influence of plasmacytoma RNA on surface immunoglobulins of lymphocytes. *Blood 39*:465-471.
16. Cohen, E. P. and J. J. Parks (1964). Antibody production by non-immune spleen cells incubated with RNA from immunized mice. *Science 144*:1012-1013.
17. Deckers, P. J. and Y. H. Pilch (1972). Mediation of immunity to tumor-specific transplantation antigens by RNA inhibition of isograft growth in rats. *Cancer Research 32*:839-846.

18. Likhite, V. and A. Sehon (1972). Cell-mediated tumor allograft immunity: *In vitro* transfer with RNA. *Science* 175:204-205.

19. Mannick, J. A. and R. H. Egdahl (1962). Transformation of non-immune lymph node cells to a state of transplantation immunity by RNA. *Ann. Surg.* 156:356.

20. Paque, R. E. and S. Dray (1970). Interspecies "transfer" of delayed hypersensitivity *in vitro* with RNA extracts. *J. Immunol.* 105:1334-1338.

21. Schlager, S. I., S. Dray and R. E. Paque (1974). Atomic spectroscopic evidence for the absence of a low-molecular weight (486) antigen in RNA extracts shown to transfer delayed-type hypersensitivity *in vitro*. *Cell. Immunol.* 14:104-122.

22. Deckers, P. J., K. P. Ramming and Y. H. Pilch (1973). The transfer of tumor immunity with syngeneic RNA. *Ann. N. Y. Acad. Sci.* 207:442-453.

23. Kennedy, C. T. C., D. B. Cater and F. Hartveit (1969). Protection of C_3H mice against BP-8 tumor by RNA extracted from lymph nodes and spleens of specifically sensitized mice. *Acta Pathol. Microbiol. Scand.* 77:169-200.

24. Ohno, R., K. Esaki, Y. Kodera, H. Shiku and K. Yamada (1973). Experimental models and the role of RNA in immunotherapy of leukemia. *Ann. N. Y. Acad. Sci.* 207:430-441.

25. Pilch, Y. H., K. P. Ramming and P. J. Deckers (1973). Induction of anti-cancer immunity with RNA. *Ann. N. Y. Acad. Sci.* 207:409-427.

26. Alexander, P., E. J. Delorme, L. D. C. Hamilton and J. G. Hall (1967). Effect of nucleic acids from immune lymphocytes on rat sarcomata. *Nature* 213:569-572.

27. Alexander, P., E. J. Delorme, L. D. G. Hamilton and J. G. Hall (1968). Stimulation of anti-tumor activity of the host with RNA from immune lymphocytes. *Nucleic Acids in Immunology*. O. J. Plescia and W. Braun, Eds. Springer-Verlag, New York.

28. Londner, M. V., J. C. Morini, M. T. Font and S. L. Rabasa (1968). RNA-induced immunity against a rat sarcoma. *Experientia* 24:598-599.

29. Rigby, P. G. (1969). Prolongation of survival of tumor bearing animals by transfer of "immune" RNA with dEAE Dextran. *Nature* 221:968-969.

30. Ramming, K. P. and Y. H. Pilch (1971). Transfer of tumor-specific immunity with RNA: Inhibition of growth of murine tumor isografts. *J. Natl. Cancer Inst.* 46:735-750.

31. Paque, R. E., M. S. Meltzer, B. Zbar, H. J. Rapp and S. Dray (1973). Transfer of tumor-specific delayed hypersensitivity *in vitro* to normal guinea pig peritoneal exxdate cells using RNA extracts from sensitized lymphoid tissues. *Cancer Research* 33:3165-3171.

32. Veltman, L. L., D. H. Kern and Y. H. Pilch (1974). Immune cytolysis of human tumor cells mediated by xenogeneic "immune" RNA. *Cell. Immunol.* 13:367-377.

33. Meltzer, M. S., E. J. Leonard, H. J. Rapp and T. Borsos (1971). Tumor-specific antigen solubilized by hypertonic potassium chloride. *J. Natl. Cancer Inst.* 47:703-709.

34. Meltzer, M. S., J. J. Oppenheim, B. H. Littman, E. J. Leonard and H. J. Rapp (1972). Cell-mediated tumor immunity measured *in vitro* and *in vivo* with soluble tumor antigens. *J. Natl. Cancer Inst.* 49:727-734.

35. Reisfeld, R. A. and B. D. Kahan (1970). Biological and chemical characterization of human histocompatibility antigens. *Fed. Proc.* 29:2034-2040.

36. Reisfeld, R. A., M. A. Pellegrino and B. D. Kahan (1971). Salt extraction of soluble HLA antigens. *Science* 172:1134-1136.

37. Bernstein, I. D., B. Zbar and H. J. Rapp (1972). Impaired inflammatory response in tumor bearing guinea pigs. *J. Natl. Cancer Inst.* 49:1641-1647.

38. Bell, C. and S. Dray (1973). RNA conversion of lymphoid cells to synthesize allogeneic immunoglobulins *in vivo*. *Cell. Immunol.* 6:375-393.

39. denOtter, W., B. A. Runhaar, A. Ruitenbeek and H. F. J. Dullens (1974). Site-dependent differences in rejection of tumor cells with and without preimmunization. *Eur. J. Immunol.* 4:444-446.

40. Borberg, H., H. F. Oettgen, K. Choudry and E. J. Beatty (1972). Inhibition of established transplants of chemically induced sarcomas in syngeneic mice by lymphocytes from immunized donors. *Internat. J. Cancer 10*:539-547.

41. Jureziz, R. E., D. E. Thor and S. Dray (1968). Transfer with RNA extracts of the cell migration inhibition correlate of delayed hypersensitivity in the guinea pig. *J. Immunol. 101*:823-829.

42. Jureziz, R. E., D. E. Thor and S. Dray (1970). Transfer of the delayed hypersensitivity skin reaction in the guinea pig using RNA-treated lymphoid cells. *J. Immunol. 105*:1313-1321.

43. Thor, D. E. and S. Dray (1973). Transfer of cell-mediated immunity by immune RNA assessed by migration-inhibition. *Ann. N. Y. Acad. Sci. 207*:355-368.

44. Paque, R. E. and S. Dray (1974). Transfer of delayed hypersensitivity to nonsensitive human leukocytes with Rhesus-monkey lymphoid RNA extracts. *Transpl. Proc. 6*:203-207.

45. Katz, K. H., L. Ellman, W. E. Paul, J. Green and B. Benacerraf (1972). Resistance of guinea pigs to leukemia following transfer of immunocompetent allogeneic lymphoid cells. *Cancer Res. 32*:133-140.

THE CLINICAL EXPERIENCE IN THE TREATMENT OF RENAL ADENOCARCINOMA WITH IMMUNE RNA[1]

Jean B. deKernion, M.D., Kenneth P. Ramming, M.D., Donald G. Skinner, M.D. and Yosef H. Pilch, M.D.

From the Department of Surgery
Divisions of Urology and Oncology
Veterans Administration Hospital
Sepulveda, California
and the University of California
Los Angeles, California
and the Department of Surgery
Harbor General Hospital
Torrance, California

Adenocarcinoma of the kidney is an uncommon tumor associated with a very poor prognosis. Because of the anatomic position of the kidney deep in the retroperitoneum, about one-third of patients with this tumor first present with distant metastases. Furthermore, no more than 50 percent of patients without metastases are cured by definitive surgery. Therefore, over two-thirds of patients with this malignancy will require therapy for metastatic tumor with little hope of palliation. There have been no effective chemotherapeutic agents in the treatment of renal carcinoma, and clinical trials have failed to demonstrate a significant objective response to cytotoxic drugs (1,2). On the basis of studies on the hormone-induced renal tumor in hamsters, Bloom recommended the use of progesterone (3). However, only an average of 15 percent with a brief objective response was reported in 272 cases treated with progesterone (4). In a recent randomized clinical study, hormones appeared to produce no more objective responses than did placebo (5). The need for newer approaches to therapy was underscored by the report of Mostofi (6) who reviewed 1700 cases of renal carcinoma with metastases at the time of diagnosis and reported only two survivors after 2 years regardless of therapy.

---

[1] This work has been supported by VA Grants Nos. 1797 and 7331 and NCI Grant No. CA 15372.

This tumor occasionally regresses spontaneously and this characteristic has had some influence on the interpretation of clinical therapeutic trials. The real incidence of spontaneous regression is less than popularly believed, however, having been reported in only 40 cases in the world literature, and possibly in a few other unreported instances (7). However, the very fact that spontaneous regressions do occur suggests that this tumor might be subject to host immune mechanisms. Further evidence of tumor-host immune system interrelationships is suggested by the recognized erratic behavior of renal carcinoma. Patients occasionally develop distant metastases many years after complete excision of the primary tumor, suggesting prolonged suppression of growth of the occult tumor foci. It is possible in rare instances that metastases may regress and recur repeatedly, although histologic confirmation is difficult to obtain.

The clinical impression that the host immune system is operative in the natural course of renal carcinoma finds support in laboratory studies of the immune response in patients with tumors. Using the assay of lymphocyte-mediated cytotoxicity *in vitro*, Bubenik et al. (8) demonstrated specific cytotoxicity in patients with renal carcinoma and found evidence suggesting the presence of some common tumor-associated membrane antigens among renal carcinomas. Correlation between cellular immunity and clinical status was reported by Brosman et al. (9) who documented decreased reactivity to skin test antigens and depression of monocyte chemotaxis in patients with advanced tumors. Hakala et al. (10) demonstrated a humoral complement-dependent cytotoxic factor in the sera of patients with renal carcinoma.

The clinical and laboratory evidence of significant host immunity in patients with renal carcinoma, as well as the paucity of effective therapeutic alternatives, supports the institution of clinical trials with immunotherapeutic agents. Occasional case reports employing various methods of immunotherapy have appeared in the literature (11). However, no clinical trial has been specifically concerned with immunotherapy of renal carcinoma. Evidence for the mediation of immunity by RNA extracts from the lymphoid tissues of specifically sensitized animals and a discussion of theoretical advantages of the application of this form of immunotherapy in clinical trials has been reported (12,13). The purpose of this paper is to present the results of the initial clinical trials of immunotherapy of human renal carcinoma with xenogeneic immune RNA.

MATERIALS AND METHODS

Patient Selection

Two kinds of patients presented with gross disease: those who had the primary tumor *in situ* and those who had had previous nephrectomy. Because of the rare occurrence of disappearance of distant metastases after removal of the primary tumor, surgeons have been prompted to perform nephrectomy even in the face of distant metastases with the small hope of temporary regression. Such a procedure in this group of patients, however, is associated with a prohibitive mortality rate (6-10 percent) (14), out of proportion to the temporary response rate of 0.3 percent (4). We therefore formulated a treatment plan intended to avoid imprudent surgery, but to allow the possibility of increasing the effect of immunotherapy. The patients were treated for two months with immune RNA. If, during that time, they showed arrest of growth or decrease in size of the metastases and no new lesions had appeared, a nephrectomy was performed. This was done to reduce the bulk of tumor since previous studies have demonstrated that immunotherapy is most effective in patients with minimum tumor burden (15). Patients who did not respond during the two-month initial trial of therapy were deemed poor candidates for immunotherapy and instead were offered experimental chemotherapy without nephrectomy.

Patients who developed metastases after curative nephrectomy received immunotherapy for 3 months. Therapy was continued thereafter if growth was arrested or the metastases regressed. Therapy was terminated in patients who did not respond, and they were offered alternative experimental therapy.

Those patients with a 50 percent or greater chance of having recurrence or metastases within two years were defined as having minimum residual disease. Patients who had had complete excision of isolated distant metastases were also included in this group. All of these patients will continue therapy for two years or until new lesions appear.

Preparation of Immune RNA

Adult rams were immunized with 4 weekly intramuscular injections of 4 cc of an emulsion of human renal cell carcinoma with an equal volume of Freund's complete adjuvant. Two weeks later the RNA was extracted from the spleen and lymph nodes of the rams by the hot phenol method of Ramming and Pilch (16). After precipitation in cold ethanol, the protein content was minimized by incubation with pronase and by repeated precipitations in solutions made 2M with respect to potassium acetate. The final RNA solution was dialyzed against distilled water and was sterilized by passage through micropore filters. It was lyophilized in 4-milligram aliquots in sterile containers. Bacterial and fungal cultures performed on each batch of RNA were always negative. The final protein concentration was never more than .001 milligram/milligram of RNA. Each preparation was analyzed by disc gel electrophoresis to demonstrate that the RNA was not degraded.

Separation of Human Lymphocytes

Lymphocytes were obtained from heparinized peripheral whole blood from each patient before therapy and at 2- or 3-month intervals during therapy. The lymphocytes were isolated on ficoll-isopaque gradients and frozen in RPMI 1640 tissue culture medium containing 40 percent agamma human serum and 10 percent dimethylsulfoxide (DMSO). The cells were step frozen at 1°C./minute in liquid nitrogen and stored in the vapor phase of a liquid nitrogen freezer.

Lymphocyte Cytotoxicity Assay

The assay was modified after the method of Cohen et al. (17). Specific numbers of cells suspended in tissue culture media were incubated in microtest plates for 2 hours at 37°C. The lymphocytes were then added to the microtest wells in 50:1 and 100:1 lymphocyte to target cell ratios and were incubated for 48 hours at 37°C. Media and dead cells were then aspirated, and the residual radioactivity in each well was measured with a gamma counter. The results were expressed as the Cytotoxicity Index (C.I.) calculated according to the formula:

$$C.I. = \frac{cpm\ control - cpm\ patient}{cpm\ control}$$

Significance was determined using the Student's t-test for unpaired data.

Skin Testing

Patients were tested prior to RNA therapy with both common recall antigens and DNCB. Later, patients were tested only with DNCB. Sensitizing and challenging doses were applied according to the method of Eilber et al. (18), and the skin tests were read and graded at 24 and 48 hours. Repeat skin tests were applied at various intervals during therapy.

Administration of RNA

Whenever possible RNA which had been extracted from a ram inoculated with autochthonous tumor was used. Because autochthonous tumor tissue was not always available, some patients were treated with RNA from sheep immunized with renal adenocarcinoma tissue from other patients. The lyophilized RNA was resuspended in sterile saline with preservative, yielding a pH of 7.34 and an osmolarity of 380 milliosmoles. Patients were skin tested with 100 μg of RNA prior to therapy, after which the RNA was injected intracutaneously in multiple wheals in lymph node-bearing areas. Doses ranged from 2 mg to 40 mg per week. Patients now initially receive 4 mg per week. If there is no improvement or measurable decrease of tumor size, the dosage is doubled every 6 weeks to a maximum dosage of 40 mg per week.

RESULTS

Twenty-three patients have been treated for between one and 21 months (see Table 1). Patients with C.N.S. metastases or with a life expectancy of less than 3 months were excluded. Ages ranged between 44 and 71 years. The primary tumor was not excised in 7 of the 19 patients with metastatic tumor. Four showed an initial clinical response and subsequently underwent nephrectomy. The other 3 did not respond and therefore did not undergo nephrectomy or receive immunotherapy.

TABLE 1.
Summary of Clinical Results

Status	Improvement	Stability or Possible Improvement	Failures	Indeterminate	Total
Metastases	3	5	7	3	18
MRD*	–	–	0	5**	5
Total	3	5	7	8	23

*Minimum residual disease
**Two of these patients were classified MRD after complete excision of solitary metastases.

Three patients were believed to be unevaluable or to have an undetermined response. Two of these patients have been treated for less than two months. One patient with a single pulmonary metastasis and a tumor in a solitary kidney has received RNA therapy for several months, but later refused to undergo the planned partial nephrectomy. His primary and metastatic lesion appear to be stable, but the influence of the RNA on the primary tumor is difficult to assess.

Patients Demonstrating Improvement

We believe three patients have shown definite improvement during RNA therapy.

Patient O.W., a 48-year-old male, underwent right nephrectomy in October, 1972 and in January, 1973, a large right renal fossa recurrence was re-excised and a partial colectomy was performed. He then received BCG therapy for 3 months, but in August, 1973, a large recurrence in the right flank was again noted. The mass continued to grow while he was receiving 5,000 R radiation therapy to the area. In November, 1973 this large mass was excised, along with most of the small bowel and part of the posterior abdominal muscles. There were positive microscopic tumor margins. Since January, 1974 he has received RNA therapy. The RNA used was initially prepared against autochthonous tumor and subsequently against tumor from other patients. The patient remains asymptomatic and free of disease.

Patient S.C., a 45-year-old white male, underwent thoracotomy in December, 1973. A mediastinal mass and multiple small pulmonary nodules were identified and biopsy showed histologic pattern of clear-cell carcinoma. No tumor was resected, and a subsequent IVP showed a large right renal carcinoma. The mediastinal mass increased in diameter in the ensuing 5 weeks, but there was a measurable radiographic decrease in size during the subsequent two months of RNA therapy. He therefore had a palliative right nephrectomy, and the mediastinal mass decreased in size slightly in the ensuing 3 months. The mass has remained stable for a total of 21 months and the patient is asymptomatic (Figure 1).

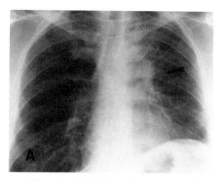 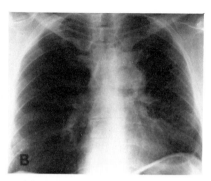

Fig. 1. a) Mediastinal mass at the onset of RNA therapy (arrows). b) After 21 months of therapy the mediastinal mass has decreased in size and there are no new lesions.

Patient G.M., a 70-year-old male, underwent a left radical nephrectomy in March, 1974. Three months later he had small bilateral pulmonary metastases and was treated with progesterone for 6 months. During progesterone therapy the lesions slowly increased in size and the hormones were discontinued. RNA immunotherapy was begun in April, 1975. The lesions remained stable for 3 months, and in the last 3 months to the time of this writing, they have decreased measurably in size (Figure 2).

Patients Demonstrating Stability or Possible Improvement

Five patients who had proven, measurable, growing metastases demonstrated either growth arrest or decrease in size of the metastatic lesions. Three of these patients (H.F., H.H., and G.P.) demonstrated no further growth of their lesions for a two-month period, after which they underwent palliative nephrectomy and continued immunotherapy. The other two patients in this group presented with metastases at varying intervals after previous nephrectomy.

Patient H.F., a 47-year-old female, demonstrated no change in the size of her pulmonary lesions before nephrectomy and in the ensuing 12 months. However, during that time, she developed a skeletal lesion and a pathologic fracture. Since then, the pulmonary lesions have slowly increased in size and a solitary brain metastasis has appeared.

Patient H.H. demonstrated growth arrest of previously rapidly growing multiple bilateral pulmonary lesions during 3 months of RNA therapy prior to nephrectomy. The growth arrest of the tumors continued for several months thereafter. To the present time, some lesions have regressed or disappeared but several large lesions have slowly increased in size. She remains asymptomatic.

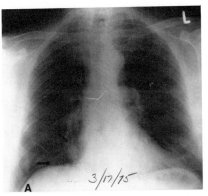 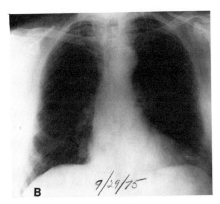

Fig. 2. a) Pre-treatment chest x-ray, March, 1975, showing multiple bilateral pulmonary metastases (arrows). b) Chest x-ray of September, 1975 shows decrease in size or disappearance of some lesions and growth arrest of others (arrows).

Patient A.M., underwent nephrectomy in May, 1975 on another urology service, although two right upper lobe lung lesions were present. These tumors grew in the interim before RNA therapy was begun in June, 1975. Since then, growth arrest has been documented and no new lesions have appeared (Figure 4). At the present time, she is a candidate for surgical excision of the pulmonary metastases.

Patient G.P. had known pulmonary metastases at the time of nephrectomy in June, 1975. Since onset of RNA therapy following the nephrectomy, the pulmonary lesions have not increased in size. However, he has developed a solitary skeletal lesion.

Patient M.S., a 50-year-old male, underwent a left radical nephrectomy for a Stage A carcinoma in October, 1972 followed by 4,200 R radiation therapy to the left renal fossa. In July, 1975, he was found to have asymptomatic pulmonary metastases and RNA therapy was begun in August, 1975. There has been slight decrease in size of the lesions during two months of therapy, but the response must be maintained for 3 months before definite improvement can be documented (Figure 5).

Treatment Failures

Seven patients failed to show any response to the therapy. Two of these had pulmonary and skeletal metastases which rapidly progressed. Three other patients with diffuse visceral metastases, including liver and brain lesions, also had no response to the therapy. These five patients died within 6 months.

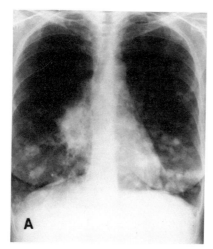
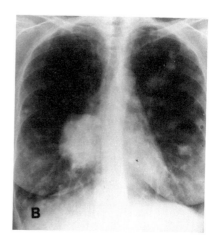

Fig. 3. a) Chest x-ray of March, 1974, prior to RNA therapy, showed a large hilar metastasis (arrow) and multiple bilateral pulmonary lesions. b) One year later, the hilar mass has increased, but there has been concomitant clearing of many smaller lesions in both lungs (arrows).

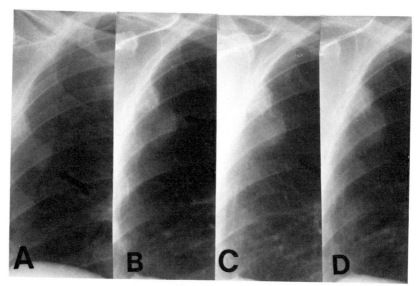

Fig. 4. a) Peripheral right upper lobe metastases before nephrectomy, May, 1975. b) Growth of the tumors seen in chest x-ray of June, 1975 prior to RNA therapy. c) Growth arrest of the tumors, July, 1975. d) Growth arrest maintained, September, 1975.

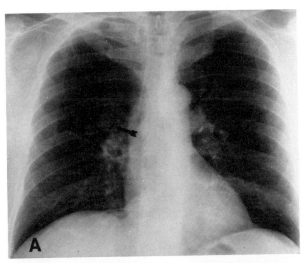

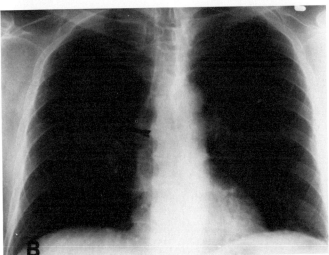

Fig. 5. *a) Two nodules near each hilum on the pre-treatment chest x-ray, July, 1975 (arrows). b) Slight decrease in diameter of metastases after 2 months of therapy (arrows).*

Another patient received only 3 weekly injections of RNA, during which time she was found to have a malignant pleural effusion and was withdrawn from therapy.

A 46-year-old male presented with 3 small pulmonary nodules two years after a nephrectomy. He has been treated with immune RNA for 4 months, during which time there has been continued slow growth of the tumors. This is the only patient with metastases confined to the pulmonary parenchyma who did not show some evidence of a response.

Minimum Residual Disease

Two patients with Stage B primary renal adenocarcinoma have received RNA therapy following nephrectomy for 3 and 20 months, respectively. One patient had a carcinoma from a solitary kidney excised in February, 1975 and has received RNA since April, 1975. Two patients have received immunotherapy after complete excision of isolated distant metastases for 12 and 6 months, respectively. All patients in this category currently have no evidence of tumor.

Lymphocyte Cytotoxicity

There appears to be a correlation between lymphocyte cytotoxicity and the patient's clinical course. Patients who failed to respond to the therapy and who manifested rapid growth of their tumor had a decrease in cytotoxicity when compared to pre-treatment levels. Patients with minimum residual disease, all of whom have shown no evidence of recurrence, have maintained their pre-treatment levels. Clinical improvement or stability has been associated with an increased cytotoxicity index in all patients tested, as in patient H.H. (Figure 6). However, it is not possible to definitely ascribe these changes in cellular immune response to the RNA therapy in all patients. The fluctuations of the cytotoxicity could be related to a change in the patient's tumor

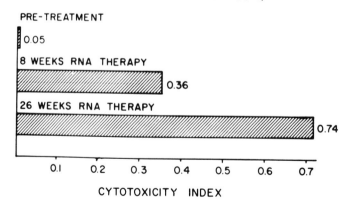

Fig. 6. *Lymphocytotoxicity of patient H.H. increased significantly during immune RNA therapy. Nephrectomy was performed after 12 weeks of therapy.*

burden and totally unrelated to the immunotherapy. Further studies will hopefully elucidate the influence of the immune RNA on the cellular immune response to tumor antigens.

DISCUSSION

It has been claimed that immunotherapy can only be expected to be effective in patients who have minimal tumor burden (15). Since this was a phase I study, primarily concerned with dosage and toxicity, therapeutic effect was a secondary consideration and patients with far-advanced tumors were included. However, it is not unreasonable to attribute the clinical responses noted in some of the patients to the immune RNA, especially in view of the natural history of this disease. In a review of over 1700 patients with metastatic renal carcinoma, Mostofi (6) reported only two survivors two years after appearance of the metastases. Middleton (19) reported no survivors after two years in patients with multiple metastases, and there were no cases of temporary arrest of growth after nephrectomy regardless of other therapy. Two of our patients have survived for 21 months, have no evidence of progression of disease and remain asymptomatic. Patient O.W. had rapid growth of residual tumor following his nephrectomy and a second local recurrence within 6 months after a complete re-excision of the tumor. Since the margins were positive following his last operation, it is surprising that he has survived almost 2 years.

Patient S.C. had a dramatic change in the growth curve of the mediastinal metastasis and the lesion has remained stable for 21 months. Furthermore, the multiple pulmonary nodules noted at the time of his thoracotomy have not been detectable on subsequent chest x-rays. In view of the natural history of this tumor, growth of these lesions would be expected over a 21-month period.

The peculiar behavior of pulmonary metastases from renal carcinoma has been previously mentioned and interpretation of changes during therapy must be made with caution. However, the 6 patients who demonstrated decrease in size of their pulmonary parenchymal metastases or arrest of growth, all demonstrated these changes at the time of onset of RNA therapy. Furthermore, it is unlikely that the changes in the metastases could be attributed solely to stimulation of host immune response related to nephrectomy, since 3 patients demonstrated stabilization of growth of their lesions before nephrectomy, and the remainder of the patients had nephrectomy months or years prior to the onset of immunotherapy.

Slow growth of some pulmonary metastases with concomitant resolution of other pulmonary lesions has been noted in several patients. Patient H.F. (see Figure 3), after initial growth arrest of all lesions, has demonstrated continued slow growth of the large hilar tumor along with regression and disappearance of other smaller tumors throughout both lungs. A similar occurrence has been noted in patient H.H., also a middle-aged female, who manifested growth arrest of the pulmonary metastases during RNA therapy for 3 months prior to nephrectomy. During the 9 months since her surgery, the two large lesions have grown very slightly, whereas smaller tumors have completely disappeared in some areas. This phenomenon makes interpretation of changes in tumor doubling time difficult. However, such diffuse pulmonary disease is usually very rapidly fatal. Prior to immunotherapy, both patients had proven rapid growth and spread of these tumors. They are alive 17 and 12 months respectively after beginning therapy and have remained asymptomatic throughout most of this time. Table 2 summarizes the current status of all patients who responded to RNA therapy.

The failure of immunotherapy to alter the course of brain and hepatic lesions is not surprising. This has been the experience in other forms of immunotherapy (20). We have also noted that skeletal metastases did not respond. The two patients who manifested skeletal metastases when treatment was begun rapidly progressed. Two other patients in whom growth of their pulmonary tumors was arrested (H.F. and R.P.) concurrently developed new solitary skeletal metastases. There is little data regarding the influence of immunotherapy on skeletal lesions, but perhaps bone metastases have a decreased response to host immune stimulation. Until further experimental and clinical data are reported, we have now excluded patients with skeletal metastases from the protocol as it is currently structured.

The effects of immune RNA on the patient's immune response require more detailed study. All patients were strongly positive to DNCB prior to beginning therapy, and this sensitivity was maintained in those who responded

TABLE 2.
Current Clinical Status of Patients who Responded to RNA Immunotherapy.

Patient	Duration of Response (mo.)	Current Status
O.W.	21	No evidence of tumor
S.C.	21	Stable mediastinal tumor
H.F.	12	Progression; skeletal metastasis
H.H.	12	Variable growth and regression
G.M.	4	Partial regression
A.M.	3	Growth arrest
M.S.	2	Growth arrest
G.P.	4	Growth arrest of lung lesions, skeletal metastasis

or remained stable. As expected, patients whose tumor growth progressed rapidly became insensitive to skin-test antigens late in the course of the disease.

The correlation between the lymphocyte cytotoxicity and the clinical course is not surprising. However, in most cases it is difficult to ascribe increases in cytotoxicity directly to the RNA therapy. Since the mode of action of immune RNA is unknown, other tests of specific aspects of the immune response may be important. Serum factors such as cytotoxic antibodies and blocking factors have not been yet studied but will be measured in the future. As in other immunotherapeutic trials, however, interpretation of the presently available tests of the immune response will probably be difficult.

All patients who had no response to therapy died within 8 months of the recognition of the metastases. Except for the presence of central nervous system or skeletal tumors, there appear to be no clinical or immunologic factors which differentiate or are predictive of those who will or will not respond. All patients in whom immunotherapy failed or who subsequently had progression of disease after an initial response received either hormonal or cytotoxic agents. None showed a response to these drugs.

It is important to emphasize the remarkable absence of toxicity associated with this form of immunotherapy. Skin rash, edema, scarification, vesical formation, infection, and pain, which often follow other forms of immunotherapy, were never encountered in any of our patients. Systemic toxicity was minimal or perhaps absent. Two patients experienced transient light-headedness and fatigue approximately 12 hours after the injection. Neither they nor any other patients developed fever, chills, gastrointestinal symptoms, neurologic symptoms, or other manifestations of allergy. Single doses up to 40 mg have been tolerated with no side effects, and some patients have received up to 200 mg total dose during therapy.

It has thus far not been possible to recognize a dose-response relationship. Patients who did not respond to lower doses were not improved while receiving higher doses. Further clinical experience must be gained, however, before dose-related response can be accurately assessed.

This preliminary trial does not conclusively establish the efficacy of immune RNA therapy in the treatment of renal carcinoma. The data suggest that treatment with xenogeneic immune RNA has altered the natural history of the tumor in a group of patients with a historically poor prognosis. Furthermore, the absence of adverse reactions and the preliminary data suggesting alteration of cellular immunity encourage further pursuit of this therapeutic modality. Until other more effective therapeutic alternatives exist, expansion of clinical trials with immune RNA is warranted and must include randomized trials in patients with minimum residual disease in whom immunotherapy has the best chance to control tumor growth.

REFERENCES

1. Talley, R. W. (1973). Chemotherapy of adenocarcinoma of the kidney. *Cancer* 32:1062-1065.
2. Lokich, J. J. and J. H. Harrison (1975). Renal cell carcinoma: Natural history and chemotherapeutic experience. *J. Urol.* 114:371-374.
3. Bloom, H. J. G., C. E. Dukes and B. C. Mitchley (1963). Hormone-dependent tumors of the kidney I. The estrogen-induced renal tumor of the Syrian hamster Hormone treatment and possible relationship to carcinoma of the kidney in man. *Br. J. Cancer* 17:611-645.
4. Bloom, H. J. G. (1973). Hormone-induced and spontaneous regression of metastatic renal cancer. *Cancer* 32:1066-1071.
5. Cox, C. E. (1975). Personal communication.
6. Mostofi, F. K. (1967). Pathology and spread of renal cell carcinoma. *Renal Neoplasia.* J. S. King, Jr., Ed. Little, Brown & Company, Inc., Boston.
7. Holland, J. M. (1973). Cancer of the kidney—natural history and staging. *Cancer* 32:1030-1042.
8. Bubenik, J., J. Jakoubkova, P. Krakora, M. Baresova, P. Helbich, V. Viklicky and V. Malaskova (1971). Cellular immunity to renal carcinomas in man. *Int. J. Cancer* 8:503-510.
9. Brosman, S., M. Hausman and S. J. Shacks (1975). Studies on the immune status of patients with renal carcinoma. *J. Urol.* 114:375-380.
10. Hakala, T. R., A. E. Castro, A. Y. Eliot and E. E. Fraley (1974). Humoral cytotoxicity in human renal cell carcinoma. *Invest. Urol.* 11:405-410.
11. Lange, P. H., T. R. Hakala and E. E. Fraley (1973). Immunobiology of genitourinary tumors. *Urology* 11:485-492.
12. Pilch, Y. H., D. Fritze, K. P. Ramming, J. B. deKernion and D. H. Kern (1976). The mediation of immune responses by I-RNA to animal and human tumor antigens. *Immune RNA in Neoplasia.* Mary A. Fink, Ed. Academic Press, New York.
13. Kern, D. H. and Y. H. Pilch (1976). Mediation of antitumor immune responses with I-RNA. *Immune RNA in Neoplasia.* Mary A. Fink, Ed. Academic Press, New York.
14. Skinner, D. G., R. B. Colvin, C. D. Vermillion, R. C. Pfister and W. F. Leadbetter (1971). Diagnosis and management of renal cell carcinoma. *Cancer* 28:1165-1172.
15. Morton, D. L. (1972). Immunotherapy of cancer. *Cancer* 30:1647-1653.
16. Ramming, K. P. and Y. H. Pilch (1971). Transfer of tumor-specific immunity with RNA: Inhibition of growth of murine tumor isografts. *J. Natl. Cancer Inst.* 46:735-750.
17. Cohen, A. M., J. F. Burdick and A. S. Ketcham (1971). Cell-mediated cytotoxicity: An assay using ^{125}I-iododeoxyuridine labeled target cells. *J. Immunol.* 107:895-898.
18. Eilber, F. R. and D. L. Morton (1970). Impaired immunologic reactivity and recurrence following cancer surgery. *Cancer* 25:362-367.
19. Middleton, R. G. (1967). The value of surgery in metastatic renal carcinoma. *Renal Neoplasia.* J. S. King, Ed. Little, Brown & Company, Inc., Boston.
20. Morton, D. L., E. C. Holmes, F. Eilber and W. Wood (1971). Immunological aspects of neoplasia: a rational basis for immunotherapy. *Ann. Intern. Med.* 74:587-604.

DISCUSSION

A. Arthur Gottlieb, Chairman

In considering the reported observations concerning "immune RNA" we seem to be faced with a set of phenomena of considerable interest whose biologic and chemical identity are still unclear.

It is important to recognize that while we think of I-RNA as being a fairly large RNA (10-18S) free of antigen, neither of these beliefs is free of objection. With regard to size, one cannot conclude on the basis of sedimentation data alone that the RNA involved is indeed a single continuous molecule; it is equally plausible that it is a group of smaller RNAs (4-5S) tied together by proteins (antigens?). If this be so, it is clear that such small RNAs cannot serve as messengers for the production of γglobulin-like subunits. The only other suggestive evidence that "immune RNA" is messenger-like is its binding (in some cases) to oligo (dT) cellulose, thus indicating that it contains stretches of poly A sequences, and the demonstration that some of these "immune RNAs" are translatable, at least in part, in cell-free systems. Nevertheless, the evidence that these "immune RNAs" are messengers is flimsy at best. Indeed, since most of the immunologic phenomena which are thought to be induced by "immune RNA" are mediated by T cells, it is difficult to know what the product specified by such a messenger ought to be. There is no real information available concerning the molecule(s), globulin-like or not, which render the T cell uniquely sensitive to a given antigen. Not knowing the product, one has real difficulty in seeing what the messenger RNA for this product ought to be.

With regard to the question of antigen, it is well to recognize that although the amounts of antigen in these preparations appears to be minute, we do not know how many molecules of antigen are present in relation to numbers of active molecules of RNA. Since the latter is unknown, and indeed the active molecules may represent only a very small fraction of the total RNA molecules present, it is conceivable that the ratio of antigens to active RNA molecules can be quite large (perhaps even 1:1). Amongst the several immunologic phenomena which appear to be mediated by "immune RNAs" — and each of these may indeed to programmed by a different RNA — only the allotype transfer experiment can be interpreted as being due to an RNA which is free of antigen (or at least, if antigen is present, it does not play a significant role). With respect to the other immunologic phenomena, we must recognize that the question of antigen contamination is unresolved.

It might be useful to go back to the molecule in macrophages which I described some years ago and which was referred to as macrophage RNP (1). You will recall that when macrophages are exposed to antigens, fragments of these antigens become associated with this RNP complex. I can't help but wonder whether some of the "immune RNA" species described at this meeting are really due to aggregates of this unusual ribonucleoprotein complex. I suppose that the only phenomenon so far described which would be completely attributable to RNA alone without antigen would be the transfer of allotypic specificity.

Since the clinical phenomena which have been described to date are largely phenomenological, and in many respects anecdotal, it would be comforting in these investigations to have (1) *an* RNA molecule, or group of RNA molecules that display the immunologic activities in question and whose activity can be reproducibly demonstrated in more than one laboratory; or (2) a clear demonstration that after lymphocytes are exposed to "immune RNA", something objectively measurable occurs in these cells – this could be the synthesis of polydeoxy or polyribonucleotides, a rise in one or another enzymatic activities, or a demonstration that different RNA molecules get into different cells which then mediate different functions.

Finally, I should like to comment on the issue of clinical trials. If I were faced with judgment of how to proceed at this point, I would choose to pursue better isolation and characterization of the RNA molecules involved rather than go forward with clinical testing of these materials. First, I do not think it wise to perform clinical trials without the purified RNAs in hand since the active materials (RNAs) may behave differently when pure than when present in crude mixtures, i.e., such mixtures may contain inhibitors of the active RNA. Consequently, if a clinical trial with the crude material were to indicate that these mixtures were not effective, one may well miss the opportunity to find any useful benefit of the purified RNA.

Secondly, I think there is a real risk, not yet fully spoken to, in administration of these RNAs. The basic protocol is immunization of sheep with human tumor cells, recovery of RNA from the sheep spleens, and injection of these RNAs into human recipients. Since the human tumor may contain putative viral sequences or incomplete viral structures which may be further replicated in the spleen of the sheep and since RNA is the basic genetic material of tumor viruses, there is a possibility that such putative viral RNA sequences might be transmitted from patient to patient. In view of the sophisticated technology available to detect murine viral RNA sequences in human cancer cells it would, I think, be incumbent on investigators of "immune RNA" who propose to put this into patients to clearly prove that the murine viral sequences are not present.

REFERENCE

1. Gottlieb, A. A. and D. S. Straus (1969). Physical studies on the light density ribonucleoprotein complex of macrophage cells. *J. Biol. Chem.* 244:3324-3329.

SESSION IV
Basic Concepts: How Does I-RNA Work?

Chairman, Robert J. Crouch

DISCUSSION

Robert J. Crouch, Chairman

Previous sessions in this workshop laid the groundwork for the material discussed in this session and much of the interchange that follows relies on the information presented.

Most of the molecular biologists present at this meeting had no experience in performing experiments with immune RNA. However, three participants related some of their attempts to become involved in the area of immune RNA. Dr. Ulrich Loening of the University of Edinburgh presented some results obtained in his laboratory by Dr. Fraser. Utilizing MOPC 47A tumors as their RNA source, they were unable to demonstrate production of free IgA with rabbit anti IgA but were able to detect surface marker formation. Dr. Sidney Pestka of the Roche Institute of Molecular Biology reiterated his findings that purified messenger RNAs for heavy and light chain proteins of immunoglobulins fail to form active antibody in a cell-free protein synthesizing system. Dr. Koch, also from the Roche Institute of Molecular Biology, presented his information concerning the cell-free translation of immune RNA, isolated by Dr. Fishman, to form anti-T2 antibody. These interesting endeavors reflected three entirely different approaches to the problems perceived by these investigators. Basically Dr. Loening's experiments reflect attempts to repeat some of the earlier findings described by many participants in this meeting while the involvement of messenger RNA for immunoglobulins in immune RNA was clearly the question approached by Dr. Pestka. Dr. Koch's collaborative efforts with Drs. Fishman and Adler clearly reflect an attempt to advance the understanding of immune RNA and its relationship to messenger RNA.

Dr. Pestka made a series of comments on the study of immune RNA and the interpretations that are available. He indicated that the results demonstrating RNA transfer are exciting and valid but there appear to be many diverse effects which can be explained by RNA antigen complexes or perhaps information transfer due to RNA alone; and that transfer of allotypic characteristics by immune RNA raises a number of questions concerning the molecular composition of immune RNA. One consistent result presented at this meeting was the size distribution of immune RNA on neutral sucrose gradients, a size ranging from 8-16S. Pestka pointed out that 16S RNA is barely sufficient to contain RNA species large enough to correspond to heavy chain message and most reports of the size of immune RNA tended to be of lower values. The problem of how this size RNA can contain the messages for light, two heavy (μ and γ), and J chain immunoglobulin proteins, as must be the case for the diverse effects of immune RNA, remains to be resolved. A complex of RNA or a distribution of all of the required RNA species might give rise to the observed S values of immune RNA. Pestka suggested that a better understanding of the molecular composition of immune RNA awaits purification, characterization and reconstitution of the active material.

Methodology remains a critical problem in the eyes of many participants at this meeting, and to alleviate the subjectivity inherent in some procedures Pestka recommended that double-blind studies be routine procedures. No conclusions should be considered valid and no experiments published by reviewers until double-blind studies are performed. Furthermore, exchange of materials should be carried out when discrepancies arise. There is clear need for communication and collaboration of biochemists, immunologists and immunotherapists.

In his discussion, Dr. Loening presented the procedures commonly employed in his laboratory for RNA isolation detailing the profiles obtained on polyacrylamide electrophoresis of the RNA. He described RNAs which are products of ribosomal RNA catabolism with mobilities slightly smaller than 28S and 18S ribosomal RNA. Paque reported that he had devoted the better part of a summer isolating immune RNA by a variety of techniques; most, if not all, proved satisfactory. Thus, different RNA isolation techniques all yield active immune RNA.

Pestka and Koch pointed out that examples of the uptake and functioning of RNA in cells from eukaryotic sources is well established. Pestka mentioned his ability to reproduce the work demonstrating interferon production by cells exposed to interferon messenger RNA; Koch reminded us that polio RNA is infectious. However, no results have yet been published that demonstrate uptake and utilization of immunoglobulin messenger RNA. But such a report would not be greatly disturbing to most molecular biologists since many messenger RNAs have been successfully translated in amphibian oocytes. If the uptake and translation of immunoglobulin messenger RNA were to result in the production of proteins that can be considered antigenic, the relationship of such RNA to immune RNA would be interesting indeed.

Translation of messenger RNAs injected into amphibian oocytes results in an overall increase in protein synthesis within the oocyte and seems to reflect an excess of protein synthetic machinery above that being used for translation of the endogenous messenger RNAs. In light of the fact that endogenous protein synthesis continues in the presence of exogenously supplied mRNA in oocytes, it seems of interest to learn of the fate of endogenous synthesis in cells challenged with immune RNA. There are actually two aspects to this problem. First, is any RNA able to enter spleen cells and flow into the translational apparatus or is there a special recognition of immune RNA in this system? None of the participants of the meeting had any information on this question. Secondly, once the immune RNA has entered the cell, is there equal opportunity for the production of immunoglobulins of allotypes of the donor and the recipient cell types?

Dr. Fishman presented some interesting preliminary data which suggests that endogenous allotype expression is strongly suppressed when one observes the phenomena induced by immune RNA. By following the allotypic characteristics of the spleen cells, Fishman noted a loss of the allotype produced in the cells before exposure to immune RNA and an increase in the allotype corresponding to that of the cell from which the immune RNA was isolated. Also, presentation of two different allotypically distinct immune RNA preparations to spleen cells resulted in partial expression of both types of allotypic markers of the donor RNA to the exclusion of that of the recipient spleen cell. The two donor immune RNAs were from cells exposed to either T2 phage or SP82 phage. If immune RNA from cells exposed to T2 phage was given to spleen cells, the antibody activity produced was amplified in the presence of antiallotype sera of the donor cell type but not the recipient. When both immune RNA from cells of two distinct allotypes which were exposed separately to T2 phage or SP82 phage were given simultaneously to spleen cells, exclusion of the recipient allotype was noted and amplification was again observed with antiallotype sera of the T2 phage exposed cells. Surprisingly, antiallotype sera of the SP82 exposed cells also amplified anti T2 antibody activity. Whatever occurs in this situation is certainly not simple, and one must ask how the recipient cell loses the capacity to continue the production of antibody of its own allotype. Competition between the RNA of the recipient and donor appears reasonable, but there must be concern for messenger RNAs of types other than immunoglobulins that need to function in the cell. Inhibition of nonimmunoglobulin mRNAs should lead to drastic changes in the cell, if not cell death.

For several years now Dr. D. Jachertz of the University of Berne has been publishing a number of papers describing the complete production of antibody in a cell-free system (see this volume). Many immunologists are reluctant to accept Jachertz's proposals since they contradict currently held theories of immunology. This workshop, bringing together molecular biologists and immunologists, provided an opportunity to examine Jachertz's system in biochemical terms (i.e., does the cell-free system described by Jachertz fit with the current knowledge of the biochemistry known about other cell-free systems?). The basic conclusion is that some major problems exist at the biochemical level prohibiting ready acceptance of the system utilized by Jachertz. Some calculations follow that point out the major areas in question.

One of the striking features of Jachertz's system is the low concentrations of materials used (2×10^{-4} µg/ml DNA, an S-100 containing 0.3 µg nitrogen/ml, 2 µM ribonucleoside triphosphates, and 10^5 antigen molecules). Dr. Roeder estimated that of the 5-6 µg protein present in the reaction mixture there would be no more than 500-600 picograms of DNA dependent

RNA polymerase and that any synthesis of RNA under these conditions would be unlikely. However, if one allows for DNA dependent RNA synthesis under these conditions, some further calculations can be made. Immunoglobulin genes are thought to represent approximately one millionth the total cellular DNA; or in Jachertz's system, 2×10^{-10} μg/ml DNA. Assuming the rate of transcription to be similar to that found in *E. coli* of 50 nucleotides per second per RNA polymerase (an estimated rate that must be too high for Jachertz's system considering the dilute nature of *both* protein and ribonucleoside triphosphates) and assuming packing of the RNA polymerase molecules similar to that observed for transcription of ribosomal RNA genes in eukaryotic systems (a packing some 10-fold higher than seen for messenger transcription), the amount of RNA formed by this process would be approximately 150 times the amount of DNA present or 150×10^{-10} μg RNA.

The calculation is as follows: Assume a gene length of 1000 nucleotides. One molecule of RNA polymerase would transverse this sequence in 20 seconds. During the five minute incubation used by Jachertz, one molecule of RNA polymerase would synthesize 15 RNA molecules. If 10 RNA polymerase molecules can be functioning simultaneously of the same piece of DNA then 10×15 or 150 molecules of RNA would be generated. Due to asymmetric transcription only 50% or 1×10^{-10} μg of DNA would be generating RNA or 150 molecules of RNA · 1×10^{-10} μg DNA = 150×10^{-10} μg of RNA = 1.5×10^{-14} g RNA nucleotides.

Jachertz generates

$\sim 1.58 \times 10^{-8}$ g RNA nucleotides
10^{-10} moles of ^{3}H UTP 20 c/mmole
10^{-9} moles of cold UTP
So S.A. $\sim$ 2 Ci/mmole
    2 Ci = $2.2 \times 10^{12} \times 2 = 4.4 \times 10^{12}$ dpm/mmole
4.4×10^9 dpm/μmole
4.4×10^3 dpm/pmole
0.1 ml get $\sim$10,000 dpm incorporated in 5 min. or 5×10^4 dpm/0.5 ml mix

$$\frac{5 \times 10^4 \text{ dpm inc.}}{4.4 \times 10^3 \text{ dpm/pmole}} = 11.3 \text{ pmole}$$

If base ratio is 0.25 UMP then,
    $4 \times 11.3 = 45.2$ pmole RNA nucleotides
    1 pmole = 350 pg
    45.2 pmole = 15.8×10^3 pg RNA nucleotides
              = 1.58×10^{-8} g RNA nucleotides

a value 10^6 higher than can be accounted for by DNA dependent RNA polymerase.

Jachertz suggests that the remainder of the RNA generated is due to an RNA dependent RNA polymerase. In order to generate the amount of RNA observed, this putative RNA dependent RNA polymerase would have to pack more tightly and synthesize RNA at fantastic rates – something on the order of 10^3 to 10^4 the rate observed *in vivo* for RNA synthesis in *E. coli*. These calculations are based on the observation of Jachertz that the incorporation of ^{3}H UTP is absolutely dependent on antigen and, presumably, all of the transcription represents transcription of one, or a few, genes. Another way to view the magnitude of this effect is to account for all of the synthesis observed by assuming that all of the DNA is transcribed and this would, under maximal conditions of transcription, generate the amount of incorporation detected. Dependence of this transcription on antigen would require the entire complement of cellular DNA to be directly involved in production of immunoglobulins. The likelihood of the majority of the cellular DNA being devoted to immunoglobulin genes seems improbable and the biochemistry suggests that the amount of RNA formed with a limited number of genes would require enzymes with rates of synthesis much, much higher than any known enzyme.

It seems fitting that this session on immune RNA concluded with a discussion of the possibility of antigen contamination in the immune RNA preparations accounting for the antibody formation attributed to RNA. It seemed to be generally agreed that tests for antigen contamination or "superantigen" are not going to solve this problem. The experiments performed by Schlager *et al.* using the ARS-NAT antigen permitted them to set a very low level of contamination, but in terms of molecules of antigen – not moles – there is still room for debate about antigen contamination. Reconstruction experiments are possible in which the minimal amount of detectable antigen – in the case of ARS-NAT < 0.0000065% by weight of immune RNA – is added along with immune RNA not specific for the antigen to ask if that amount of antigen is sufficient to account for the observed response. Fishman has reported such experiments for T2 phage and suggests that free antigen is insufficient. An interesting approach was suggested by Dr. Peters. He suggested the use of an antigen that would be chemically or thermolabile and would be totally destroyed upon appropriate treatment. To exclude superantigen (complexes of antigen or modified antigen and RNA) on any chemical basis is almost impossible since the levels are not measurable, as demonstrated in the ARS-NAT situation; and any modification of antigen leaves the investigator guessing as to the molecule that he is trying to detect.

The strongest support for non-involvement of antigen in immune RNA is based on the allotype transfer experiments. At this meeting it was pointed out that new allotypes have been discovered in animals which were previously thought to be absent. The suggestion that immune RNA selectively stimulates the formation of the donor allotype in the recipient cell does not necessarily rely on messenger RNA transfer but does suggest some information transfer. Again the antigen contamination becomes critical. Finding unusual allotypes in homologous system transfers, because of the demonstration of spontaneously arising allotypes, lessens the strength of arguments of "genetic transfer" by immune RNA. Even though the allotype transfer experiments are useful in demonstrating transfer of information, the lack of reproducibility in the hands of many workers as well as the problem of spontaneous allotype induction suggest that a more useful approach is to use the major changes among differing species to perform experiments similar to the allotype transfer experiments. It was well documented in papers presented at this meeting that effects of immune RNA can be seen in cross-species transfers. Such xenogeneic transfers can be used to test, in a clear manner, the genetic origins of the immunoglobulins produced by stimulation with immune RNA.

SESSION V
Applications to Tumor Immunology and Immunotherapy

Chairman, Peter Alexander

DISCUSSION

Peter Alexander, Chairman

RNA-rich extracts prepared from lymphoid tissue of suitably immunized animals have been found in several different laboratories to retard in an immunologically specific manner the growth of some autochthonous chemically-induced tumors and of syngeneically transplanted tumors in experimental animals. The group discussed three aspects of this phenomenon: (a) the mechanisms of action; (b) the nature of the lymphoid cells which are altered by RNA, and the nature of the lymphoid cells which are the source of the active principle; and (c) the persistence of RNA in body fluids in the face of plasma RNAse activity.

Mechanism of action

The suggestion that the injected RNA enters the lymphoid cells of the host and causes these to synthesize immunologically specific products directed against the tumor-specific antigens of the tumor (i.e. antibodies and/or antigen-specific receptors on cytotoxic lymphoid cells) has frequently been made. The chairman of this session felt that alternative mechanisms should be considered for the following reasons: (a) In experiments using well characterized messenger RNA (e.g., for hemoglobin), which effectively translate for the synthesis of the specific proteins in cell-free systems, incubation of whole cells in solution with such mRNA did not result in the synthesis of the specific protein by the treated cells. Also, A. Grassman reported that defined mRNA was transplanted by cells when introduced by micro-inoculation, but that no translation occurred if the cells were exposed to the same mRNA in the medium. (b) In experiments carried out in Alexander's laboratory (1,2), an immunologically specific antitumor effect was obtained with extracts containing the RNA of immunoblasts produced by stimulation with tumor; but on further purification, which removed much of the contaminating protein and DNA, the extract lost antitumor activity. c) RNA-rich extracts of the type that showed antitumor activity when obtained from immunoblasts directed against a variety of bacterial antigens did not induce the formation of antibacterial antibodies when injected into heterospecies.

Alexander concluded from these experiments that the antitumor action has to be ascribed to an effect other than the transcription by the recipient cells of a protein with immune-specific properties. He suggested that the observed antitumor effect may be multifactorial and may depend on the cooperative action of several reactions produced by the ingested RNA-containing material. The obvious alternate mechanism to information transfer is that the active material is "super antigen", an entity which has been well described in other systems, in which the antigenic site is covalently linked to a special RNA. There is also much evidence that RNA can act as an adjuvant without being covalently linked to antigen and this applies particularly when the protein is very firmly associated with a particular type of RNA. This is most strikingly demonstrated in immunization against mycobacteria (3), where a 16s particle containing 83% RNA was highly immunogenic for the tightly associated protein. 0.06µg of RNA-associated protein was sufficient to immunize mice against mycobacteria. The observation by Pilch and his colleagues that the antitumor activity is abolished by RNAse and not by proteolytic enzymes is not incompatible with either of these mechanisms since firm association of protein with nucleic acids protects the former against enzymatic attack.

The following other mechanisms by which RNA may augment the well established immunotherapeutic effect (4) produced by stimulating distant lymphoid tissue of the tumor bearing host with antigen were referred to by Alexander: (a) Double stranded RNA has a non-specific tumor destructive effect in which the immune reactivity of the host against the tumor contributes (5). It is possible, therefore, that amounts of dsRNA which are insufficient to affect the tumor by themselves may do so in conjunction with specific immunization with antigen. (b) McCullagh (6) has shown that high dose tolerance of rats to sheep red cells can be abolished if the thoracic duct cells of the tolerant rat are incubated *in vitro* with RNA derived from reactive nodes as well as the specific antigen. When these thoracic duct cells are reinjected, the rats make antibody to the sheep red cells. (c) Jones and Lafferty (7) find that a lymphocyte transfer reaction characterized by an intense host response can be mimicked by injecting RNA from allogeneic lymphocytes. While there is no evidence that one or more of these processes referred to above are involved in the antitumor action of the RNA extracts, it is important to realize that there are alternatives to the immune mRNA hypothesis for the antitumor action of RNA-containing extracts.

Dr. Pilch suggested that there may be receptors for RNA on the surface of T-cells and that binding of specific RNA to them may induce the synthesis of proteins involved in specific T-cell killing by derepression. However, Dr. Pilch also felt that he had preliminary data which supported the mRNA hypothesis in that injection of the antitumor RNA caused the formation of antibodies in rats. The work, however, was not completed and he had no data which indicated whether the immunoglobulin was of host type or of the phenotype of the RNA donor. If injection of sheep RNA into rats results in the formation of sheep immunoglobulin with anti-rat tumor activity, this would constitute proof for the mRNA hypothesis. Dr. Pilch stated that such experiments were now in progress in his laboratory.

Nature of Lymphoid Cells

Dr. Kern reported preliminary data on RNA-induced cytotoxicity using the system previously described by the Pilch group in which spleen cells from normal rats were incubated with "immune RNA." Some of these spleen cells became specifically cytotoxic to the sarcoma which had been used to immunize the rats from which the "immune RNA" was extracted. Dr. Kern found that this specific cytotoxicity resided in the non-adherent cells of the treated spleen and he had indications that the cytotoxic cells were T-cells. A lively discussion ensued on the nature of the *in vitro* test used by the Pilch group. Dr. Sell and others questioned whether this really measured cytotoxicity and several suggestions were made for examining more precisely the interaction between the cells treated with "immune RNA" and the tumor cells. This is clearly a key problem, since the most compelling evidence that the *in vivo* antitumor action of RNA results from the transformation of normal lymphoid cells into specific cytotoxic cells relies on this test. Dr. Kern further presented very recent experiments which indicated that RNA, active in this *in vitro* transformation of spleen cells, was derived from the T-cells of the immunized donor rats and that the cells of the mononuclear phagocytic series do not yield active RNA. In this respect this system appears to differ from that of Fishman and Adler, where macrophages provide the RNA that in their experiments induces specific antibody synthesis in normal lymphoid cells. It became clear that a question which needs resolution in all of the experiments in which transfer of immunological information by RNA is found is the nature of the cells which provide the RNA that is active.

Persistence of RNA in tissue fluids

If immune RNA is to act *in vivo* it has to persist in tissue fluids long enough to be bound and taken up by the target cells. Plasma from lymph and blood contains a variety of RNAses that degrade both single and double stranded RNA. The possibility that there may be serum proteins which protect RNA was raised by Dr. Heller who found that a serum α-globulin protected the information transfer activity of RNA derived from myeloma cells. This RNA was found by him to convert normal lymphocytes into lymphocytes which have the specific myeloma protein in their plasma membrane. There was no evidence, however, that this protection by a serum protein is a phenomenon that applies to all types of RNA. It became clear from the comments of the biochemists in the group that much work needs to be done to determine the nature of the plasma RNAses. Especially, experiments need to be carried out to establish the rate at which the activity of putative immune RNA is lost on incubation in serum. Dr. Robertson pointed out the interesting possibility that serum RNAses may be restricted in their actions to RNA having particular configurations and that immune RNA may be resistant to RNAses which degrade other types of RNA. This hypothesis is, of course, amenable to test.

Dr. Jachertz emphasized that in all of his experiments transfer of immunological information by RNA only occurred when the cells were exposed to optimal concentrations of active RNA. For example, in several of his systems 5 pico grams of RNA were active, but both 25 and 1 pico grams were inactive. He stressed that the amount of RNA which was used to induce regressions of tumors was orders of magnitude greater than the concentrations at which he observed activity. This raises the possibility that destruction of much of the ingested RNA by the RNAses of the body may be a prerequisite for obtaining information transfer *in vivo*.

REFERENCES

1. Alexander, P., E. J. Delorme, L. D. G. Hamilton and J. G. Hall (1967). Effect of nucleic acids from immune lymphocytes on rat sarcomata. *Nature. 213*:569-572.
2. Alexander, P., E. J. Delorme, L. D. Hamilton and J. G. Hall (1967). Stimulation of anti-tumor activity of the host with RNA from immune lymphocytes. Nucleic Acids In Immunology, (proceedings of a symposium held at the Institute of Microbiology of Rutgers University). O. J. Plescia and W. Braun, Eds.
3. Loge, R. V., W. E. Hill, R. E. Baker and C. L. Larson (1974). Delayed hypersensitivity reactions provoked by ribosomes from acid-fast bacilli: physical characteristics and immunological aspects of core ribosomal proteins from *Mycobacterium smegmatis*. *Infection and Immunity. 9*:489-496.
4. Haddow, A. and P. Alexander (1964). An immunological method of increasing the sensitivity of primary sarcomas to local irradiation with X-rays. *Lancet 1*:452-457.
5. Alexander, P., I. Parr and E. Wheeler (1973). Similarities of the anti-tumor actions of endotoxin, lipid A and double-stranded RNA. *Br. J. Cancer 27*:370-389.
6. McCullagh, P. (1973). The abrogation of immunological tolerance of sheep erythrocytes by means of sub-cellular extracts. *Aust. Exp. Biol. Medic Sc. 51*:783-791.
7. Jones, M. A. S. and K. J. Lafferty (1966). The dermal reaction induced in sheep by homologous lymphocytes and an RNA fraction extracted from homologous lymphocytes. *11th Congress Int. Society Blood Transfusion*, Sydney, 635-644.

GENERAL DISCUSSION

IMMUNE RNA – QUESTIONS NOT ANSWERED

Stewart Sell, M.D.

Department of Pathology
School of Medicine
University of California, San Diego
La Jolla, California 92093

The Woods Hole workshop on "immune RNA" was productive more in the questions raised than in the answers provided. The term "immune RNA" was used in a general way at the conference to refer to phenol-extracted preparations from the lymphoid tissues of immunized donors that could specifically transfer immune reactivity *in vitro* to non-sensitized recipient lymphoid cell preparations or *in vivo* to individuals not previously immunized. Two general questions became apparent: (a) Are the data interpreted as demonstrating the transfer of specific information to recipient lymphoid cells by "immune RNA" reproducible and valid? (b) If the phenomena are real, what are the theoretical explanations and practical applications? To approach these two general questions I have organized this overview of the conference by asking five more specific questions about immune RNA: (a) Where does it come from? (b) What is it? (c) Where does it go? (d) What does it do? and (e) How does it work?

Where does it come from? The general assumption, mainly based on earlier published work, is that the extracted materials used in most transfer experiments described in this volume are derived from macrophages. Alexander presented old observations that phenol extracts from cell suspensions containing macrophages were effective in inhibiting the *in vivo* growth of a mouse fibrosarcoma, whereas extracts from activated lymphoblasts from the draining lymph nodes of an immunized donor did not. The earlier studies of Fishman and Adler also suggest that the macrophage is the source of the biologically active extract. However, as the experimental protocols were described, it became clear that this assumption could not be made for most of the systems studied. The phenol extracted material used for transfer is usually obtained from mixed lymphoid cell populations, i.e., lymph node, spleen or peritoneal exudate cells. Although reasonable data were presented to support the notion that the transferred material is derived from cytoplasm and not nuclei, the cell type from which it is extracted is not clearly defined. Therefore, one approach for further study would be to use purified lymphoid cell populations from immunized donors to define the cellular source of "immune RNA."

What is it? A major effort of the conference went into discussion of what was present in the phenol extractable material. In short, the composition of the active material is not known. The most likely possibilities are that it is RNA, RNA-antigen complex or RNA fragments. This conclusion is based on the fact that RNAase treatment eliminates activity. Even this point is not clear, however, because in some systems studied the RNAase added to the immune RNA is not removed prior to transfer to recipient cells and therefore may actually be acting directly on effector cells.

Assuming that RNA molecules of some form are required, a number of inconsistancies became apparent during the discussion:

(a) The size of the RNA in phenol extracted material was found by several workers to be between 8-16S after fractionation on sucrose gradients. RNA of this size is not big enough to code for a complete antibody molecule, but it is enough for an Ig light chain or some nuclear proteins. As the results of different experiments were presented, one became overwhelmed at the amount of specific information that was transferable by various "RNA" preparations. No one was able to identify an RNA molecular form which would clearly fulfill the requirements for such information transfer.

(b) The phenol extraction methods used most likely produce RNA fragments as well as intact RNA. The question was raised as to how such fragments could transfer information. The studies on viroid RNA described by Hugh Robertson suggested the possibility that RNA fragments might be incorporated into new sequences or function in a previously unexpected way. This virus-like RNA has no messenger function, but appears to have an "initiator" function that produces drastic effects on cells in that they make many more RNA molecules.

(c) The active phenol extracts of RNA are usually crude extracts that contain a variety of species of RNA as well as other materials. Several methods for extracting more purified RNA were presented. However, it seems that the transfer activity of the phenol extracts required "dirty" preparations. It was stated by several workers that more refined extraction procedures resulted in inactive material.

(d) An interesting historical point brought out was that the phenol extraction method was originally designed for preparation of endotoxin. This raised the possibility that some effects attributed to "immune RNA" might be caused by endotoxin. The specificity found in many transfer experiments argues against some of the non-specific effects that endotoxin might produce, but a specific response could be explained by the simultaneous presence of endotoxin and small amounts of antigen (see below). The presence or absence of endotoxin in the various phenol extracts was not tested directly; but in at least some experiments done after gradient fraction of RNA, endotoxin should have been removed.

(e) The presence of antigen in phenol extracts and the possible function of RNA-antigen complexes as "super-antigens" was discussed in detail. In fact, there is a long-standing disagreement as to whether or not antigen is present in the material being transferred (1,2). The presence of antigen could not be conclusively ruled out in most of the preparations described. The most pertinent studies on this point were those of Schlager et al. who were unable to detect a known antigen potentially measurable in very small amounts in an active "immune RNA" preparation (3) and that of Fishman and Adler who treated RNA-Ag with antibody, presumably precipitated out RNA-Ag complexes, and were still able to transfer specific antibody formation (4). However, these studies are not conclusive as antigen might be highly active when complexed to RNA even in undetectable amounts. Crouch calculated that at the level of detection available to Schlager et al. (this volume) there could still be 1 molecule of antigen for each 1,000 molecules of "immune RNA." Accepting the assumption of Crouch that 1 mg of total RNA contains only 50 μg of "immune RNA" due to the fact that most of the RNA is ribosomal or transfer RNA and not active in transferring the capacity to produce antibody, 1 mg of total RNA could contain 10^9 molecules of antigen. We have found that as little as 10 pg of TNP-hemocyanin (KT) is sufficient to stimulate an antibody response by rabbit or mouse spleen cell cultures (Redelman and Trefts, unpublished data). Assuming a molecular weight of 8×10^6 for KT, this represents $10^7 - 10^9$ molecules of antigen.

The above calculations illustrate that very small amounts of antigen may be immunogeneic when administered in the proper form. Therefore, any experiments designed to rule out the presence of minute amounts of immunogen in RNA preparations must be particularly well-controlled to be interpreted. The attempts to remove antigen by precipitation with antibody mentioned above fall into this category. It may be impossible to eliminate the functional presence of antigen by precipitation with antibody. Antibody precipitation of antigen in solution almost always results in the production of some soluble complexes and these may be highly immunogenic. The antibody precipitation experiments of Fishman and Adler (4) should be repeated using solid antibody immunoabsorbant columns with marker radiolabeled antigen. Clearly, more definitive studies are needed to rule out antigen carry over.

Where does it go? In order for "immune RNA" to exert a function on effector cells, it is presumed necessary that it enter the effector cells. However, the data on where "immune RNA" goes are almost nil. Few, if any, attempts have been made to determine the fate of immune RNA *in vitro* and none *in vivo*. Labeled macrophage RNA can enter lymphocytes (5). However, penetration of an active fraction of the "immune RNA" into the cytoplasm of effector cells has not been clearly demonstrated. The fate of "immune RNA" might be traced by labeling the presumptive "immune RNA" in the 8-16S fractions of RNA preparations and determining its cellular distribution using autoradiography.

The role of ribonuclease in determining the fate of transferred RNA preparations was also discussed in some detail. Most investigators use some method to try to prevent digestion of their RNA preparations. This is feasible for *in vitro* experiments, but little protection seems possible *in vivo*. It was shown that RNA added to serum is rapidly—perhaps within seconds—degraded. The question was raised as to how RNA could get to effector cells *in vivo* if there was such a rapid degradation of RNA. The answer supplied was that the large amounts of RNA used (on the order of milligrams/animal) would permit a small fraction of the transferred material to survive and reach a functional destination.

The route of injection of RNA preparations into recipient animals is an important factor in determining *in vivo* distribution. For most transfer studies, RNA preparations are injected intradermally and not systemically. This route is classically that used for active immunization by antigen and is not prone to provide rapid dissemination of injected material. The route of injection achieves even greater significance in view of the fact that polynucleotides, including double-stranded RNA, and endotoxin, are known to function effectively as adjuvants in induction of immune response particularly after intradermal injection. In summary, the biological distribution and the eventual functional site of activity of "immune RNA" preparations are unknown.

What does it do? One of the key questions regarding evaluation of the functional activity of phenol extracted immune material is whether or not the immune assays used for determining activity are valid. The systems described in this symposium included *in vitro* and *in vivo* tumor inhibition, transfer of delayed hypersensitivity *in vitro*, transfer of specific antibody production and formation of donor allotypic markers *in vitro*. These studies almost always demonstrate immune specificity in that phenol extracts from nonimmunized donors do not work. Taken at face value, these studies are extensive, well-controlled and apparently specific. However, at the Woods Hole meeting, details of specific methods and techniques were not always presented and results were usually given as ratios, indicies or plus-minus without the raw data being available. A disturbing feature was the tendency to present the results of one positive experiment without indicating the percent of total experiments performed that were positive. It became clear in a few presentations that negative results were, in fact, not uncommon. Independent verification of most of the phenomena studied is needed.

Repeated arguments were made for the validity of the effects reported. I would like to discuss these under four headings: (a) tumor inhibition, (b) antibody production, (c) allotype transfer and (d) transfer of species specific immunoglobulin production.

(a) Tumor Inhibition — From the bulk of the data presented it seems that the most reproducible finding is that "immune RNA" can retard or reverse the growth of an experimental tumor *in vivo*. That delayed hypersensitivity to more classic antigens such as PPD can be transferred *in vitro* also seems valid from the results of Paque. But the transfer of delayed hypersensitivity to PPD *in vivo* seems not to have been possible and the transfer of tumor killing effects of "immune RNA" *in vitro* depends upon the interpretation of an inadequately characterized "cytotoxicity" assay. This assay as used by the Pilch group measures tumor cell adherence. There was not clear evidence given that the nonadherent cells were killed or even damaged. The ability of this assay to measure an immune event or even a killing event must be firmly established. The clinical studies on tumor inhibition described at the conference were clearly negative. In fact, the design of the study, based on the assumptions that renal cell carcinoma consistently grows and that metastases do not have periods of stable growth or regression, is inadequate. Metastases from renal cell carcinomas may remain quiescent for many years (6) and have been known to regress spontaneously (7).

(b) Antibody Production — The *in vitro* passive transfer of antibody production by phenol extracts has been duplicated in a number of laboratories. Although the rise in titers or increase in plaque-forming cells is not large, the data indicate a high degree of specificity and sufficient controls are included to justify confidence that specific antibody formation can be transferred by phenol extracts of lymphoid tissue from immunized donors (immune RNA).

(c) Immunoglobulin Allotype Transfer — Allotypes are immunochemically and structurally (amino acid sequence) recognizable variants of immunoglobulin molecules that are genetically determined and inherited according to Mendelian laws. The ability of "immune RNA" to transfer specific genetically controlled allotypic information presumably not present in the genotype of the cells of the recipient is of much greater theoretical importance than transfer of specific delayed hypersensitivity tumor inhibition of antibody production. The early experiments of Adler *et al.* (2) employed one of the most sensitive assays for antibody available, neutralization of T_2 phage. The identification of allotypic transfer was made by a loss of neutralizing activity in culture supernates of lymph node cells by addition of anti-allotypic serum of the donor allotype and assayed by the percentage of viable phage remaining. This observation should be verified by another technique because the reduction of phage killing as an assay for allotype specificity is only a qualitative measurement and not clearly interpretable. Antibody activity to other allotypic markers not identified at the time these experiments were done or other non-specific effects might be operative. The possibility also remains that immunoglobulin carryover could account for the finding of donor allotype.

An attempt at verification was made by Bell and Dray using enhancement of plaque-forming cells by specific anti-allotype serum. At face value, the data of Bell and Dray (9-11) are most impressive and convincing. However, the question in regard to their findings is reproducibility. The incidence of negative experiments is not reported by Bell 'and Dray. From comments made at the meeting it seems well-known that it has been difficult if not impossible to duplicate Bell and Dray's results in other laboratories, or in fact by others in their laboratory. It is indeed unfortunate that these potentially important findings have not as yet been reproduced. Until this is done, allotype transfer by immune RNA cannot be accepted as a reproducible phenomenon.

The transfer of a myeloma idiotype to non-neoplastic lymphocytes by an RNA fraction from the cytoplasma or plasma is another complex observation which defies a conventional explanation at this time (see papers by Giacomoni and by Heller, this volume). The transferred myeloma idiotype appears on recipient cells more rapidly than immunoglobulin can be synthesized by plasma cells and the number of cells acquiring the transferred idiotype is enormous (12,13,14). It is possible that the rosetting techniques used may not be specific; the data presented at the conference were insufficient to evaluate this fully. Because the RNA preparations are of tumor origin, it is possible that the phenomenon does not occur in non-tumor situations. The entrance of transfer RNA into xenogeneic cells has been demonstrated in other systems. Plasma cell derived RNA could enter other differentiated plasma cells and lead to production of donor-type Ig. The biological significance of this type of transfer would be quite different from the transfer of specific immune information into uncommitted immune cells. Much mention was made of the presence of A-type particles in the tumor cells, but the role of these in the transfer phenomena was not further clarified. Further understanding and documentation of this type of phenomena is required before its significance can be evaluated.

(d) *Species Specific Immunoglobulins* — Many of the experimental protocols employed phenol extracts of the lymphoid tissue of one species (i.e., sheep) injected into another species (guinea pigs, mouse, rat, man). The most critical result demonstrating information transfer would be to find donor immunoglobulin produced by the recipient cultures or individual. This observation has not been made to date.

How does it work? One of the most interesting conclusions of the meeting was that it is not necessary to accept the reality of a phenomenon in order to speculate on its mechanism. Because of the varied nature of the systems studied, possible mechanisms for the different phenomena will be discussed separately. The mechanisms of effect of phenol extracts generally can be divided in those in which antigen may be present (RNA-Ag) and those in which antigen is not present (informational RNA or INA) as emphasized by Fishman at this conference. The phenomena included in this discussion of mechanisms include (a) tumor regression, (b) antibody transfer and (c) allotype transfer.

a) *Tumor Regression* — Peter Alexander stressed in his presentation that induction of tumor regression *in vivo* must be due to a combination of mechanisms. For most *in vivo* tumor systems a combination of immune specific and non-specific events may occur. Double-stranded RNA derived from a fungus can inhibit fibrosarcoma growth in a mouse model system, presumably through pharmacological effects. Polynucleotides are known to stimulate interferon production *in vivo* and *in vitro* and interferon may prevent or inhibit growth of tumor cells. The role of antigen or an RNA-antigen complex cannot be evaluated until a system using a well-defined tumor-specific transplantation antigen is used. Clearly, the tumor regression effect of RNA extracts has been documented in several laboratories using a variety of experimental tumors. More models are far less needed than imaginative protocols designed to determine the mechanisms — in other words, protocols to answer any of the five questions posed.

b) Antibody Production — One explanation for transfer of antibody remains the transfer of antigen in a highly immunogenic form. If information is actually passed from RNA (presumably macrophage RNA) to a lymphocyte precursor (perhaps B cell) then the cell involved in specific antigen recognition is not the lymphocyte (T or B cell) but the macrophage. This is inconsistent with most current concepts of immune recognition. There is no evidence that macrophages either produce immunoglobulin or produce information (perhaps messenger RNA) for Ig production. Hybridization studies of macrophage RNA with plasma cell DNA applied to the Heller system might be a technique that could be applied to this question. If immune RNA is being produced, a DNA probe could be used to detect it. Therefore, in spite of the experiments mentioned above in which antigen could not be detected in RNA preparations used to transfer antibody production and attempts to remove any antigen were performed, the presence of a "super-antigen" in the form of an RNA-antigen remains a plausible hypothesis. This complex may be more stable than antigen alone or more able to activate specific immunocompetent cells. The fact that RNAase destroys this effect while pronase does not can be explained by a requirement for some form of RNA for the effect and protection from pronase of the antigenic structure in the RNA-Ag complex. It is also possible that RNAase inhibits, not be affecting the RNA extracts, but by direct effect of RNAase on the recipient cells. Endotoxin is also a potent adjuvant and could be present in many "immune RNA" preparations.

(c) Allotype Transfer — Allotype transfer cannot be explained by the presence of antigen. The seemingly inescapable conclusion is that if allotype transfer does occur some transfer of information must take place. This information could be structural or gene controlling in nature. Such information could be in the form of messenger RNA which redirects plasma cells to make a new amino acid sequence, but the size of the RNA required for this is too large to be supplied by "immune RNA." Fragments of RNA might be inserted into host messenger and provide for information for allotypic structure. Immune RNA might act through reverse transcriptase to insert new DNA sequences into the host genone. The function of immune RNA might be in directing synthesis of nuclear proteins controlling gene translation. This latter possibility is supported by the recent observations of unexpected third allotypes in genetically heterozygous rabbits (15,16). Thus, control of expression of a genetically determined structure may be involved rather than actual information transfer.

None of the mechanisms proposed above seems likely. Because of the lack of reproducibility of allotype transfer by immune RNA, speculations on its mechanism may be unwarranted. In short, the transfer of allotypic specificity by "immune RNA" is not a consistently reproducible phenomenon. The clear reproducible demonstration of incorporation of labeled amino acids into immunoglobulin of donor allotype following treatment of recipient lymphoid cell cultures with "immune RNA" is required to resolve this important question. The accumulation of a protein in tissue culture medium does not prove synthesis. As a rule, it takes many negative experiments to counter-balance a few positive ones. However, the documentation of an inexplicable phenomenon has to be good. Only time will resolve this dilemna. If it is valid, definitive reproducible results should eventually be obtained.

CONCLUSIONS

What then are the answers to the two major questions posed? Are "immune RNA" transfer experiments valid; and, if so, what are the theoretical implications and practical applications? Not all of the data suggesting transfer of immune information by phenol extracts are convincing and reproducible; those that are might be explained by transfer of antigen complexed to RNA or by mechanisms that do not involve transfer of specific immune information. More definitive, reproducible and convincing experiments are thus required before the concept of transfer of specific information by immune RNA can be generally accepted. However, the repeated observation of inhibition or reversal of tumor growth in experimental animals by passive transfer of phenol extracts of lymphoid tissue is an exciting and potentially important finding. Because of the possible applicability of this phenomenon to treatment of cancer in humans, every effort should be made to define the mechanism of action and to determine the conditions which will provide the most efficacious clinical use. For the future, new approaches that will document the validity of the phenomena reported and clearly define the mechanism of transfer should receive enthusiastic support. But should the clinical application of the transfer of phenol extracts to humans with tumors be done until more is known about their effects in experimental animals so that meaningful protocols can be designed? In summary, we don't know the cellular origin of "immune RNA." We don't know what it is, where it goes, or how it works, but it seems to do some very interesting and potentially important things.

ACKNOWLEDGEMENTS

I would like to thank Mary Fink for asking me to write this summary of the Woods Hole workshop on "immune RNA" and the following, who read and contributed to this manuscript: Richard Dutton, Ruben Falcoff, Doug Redelman, Susan Swain and Park Trefts.

REFERENCES

1. Gottlieb, A. A. and R. H. Schwartz (1972). Review: Antigen-RNA interactions. *Cell. Immunol. 5*:363-365.
2. Adler, F. L. (1972). Meeting Report: RNA in the immune response. *Cell. Immunol. 5*:363-365.
3. Schlager, S. I., S. Dray and R. E. Paque (1974). Atomic spectroscopic evidence for the absence of a low-molecular-weight (486) antigen in RNA extracts shown to transfer delayed hypersensitivity. *Cell. Immunol. 14*:104-122.
4. Fishman, M. and F. L. Adler (1967). The role of macrophage-RNA in the immune response. *Cold Spring Harbor Symp. Quant. Biol. 32*:343-348.
5. Fishman, M., R. A. Hammerstrom and V. P. Bond. *In vitro* transfer of macrophage RNA to lymph node cells. *Nature 198*:549.
6. Starr, A. and G. M. Miller (1952). Solitary jejunal metastases twenty years after removal of a renal cell carcinoma. Report of a case. *N.E.J. Med. 246*:250-251.
7. Mann, L. T. (1948). Spontaneous disappearance of pulmonary metastases after nephrectomy for hypernephroma. Four year follow-up. *J. Urology 59*:564-566.
8. Adler, F. L., M. Fuhman and S. Dray (1966). Antibody formation initiated *in vitro*. III. Antibody formation and allotypic specificity directed by ribonuclei acid from peritoneal exudate cells. *J. Immunol. 97*:554-558.
9. Bell, C. and S. Dray (1969). Conversion of non-immune spleen cells by ribonucleic acid of lymphoid cells from an immunized rabbit to produce γ M antibody of foreign light chain allotype. *J. Immunol. 103*:1196-1211.
10. Bell, C. and S. Dray (1970). Conversion of non-immune rabbit spleen cells by ribonucleic acid of lymphoid cells from an immunized rabbit to produce IgG antibody of foreign light chain allotype. *J. Immunol 105*:441-556.
11. Bell, C. and S. Dray (1971). Conversion of non-immune rabbit spleen cells by ribonucleic acid of lymphoid cells from an immunized rabbit to produce IgM and IgG antibody of foreign heavy chain allotype. *J. Immunol. 107*:83-95.
12. Yakulis, V., N. Bhoopalam, S. Schade and P. Heller (1972). Surface immunoglobulins of circulating lymphocyte in mouse plasmacytoma. I. Characteristics of lymphocyte surface immunoglobulins. *Blood 39*:453-464.
13. Bhoopalam, N., V. Yakulis, N. Costea and P. Heller (1972). Surface immunoglobulins of circulating lymphocytes in mouse plasmacytoma. II. The influence of plasmacytoma RNA on surface immunoglobulins of blood lymphocytes. *Blood 39*.465-473.
14. Giacomoni, D., V. Yakulis, S. R. Wang, A. Cooke, S. Dray and P. Heller (1974). *In vitro* conversion of normal mouse lymphocytes by plasmacytoma RNA to express idiotypic specificities on their surface characteristics of the plasmacytoma immunoglobulin. *Cell. Immunol. 11*:389-400.
15. Strosberg, A. D., C. Hamers-Casterman, W. Van Der Loo and R. Hamers (1974). A rabbit with the allotypic phenotype: a1a2a3 b4b5b6. *J. Immunol. 113*:1313-1325.
16. Mudgett, M., B. A. Fraser and T. J. Kindt (1975). Non-allelic behavior of rabbit variable-region allotypes. *J. Exp. Med. 141*:1448-1452.

MICRODETECTION AND DIFFERENTIATION OF FREE AND INTRAVIRAL NUCLEIC ACIDS USING LASER FLUORESCENCE MICROSCOPY AND FLUOROCHROME MIXTURE STAINING[1]

M. A. Apple,* M. Hercher,+
T. Hirschfeld+ and T. Keefe+

*University California Medical School
San Francisco, California 94143
+Block Engineering Inc.
Cambridge, Massachusetts 02139

INTRODUCTION

Data supporting the concept that RNA can transfer specific immunity from antigen-exposed cells to cells apparently not antigen exposed has appeared (1-11). The verification that an RNA and only an RNA carries specific new information from one cell to another as part of a normal process of antigen information processing would be of immense significance.

The ability to transfer, by a selected RNA, a specific antigen responsiveness in man could create an important new modality of cancer therapy (12-14) and protection against a myriad of infectious diseases. This verification could lead to selective interference with immune responsiveness, such as by new drugs (15) which could selectively reverse transplant rejection, graft-vs-host reactions and autoimmune diseases.

The key question is: Can we verify or have we ascertained that a specific RNA receives specific antigen information from a specific cell and carries that information independently to a new cell where that specific information is

---

[1] Supported in part by NIH Grant No. CA 14894.

expressed? There has been no unequivocal demonstration of the entire paradigm even after close scrutiny:

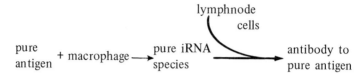

A careful examination suggests that possibly none of the individual steps has been demonstrated unequivocally, although a large amount of evidence is now available which supports and is consistent with both the individual steps and even the entire paradigm. Unequivocal evidence that a specific macrophage is *the* cell processing information from a pure antigen into a novel intercell RNA code form has not become available. The task might be simplified if experiments could be conducted on single cells. Unequivocal evidence that a specific information-containing RNA uniquely carrying information provided by a pure antigen exists and can transfer this information by a decoded expression of its base sequence in another cell also has not become available. Major difficulties in demonstrating if this step of the paradigm occurs are provided by limitations in the current methods available for separation, analysis and identification of this putative "immune RNA".

There are needs for new methods for the detection, identification, quantitation and analysis of nucleic acids in ultramicro quantities, where such methods could be rapidly applied nondestructively in solution. We report here the preliminary application of recent technology to approach this set of problems. Using laser fluorescence microscopy with attenuated total reflection illumination intensity fluctuation spectroscopy, a double fluorochrome stain and special computerized data handling techniques (16-23), we have been able to distinguish the sizes and amounts of molecules in the 200-3500A (IgG to virus size) range in 0.1 microliter samples in solution in a few minutes, distinguish RNA from DNA and distinguish forms of nucleic acid which are all singlestranded from those containing doublestranded regions. This technique presently requires 10^{-19} moles of nucleic acid to clarify these points, but the physical limit is about two orders of magnitude below this or less than 100 molecules.

METHODS

A. *Brownian size determination.* The instrument we describe (developed by Block Engineering, Inc., Cambridge, Mass.) illuminates the sample with an argon laser beam focused (at an angle beyond the critical one of total reflection) on the horizontal surface of a high refractive index prism that acts as a support for the stained sample solution. The sample's fluorescence is observed with an NA 1.25, 100X objective in a fluorescence microscope with appropriate filters and a trinocular head with an S-20 photomultiplier in its vertical arm. The electronic system then calculates the log-log power spectrum of the signal's autocorrelation function, a straight line whose slope multiplied by a calibration constant gives the log of the particle's Brownian diameter (16-18,22,23).

The typical staining procedure uses a solution containing 0.2M pH7.2 phosphate buffer, 2% bovine serum albumin with either 10^{-5}M ethidium bromide or a mixture of $6.7.10^{-6}$M ethidium bromide and $6.7.10^{-7}$M acridine orange added (20,27). The corresponding emission bands at 520 and 590_{nm} are isolated by appropriate filters.

Where viral samples were used, the individual viruses could be observed moving randomly through the field. Natural backgrounds, as for instance human serum, show no observable particles under these conditions and do not appear to produce significant background problems.

B. *Nucleic acid content per particle and concentration determination.* The output fluctuations of the laser fluorescent microscopy instrument contain both the background shot noise (simply related to the measured steady state background), the signal shot noise (simply related to the measured integrated fluctuating signal), and the concentration fluctuations of the sample. Subtracting the known noise sources from the measured fluctuations then gives the concentration fluctuations themselves, which are simply related to the concentration. This in turn can be combined with the total measured signal to give the signal per individual particle and thus the particle's nucleic acid content (19).

C. *Nucleic acid type determination with the fluorochrome stain mixture.* These can be made in the instrument by viewing particles stained with the combination stain with the filters corresponding to each of the spectral emission peaks obtained. The nucleic acid can then be classified by the intensity ratios at these peaks. If somewhat larger sample quantities are available, the measurements may be done macroscopically in a spectrofluorimeter, as is done for the pure chemical samples described, which contained about 10^{-10} mol/μl of nucleic acid. A 4765Å laser is used for excitation.

D. *Materials.* Nucleic acids and viruses were obtained from standard commercial sources or prepared by previously published methods (24-26).

RESULTS

A. *DNA oligonucleotide fluorescence.* The intensity of blue laser stimulated fluorescence at each wavelength over the range of 485-645 nm was plotted for oligo-dT (12-18 bases) and oligo-dG (12-18 bases). A sharp peak occurred in both cases (Figure 1 and Figure 2) at approximately 522nm with a shoulder at higher wavelengths for both these single stranded DNAs.

B. *Single stranded RNA fluorescence.* The fluorescence intensity of poly-rC, a single stranded RNA, showed a sharp peak at 520nm and a lesser peak at about 560nm with an additional shoulder at higher wavelengths (Figure 3).

C. *Soluble RNA fluorescence.* The fluorescence spectrum emitted by 4-20S RNA (some single, mostly double-stranded) in response to excitation at 4765Å shows a major departure from the pure single strand RNA; the maximum intensity is at 590nm and a second small peak is at 520nm. The 590nm peak appears associated with the formation of a two-stranded duplex (Figure 4).

D. *Fluorescence pattern of RNA:DNA duplexes.* RNA:DNA duplexes, such as those formed by transcription from RNA or from DNA, show a fluorescence pattern with a major peak at 590nm, a shoulder at about 615nm and a minor peak at 520nm. The relative height of the 520nm peak appears to be dependent on the base composition of poly-purine:poly-pyrimidine duplexes (Figure 5 and Figure 6).

E. *Fluorescence of double-stranded DNA.* The fluorescence emission spectrum of homopolymer double stranded DNA shows a major peak at 590nm (Figure 7 and Figure 8), a shoulder at about 607nm and a minor peak at about 518-520nm. High molecular weight thymus DNA has about the same pattern (Figure 9) with a less sharp shoulder. The 520nm peak was relatively lower with the poly dA:poly-dT pair than with the poly dG:dC pair.

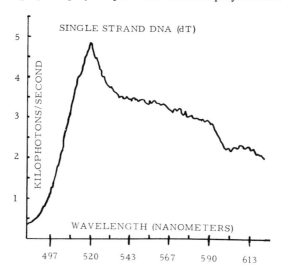

Fig. 1. *Laser fluorescence microscopy emission spectrum of oligo-dT.*

F. *590nm/520nm fluorescence ratio of nucleic acids.* If we compare the two major fluorescence emission peak sizes by making a ratio of the 590nm to the 520nm fluorescence following a 4765A excitation, we see that the ratio is below 1.0 for pure single stranded RNA or DNA but greater than 2.0 for either double stranded nucleic acid (Table 1). The double stranded DNA of either A:T or G:C base composition have over twice as high a ratio as the corresponding base composition of an RNA:DNA hybrid. Thus, it is possible to distinguish strandedness, hybrid formation, and the base pair composition. If the latter either varies between narrow limits or its effect alone can be measured by reference compounds, strandedness and hybridization can be measured by themselves.

G. *590nm/520nm fluorescence and nucleic acid type.* With a series of viruses of known nucleic acid type (RNA or DNA), the 590nm/520nm fluorescence intensity ratio (Table 2) appeared to clearly distinguish them (20). There are clear differences between the 590nm/520nm ratio in RNA viruses as compared to free RNA.

H. *Brownian motion size analysis.* Using polystyrene particles of known size to provide initial calibration by comparing the log power spectrum of the emitted fluorescence fluctuation autocorrelation to log frequency, a method of Brownian motion analysis was derived that could be extended to molecules the size of proteins (23). This analysis for molecules ranging in size from an influenza virus down to IgG correlated well with that given by direct electron microscopy (Figure 10).

TABLE 1.

Fluorescence Intensity Ratio at Two Wavelengths with Free Nucleic Acids: Presence or Absence of Double Strandedness

Nucleic Acid	Type	590nm / 520nm Fluorescence Intensity ratio
p-dA:p-dT	ds-DNA	5.8
p-rA:p-dT	ds-RNA:DNA hybrid	2.6
p-dG:p-dC	ds-DNA	2.8
p-rG:p-dC	ds-RNA:DNA hybrid	1.2
p-dA/dT/dC/dG	ds-DNA (thymus)	2.2
p-dA/dT/dC/dG	ds/ss-DNA (sperm)	1.4
dT_{12-18}	ss-DNA	0.6
dG_{12-18}	ss-DNA	0.4
p-rA/rU/rC/rG	ds/ss-RNA (sol. RNA)	2.2
p-rC	ss-RNA	0.7

Laser microscopy of microliter samples of polynucleotides in solution with a mixture of fluorochrome stains showed a 15-fold range of 590/520nm fluorescence.

TABLE 2.

Fluorescence Intensity Ratio at Two Wavelengths with Viruses: DNA/RNA Nucleic Acid Type

Virus	Nucleic Acid Type	590nm / 520nm Average Fluorescence
f2	RNA	16.5
MS2	RNA	16
Qβ	RNA	12.5
T2	DNA	2.45
T4	DNA	1.25
ΦX-174	DNA	1.00

Within the limited number of examples here examined, the ratio of average fluorescence intensity at 590nm-vs-520nm is much higher for RNA viruses than for DNA viruses; the RNA virus ratio is much higher than that found for free RNA (22).

DISCUSSION

Under the conditions of fluorescent spectral analysis of blue laser excited solutions of combination fluorochrome stained polynucleotides, several features of analytical importance emerge. Single stranded polynucleotides (Figures 1-3) show a single sharp peak at about 520_{nm} with the base composition (G, T, C) or nucleic acid type (DNA, RNA) not significantly altering the wavelength of this maximum; the shoulder peaks at higher wavelengths are influenced by base composition (e.g., Figure 1-vs-Figure 2). Any mechanism of producing double strandedness, whether the loosely associated RNA:DNA hybrid (Figures 5 and 6) or the tightly wound DNA:DNA duplex (Figure 9), creates a major shift in the fluorescence emission ratios. This changed emission ratio shows as a peak around 590_{nm} and is higher than the 520_{nm} peak when the polynucleotide is double stranded. It is evident from inspection of Figures 1-9 that one could readily detect the presence of small amounts of a double stranded region in a polynucleotide preparation.

Under properly modified conditions, the kinetics of double strand formation from single strand (e.g., replication, transcription) could be readily studied, providing a possibly whole new realm of information on the kinetics of polymerization of nucleic acids. Importantly, the shift of single strand into hybrid DNA:RNA duplexes (transcription, reverse transcription) could possibly be carefully examined kinetically or applied for analysis (Table 1), for example, to detect oncorna viruses in very small quantities. A circumstantial case implicating viruses as human carcinogens is based on epidemio-

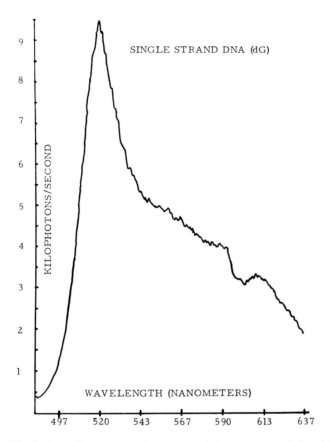

Fig. 2. *Laser fluorescence microscopy emission spectrum of oligo-dG.*

logic association (28), verification in a variety of other mammals (29-32), finding of new virus-like genetic information in human cancers (33-34) and identification of viruses which transform human cells into cancer *in vitro* (35-36). The reverse transcribing viruses have been specifically shown to be a contagious cause of mammalian cancers (37-38). To identify human viruses which convert single strands of RNA (with a primer) into RNA:DNA hybrids would allow possible identification of putative human cancer viruses. Sensitive new techniques of measuring this conversion, such as those described here, are needed to test this hypothesis of association. Among the ways RNA can transmit heritable information to cells is by reverse transcription (15). The ability to detect activation of a reverse transcribing apparatus in possibly a single cell would be an area of future application of this laser microscopy technique which is still unexplored.

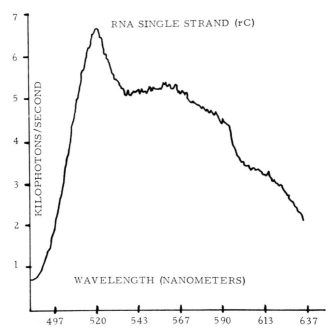

Fig. 3. *Laser fluorescence microscopy emission spectrum of poly-rC.*

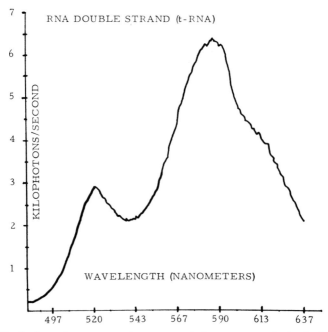

Fig. 4. *Laser fluorescence microscopy emission spectrum of mixed t-RNAs.*

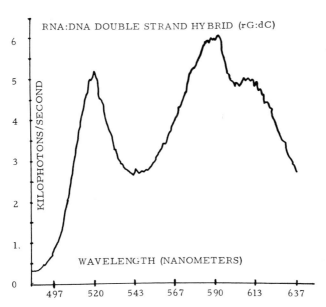

Fig. 5. *Laser fluorescence microscopy emission spectrum of poly r-G: poly-dC.*

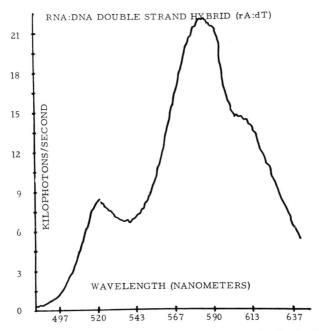

Fig. 6. *Laser fluorescence microscopy emission spectrum of poly-rA: oligo-dT.*

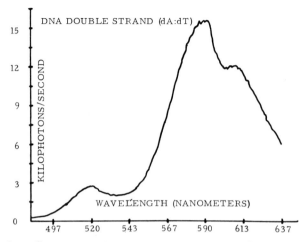

Fig. 7. *Laser fluorescence microscopy emission spectrum of poly-dA: poly-dT.*

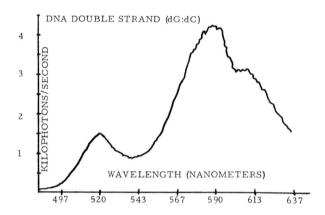

Fig. 8. *Laser fluorescence microscopy emission spectrum of poly d-G: poly-dC.*

Identification of RNA as free or RNA associated with proteins might be distinguished more readily either with modified fluorochromes or by extending the measuring of the shift of emission spectrum from the monochromatic input as the RNA combines with another moiety.

Additional analytical capabilities arise from the sizing and the nucleic acid content determination capabilities discussed above. Dyes specific to other compounds can be used to detect their quantity per particle. The instrument is electronically set to about a 200Å lower size cut off. Even without resetting, this can easily distinguish IgG from IgM or free and bound antibodies under defined conditions (Figure 10). The present detection limit of 10^{-19} mol, determined by the difficulty of handling sample volumes $<0.1\mu l$, could be decreased to 10^{-21} mol or about 10 molecules before any other limitations become significant (in practice only 1nl of sample is observed anyway). The possibility of unique new types of study of drug-receptor interactions is opened by extension of these observations.

Additional analysis is provided by using the laser fluorescence technique to distinguish RNA from DNA viruses, as shown earlier (Table 2), because the 590_{nm} peak height ratio to the 520_{nm} peak height is very much higher. The limits of sensitivity of this distinction are at present unexplored but offer promise of widespread future applicability for identification of specific viruses, determination of the homogeneity of specific virus preparations such as vaccines, and coupled with the other techniques reported here offer a variety of new, sensitive, rapid, and specific exploratory tools for the study of small molecules and nucleic acids in minute quantities.

In spite of the sensitivity of the laser microscopy method, it cannot at present pick out and identify a single subspecies of RNA from among thousands of other similar RNAs. This could be possible only when the sought after species is characterized as dissimilar in a very specific way. However, to solve some problems such as the role of a putative immune RNA, this RNA will have to be isolated in totally pure form and characterized, perhaps by the methods described herein; and it will have to be reproducibly able to create some measurable, specific immune response to an antigen not experienced by the responding cell. With the methods described here, only a few cells and a few molecules of a pure RNA species could be studied. Effects of contaminants of the order of a few parts in 10,000 would thus be an avoidable source of error.

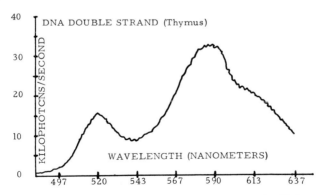

Fig. 9. *Laser fluorescence microscopy emission spectrum of calf thymus high M.W. DNA.*

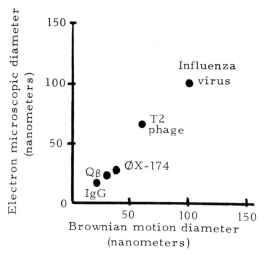

Fig. 10. Rapid size estimation of particles in aqueous media using a Laser Fluorescence Flunctuation Brownian Motion measure. Brownian motion analysis correlates well with the more laborious electron microscopic methods. The Brownian motion method was calibrated using the log power spectrum of the emitted fluorescence fluctuation autocorrelation -vs- log frequency as a function of the size of polystyrene particles (23).

REFERENCES

1. Fishman, M. (1961). Antibody formation in vitro. J. Exp. Med. 114:837-856.
2. Fishman, M. and F. Adler (1963). Antibody formation initiated in vitro: II. Antibody synthesis in x-irradiated recipients of diffusion chambers containing nucleic acid derived from macrophages incubated with antigen. J. Exp. Med. 117:595-602.
3. Braun, W. (1973). RNAs as amplifiers of specific signals in immunity. Ann. N. Y. Acad. Sci. 207:17-28.
4. Schlager, S., S. Dray and R. E. Paque (1974). Atomic spectroscopic evidence for the absence of a low-molecular-weight (486) antigen in RNA extracts shown to transfer delayed type hypersensitivity in vitro. Cell. Immunol. 14:104-122.
5. Giacomoni, D., V. Yakulis, S. R. Wang, A. Cooke, S. Dray and P. Heller (1974). In vitro conversion of normal mouse lymphocytes by plasmacytoma RNA to express idiotypic specificities on their surface characteristic of plasmacytoma immunoglobulin. Cell. Immunol. 11:389-400.
6. Adler, F. and M. Fishman (1975). In vitro studies on information transfer in cells from allotype-suppressed rabbits. J. Immunol. 114:129-134.
7. Yakulis, V., V. Cabana, D. Giacomoni and P. Heller (1973). Surface immunoglobulins of circulating lymphocytes in mouse plasmacytoma: III. The effect of plasmacytoma RNA on the immune response. Immunol. Comm. 2:129-139.
8. Saito, K., S. Kurashige and S. Mitsuhashi (1969). Serial transfers of immunity through immune RNA. Jap. J. Microbiol. 13:122-124.

9. Bell, C. and S. Dray (1973). Lymphoid cells converted by lymphoid RNA extracts *in vitro* and *in vivo* to synthesize allogenic immunoglobins. *Ann. N. Y. Acad. Sci.* *207*:200-224.

10. Adler, F., M. Fishman and S. Dray (1966). Antibody formation initiated *in vitro. J. Immunol. 97*:554-558.

11. Bell, C. and S. Dray (1969). Conversion of non-immune spleen cells by ribonucleic acid of lymphoid cells from an immunized rabbit to produce M antibody of foreign light chain allotype. *J. Immunol. 103*:1196-1211.

12. Ohno, R., K. Esaki, Y. Kodera, H. Shiku and K. Yamada (1973). Experimental models and the role of RNA in immunotherapy of leukemia. *Ann. N. Y. Acad. Sci.* *207*:430-441.

13. Pilch, Y., K. P. Ramming and P. J. Deckers (1973). Induction of anti-cancer immunity with RNA. *Ann. N. Y. Acad. Sci. 207*:409-429.

14. Dodd, M., M. E. Scheetz and J. L. Rossio (1973). Immunogenic RNA in the immunotherapy of cancer: The transfer of antitumor cytotoxic activity and tuberculin sensitivity to human lymphocytes using xenogenic ribonucleic acid. *Ann. N. Y. Acad. Sci.* *207*:454-467.

15. Apple, M. (1973). Review: Reverse transcription and its inhibitors. *Ann. Rep. Med. Chem. 8*:251-261.

16. Hirschfeld, T. (Sept. 14, 1971). U. S. Patent #3,604,927.

17. Hirschfeld, T. (March 18, 1975). U. S. Patent #3,872,312.

18. Block, M. J. (Sept. 27, 1973). Application for Letters Patent.

19. Hirschfeld, T. (Jan. 7, 1975). U. S. Patent #3,859,526.

20. Hirschfeld, T. (Aug. 12, 1975). U. S. Patent #3,899,297.

21. Hirschfeld, T. (June 3, 1975). U. S. Patent #3,887,812.

22. Hirschfeld, T., M. J. Block and W. Mueller. Oral presentation, *Ann. Meeting Opt. Soc. Am.,* Boston, Massachusetts, October, 1975.

23. Hirschfeld, T., M. J. Block and W. Mueller. An optical instrument for visual observation, measurement and classification of free virion. In publication.

24. Collaborative Research, Inc., Waltham, Mass. 02154.

25. P-L Biochemicals Inc., Milwaukee, Wisc. 53205.

26. Holley, R., J. Apgar, B. P. Doctor, J. Farrow, M. A. Marini and S. Merrill (1961). A simplified procedure for the preparation of tyrosine- and valine-acceptor fractions of yeast "Soluble ribonucleic acid." *J. of Biol. Chem. 236*:200-202.

27. Gafni, A., J. Schlessinger and I. Z. Steinberg (1973). Fluorescence and optical activity of acridine dyes bound to poly-A and DNA. *Israel J. of Chem. Vol. II,* Nos. 2-3:423-434.

28. Klein, G., G. Pearson, J. S. Nadkarni, J. J. Nadkarni, E. Klein, G. Henle, W. Henle and P. Clifford (1968). Relation between Epstein-Barr viral and cell membrane immunofluorescence of Burkitt tumor cells: I. Dependence of cell membrane immunoflourescence on presence of EB virus. II. Comparison of cells and sera from patients with Burkitt's lymphoma and infectious mononucleosis. *J. Exp. Med. 128*:1011-1030.

29. Simons, P. and D. McCully (1970). Pathologic and virologic studies of tumors induced in mice by two strains of murine sarcoma virus. *J. Nat. Canc. Inst.* *44*:1289-1303.

30. Lyons, M. J. and D. H. Moore (1965). Isolation of the mouse mammary tumor virus: Chemical and morphological studies. *J. Nat. Canc. Inst. 35*:549-565.

31. Noyes, W. (1959). Studies on the shope rabbit papilloma virus: The location of infective virus in papillomas of the cottontail rabbit. *J. Exp. Med. 109*:423-428.

32. Oroszlan, S., L. Johns and M. Rich (1965). Ultracentrifugation of a murine leukemia virus in polymer density gradients. *Virology 26*:638-645.

33. Baxt, W., J. W. Yates, H. J. Wallace, J. F. Holland and S. Spiegelman (1973). Leukemia-specific DNA sequences in leukocytes of the leukemic member of identical twins. *Proc. Nat. Acad. Sci.* 70:2629-2632.

34. Baxt, W. and S. Spiegelman (1972). Nuclear DNA sequences present in human leukemic cells and absent in normal leukocytes. *Proc. Nat. Acad. Sci.* 69:3737-3741.

35. Aaronson, S. and G. Todaro (1970). Transformation and virus growth by murine sarcoma viruses in human cells. *Nature* 225:458-459.

36. Gerber, P. (1972). Is Epstein-Barr virus a human tumor virus? *NCI Monograph* 36:65-71.

37. Jarrett, W., O. Jarrett, L. Mackey, H. Laird, Wm. Hardy, Jr. and M. Essex (1973). Horizontal transmission of leukemia virus in the cat. *J. Nat. Canc. Inst.* 51:833-841.

38. Jarrett, W. (1971). Feline leukemia. *Inst. Rev. Exp. Pathol.* 10:243-263.